Gravity's Ghost and Big Dog

Gravity's Ghost and Big Dog

Scientific Discovery and Social Analysis in the Twenty-first Century

HARRY COLLINS

The University of Chicago Press
Chicago and London

HARRY COLLINS is the Distinguished Research Professor of Sociology and director of the Centre for the Study of Knowledge, Expertise, and Science at Cardiff University, and a fellow of the British Academy.

The University of Chicago Press, Chicago 60637
The University of Chicago Press, Ltd., London

Printed in the United States of America

22 21 20 19 18 17 16 15 14 13   1 2 3 4 5

ISBN-13: 978-0-226-05229-8 (paper)
ISBN-13: 978-0-226-05232-8 (e-book)

Library of Congress Cataloging-in-Publication Data

Collins, H. M. (Harry M.), 1943–author.
Gravity's ghost ; and, Big dog : scientific discovery and social analysis in the twenty-first century / Harry Collins.—Enlarged edition.
pages cm
"Among other things, this volume contains the paperback edition of Gravity's ghost, which is included unaltered as the first part of the book. The original pages of the hardcover are reproduced with the same page numbers, though the acknowledgments, references, and index have been shifted to the end of the volume. The new edition is twice as long as the original Gravity's ghost. . . . The new material was initially written as a separate book, but the story is so closely related to Gravity's ghost that it seemd more sensible to bundle the two together. The additional material, the second section of this edition is entitled Big dog"—Preface to the Enlarged edition.
Includes bibliographical references and index.
ISBN 978-0-226-05229-8 (paperback : alkaline paper)—ISBN (invalid) 978-0-226-05232-8 (e-book) 1. Gravitational waves—Research—History. 2. Gravitational waves—Experiments. 3. Science—Social aspects. I. Title. II. Title: Big dog.
QC179.C645 2013
539.7'54—dc23

2012048347

♾ This paper meets the requirements of ANSI/NISO Z39.48–1992 (Permanence of Paper).

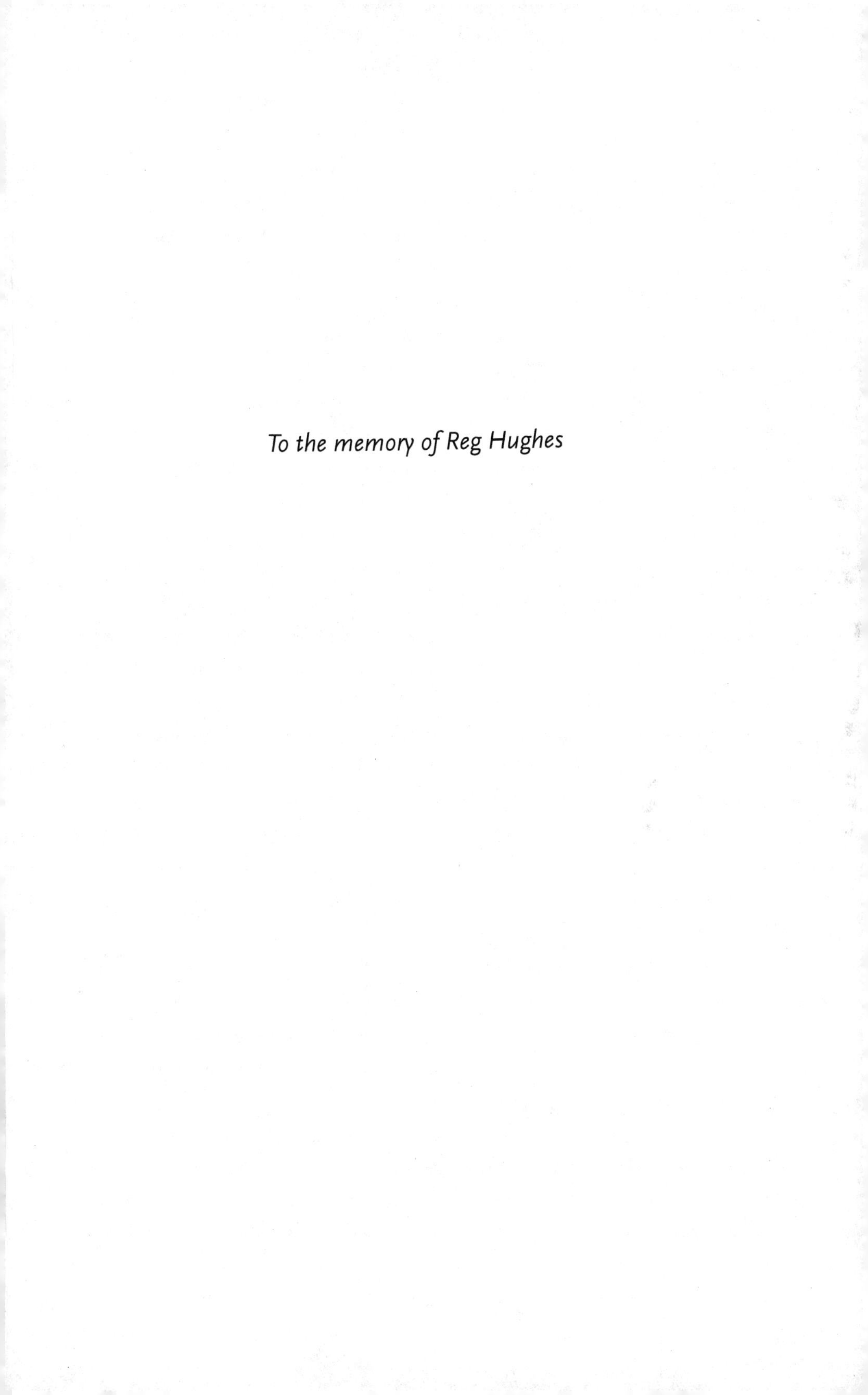

To the memory of Reg Hughes

CONTENTS

PREFACE TO THE ENLARGED EDITION

Among other things, this volume contains the paperback edition of *Gravity's Ghost*, which is included unaltered as the first part of the book. The original pages of the hardcover are reproduced with the same page numbers, though the acknowledgments, references, and index have been shifted to the end of the volume. This new edition is twice as long as the original *Gravity's Ghost*. More physics has come along and it has been too interesting to not write up. The new material was initially written as a separate book, but the story is so closely related to *Gravity's Ghost* that it seemed more sensible to bundle the two together. The additional material, the second section of this edition, is entitled *Big Dog*, the name the scientists gave to the gravitational-wave event that was the occasion for this new material.

Another addition to the volume is a third section, "The Trees and the Forest," that reflects on the sociological significance of the material and the lessons that can be learned from it—lessons about methodology and about the nature of expertise. In this section I also reflect on the place of the book in the landscape of science and technology studies, my home discipline. Much to my surprise, in recent years some of my work—the part known as the "third wave of science studies"—has been treated by some at the heart of this discipline as dangerously heterodox, though it has been very well received by the community as a whole. Some of the analysis found in *Big Dog* is inspired by the elements of the supposedly heterodox approach. The sociological introduction explains some of what is going on and how the two elements of this book fit into it. Even nonsociologist readers might find some of this of interest, especially as it explains how, much to their surprise—or in some cases, derision—I, a sociologist, came to disagree with some of my gravitational-wave physicist colleagues over substantive points to do with the way they were analyzing their data. In the new section I try to explain and justify this curious turn of events.

Gravity's Ghost reports the analysis and decision-making process for a weak signal that may have been a gravitational wave—this weak signal became known as the "Equinox Event"; *Big Dog* reports on the analysis and decision concerning a strong signal that may have been a gravitational wave. A rather different set of problems arose in the second case, and a rather different analysis emerges. In the case of the Equinox Event, it is a matter of whether the signal has any credibility; in the case of Big Dog it is a matter of how the event is to be presented to the world, what scientific importance will be claimed for it. The decision, as I show in chapter 14, is not a matter of calculation but of choices that are closer to philosophy than physics. We see the physicists walking a philosophical tightrope between certainty on the one hand and scientific significance on the other. The analysis includes a debate about whether the scientists achieved the right balance.

The events surrounding Big Dog also provide answers to some of the attempts to look into the future that are described in the postscript to the original volume, "Thinking after Arcadia" (pp. 153–58), and it feeds into the question about whether Initial LIGO could ever have made the putative detection, which was put forward as the *scientific* justification for its design (pp. 134–35). It is fortuitous that real events corresponding to those being imagined in 2009 should occur in just a couple of years.

The acronym for Advanced LIGO that is now most commonly used by the physicists is aLIGO. I used AdLIGO in *Gravity's Ghost*, and for consistency I have continued to use AdLIGO in *Big Dog*.

I

GRAVITY'S GHOST

The Equinox Event

INTRODUCTION

I begin to write this volume right now at 2:10 p.m. U.S. West Coast time on 19 March 2009. That's the first sentence. I am sitting in Los Angeles airport with an hour or two before my flight back to the UK. I have just come from the small California township of Arcadia, where I have been attending a meeting of the "LIGO Scientific Collaboration and Virgo." The Collaboration is six or seven hundred strong, and it has been trying to detect gravitational-waves using apparatuses costing hundreds of millions of dollars. These giant machines have been "on air" for a couple of years, and a lot of data has been collected which has only just been analyzed; it might or might not contain intimations of a gravitational-wave. The high point of the meeting was the "opening of the envelope"—an event that has kept me and most of the gravitational-wave community on tenterhooks in the eighteen months since the autumn of 2007.

The "envelope" held the secret of the "blind injections." The blind injections were the possible introduction of fake signals into the data stream of LIGO—the Laser Interferometer Gravitational-Wave Observatory. The idea was to see if the physicists could find them. Only the two people who had injected the signals knew the secret before this meeting.[1] Their brief was to inject fake

1. Or so I thought until Jay Marx, the executive director of LIGO, told me (private communication, October 2009), that he had known the secret from the beginning and had been keeping a "poker face" through the whole eighteen months—a remarkable performance.

gravitational-wave signals according to a randomized code. Both the shape and strength of the signals and the number of signals injected was to be decided at random. One possibility was that no signal at all had been injected. It could also be that one, two, or even three signals had been injected. On this depended whether anything suggestive in the data was a blind injection and whether any blind injection had given rise to anything suggestive. Last Monday, in that tense meeting, I and the rest of the community were told the truth; today is Thursday. Before you finish reading this book you too will know what was in the envelope. To get the greatest benefit from the book, however, I would suggest, don't flip to the end—follow events as I and the physicists lived them; read the book in the spirit of a "whatwosit," the physics version of a "whodunit."[2]

The story recounted in these pages is part of the history of gravitational-wave detection that I started to document in 1972.[3] Since a second period of very intense involvement began in 1994, I have been given more and more access to the field. I now work from a privileged and possibly unique position—an outsider, with no controls on what I disclose except those of good manners and good sense, who is privy to the inner discussions of a live scientific group as they struggle to make a discovery.

Only since the mid-1990s, with the invention of the unobtrusive digital voice-recorder, has it been possible to tell a story with what the actors said in real time playing the large role it does here. With such a device I can sit quietly in the corner of the rooms in which the events I report take place, typing notes into my laptop computer and recording anything that anyone says that sounds interesting. Of course, that I am allowed to do this as a matter of course is also something worthy of remark. It has been a long and slow matter of building trust and colleagueship with the members of the gravitational-wave community, something which has been not an onerous task for me but part of my reward for doing my job. Why did the scientists ever allow me to get started? After all, some of what I write might embarrass them. It is because they believe in the academic enterprise and know this is the right thing to allow, even if it does make for a less comfortable life. On the upside for the scientists, we can be sure of one thing: any group that is prepared to allow an outsider like me to listen in on their innermost

2. I respectfully ask reviewers not to give away the secret either.

3. The period from the start to around 2003 is described in my book, *Gravity's Shadow* (2004). See also http://www.cf.ac.uk/socsi/gravwave.

discussions is a group you can trust. What is strange is that I am probably the only person who is currently doing work of this kind.[4]

Over the years I have come to love gravitational-wave detection physics and the people in it. As a sociologist of science I have investigated quite a few fields but chose to do my career-long fieldwork study following gravitational-waves because I felt more at home in this science that in any of the others I looked at. The task the physicists have set themselves is nearly impossible, and it will, and then only with luck, take a lifetime to complete. Expecting little financial reward, they spend their existence encountering endless frustrations and disappointments in the hope of gaining a miniscule increase in understanding of how the world works; I find myself happy in such company. At times of despair and encounters with stupidity, the example of the gravitational-wave physics community has rekindled my faith in the worlds of both science and social science. Ironically, given what is in this book, it has helped me to continue to believe that a high standard in the search for truth is better than academic *realpolitik*, both in theory and in life. In what I call the "envoi," I try to elevate this point into a political philosophy, arguing that science done with real integrity can provide a model for how we should live and how we should judge.

The irony is that what I describe may be scientists trying too hard to achieve perfection. In the twenty-first century it may be better to allow the imperfection of *the best that can be done*—which gravitational-wave detection physics assuredly is—to be revealed, not disguised. The model that has led, or perhaps, misled, the philosophical understanding of the nature of science for too long is Newtonian physics, along with its successors, relativity, quantum physics, high-energy physics, and so forth. The gold standard for these sciences is exact quantitative prediction, triumphantly confirmed in more recent years via statements of high levels of statistical significance. Retrospective accounts of these triumphs have given us an unsupportable model of how the world can be known.

Two things have gone wrong. The first thing is that the domain to which the Newtonian model applies, though it takes up a huge proportion of that part of our imaginations that is devoted to science, is a tiny and unrepresentative corner of the scientific enterprise. Nearly all science is a mess—think about long-term weather forecasting, the science of climate,

4. Try to imagine how different certain commercially driven sciences might be if they let people like me listen to their inner deliberations. Wouldn't it be better if there were this kind of record of more major scientific enterprises, both taxpayer-funded and private?

the science of human behavior, economics, and so on and so on. The trick of the sciences of the very large, and the sciences of the very small, is that nothing much happens in outer space and nothing much happens in inner space; get away from the earth and there is not much there; get down to where it is all spaces between subatomic particles and there is not much there either. That is why the sciences of astronomy, astrophysics, and cosmology, and of quantum and high-energy physics, are so simple, and that is why they can more easily appear to match the idealized model. Down/up here, where most of us live from day to day, there is so much going on that it is almost impossible to make a secure prediction. So the Newtonian model is unrepresentative of science in a statistical sense and still less representative of the sciences of pressing concern to the citizen.

The second thing that has gone wrong is that the Newtonian model is not even a correct description of itself. The triumphal accounts are either retrospective, or refer to sciences that are so well established that they have made all their mistakes and become technologically secure—not reaching, but perhaps reaching toward, the reliability of your fridge or your car. The revealing science—the science which more readily shows us how humans wrest their understanding from a recalcitrant nature—is pioneering science, where things are being done for the first time and mistake after mistake is being made. Gravitational-wave detection physics is a true science in this respect. Here things are being done for the first time.

The contrast between the frontier sciences and the technologically well-developed sciences is a main lever of analysis in this book, and it may be that the sociologist's deliberately distanced perspective helps to bring it out. On the other hand, the views that emerge from the sociological perspective, at least insofar as they bear on the dilemmas of the science itself rather than its role in society, are not dissimilar from those of some of the members of the scientific collaboration being studied. The sociological contribution is, perhaps, simply to set out the arguments in a systematic way and relate them to wider issues.

Some of these sentiments might grate on those brought up in the tradition of science and technology studies, or "science studies," as it has been practiced since the early 1970s. The prevailing motif of the field has been the "deconstruction" of the idealized model of science. That I could argue above that the idealized model is not even a good description of the Newtonian sciences and their counterparts is a result of the new understanding that has come with this movement—a movement in which I have been involved from the beginning. The sentiments expressed above have to be understood in the light of what has been called the "three wave" model of science studies.[5] The First Wave took science to be the preeminent form of knowledge-making, and the job of philosophy of science was to tease out its logic and that of the social study of science to work out how society could best nurture it. The Second Wave used a variety of skeptical tools—from philosophical analysis to detailed empirical studies of the day-to-day life of science—to show that the predominant model of science was wrong and that the examples of scientific work used to support it were oversimplified. For example, the Michelson-Morley experiment was regularly described as having shown, in 1887, the speed of light to be a constant, whereas it took fifty years of dispute before scientists agreed that its results were empirically sound.[6] Most of this book is Wave Two science studies: it is going to reveal how difficult it is be sure of what you are finding out even in physics.

Wave Two shows, with the quasi-logical inevitably of the application of skepticism, that science has no philosophically or practically demonstrable special warrant when it comes to knowledge-making. The recently proposed Wave Three of science studies accepts this but argues that technically based decisions still have to be made in a modern society. Wave Three therefore seeks an alternative way of establishing the value of the science-driven thinking that we are almost certainly going to put at the heart of our technical judgements. The proposed alternative is the analysis of expertise.[7] Wave Three makes it explicit that in spite of the logic of Wave Two, which shows how sciences claim to true knowledge can be "deconstructed," science is still the best thing we have where knowledge about the natural world is concerned.[8] Here, the processes of science are unapologetically spoken of as the most valuable models for the making of technological knowledge, even though this cannot be proved to be the case by detailed description or logical analysis. Gravitational-wave physics, in spite of the fact, or perhaps because of the fact, that what it counts as a finding has to be wrested from the fog of uncertainty by human technical judgment, is an example of the best that humans can do and should do.

5. Collins and Evans 2002, 2007.

6. Collins and Pinch 1998a.

7. Collins and Evans 2007.

8. Another thing discussed in this book that does not mesh easily with contemporary science studies is that gravitational-wave physics can be contrasted with other sciences, such as high-energy physics, which can be treated as, in some sense, more perfectible. Even though it has been shown that even in the heartland of science there is no perfection to be found, the contrast between the two kinds of science remains a useful device, so long as it is understood in the spirit of saying that it is more certain that the sun will rise tomorrow morning than that tomorrow's weather will be like today's, even though both are subject to the problem of induction.

1 Gravitational-Wave Detection

A Brief History of Gravitational-Wave Detection

In 1993 the Nobel Prize for physics was awarded for the observation, over many years, of the slow decay of the orbit of a binary star system and the inference that the decay was consistent with the emission of gravitational-waves. Here, however, we are concerned with the detection of gravitational-waves as a result of their "direct" influence on terrestrial detectors rather than on stars. The smart money says that the first uncontested direct detection will happen six to ten years from now, almost exactly fifty years since Joseph Weber, the field's pioneer, first said he had seen them. Joe Weber's claim was not uncontested. It was one of some half-dozen contested claims to have seen the waves made since the late 1960s. All of these have been consigned, by the large majority of the physics community, to the category of "mistake."[1] The rejection of these results by the balance of the gravitational-wave

1. This volume, though it can be read independently, is the second in a series that should have (at least) three volumes—the first being *Gravity's Shadow*, my account of the sociological history of the attempt to detect gravitational-waves from its beginning. This includes the remarkable story of the contested claims and the shift from small science to the new, big-science, technique of interferometry. *Gravity's Shadow* ends around 2003. The third volume should be about the uncontested claims to come. The current book has few bibliographical references, and those who feel the need to go back to the original sources pertaining to the history of gravitational-wave detection, or the science studies background of what is argued here, should refer to *Gravity's Shadow*.

community was often ferocious, driven by the sense of shame at the field's reputation for unreliability, or flakiness, in the eyes of outsiders. Newcomers to the enterprise also had to justify spending hundreds of millions of dollars on the much larger instruments—the giant "interferometers"—that they felt would finally be able to make a sound detection and atone for past mistakes; if the old cheap technology really could see the waves, then there would be no need for the new, so the credibility of the old cheap technology had to be destroyed.

The proponents of the old technology fiercely resisted the destruction of their project, which caused both sides to dig themselves into polarized positions.[2] The consequence was that for decades the creative energy of most interferometer scientists was directed at finding flaws; the principle activity had become showing how this or that putative signal in either their rivals' or, subsequently, in their own detectors was really just noise. This is the problem of the negative mindset that is a central feature of what is to follow. In the meeting in Arcadia that bitter history stalked the corridors with an almost physical presence.

Weber and the Bars

Joe Weber was a physicist at the University of Maryland. In the 1950s he began to think about how he might detect the gravitational-waves predicted by Einstein's theories. Gravitational-waves are ripples in space-time that are caused by rapid changes in the position of masses, but they are so weak that only cosmic catastrophes such as the explosion or collision of stars or black holes can give rise to enough of the radiation to be even conceivably detectable on the surface of the earth. It would take a great leap of the imagination, a genius for experiment, and a heroic foolhardiness to try it. Weber was equipped with the right qualities, and he built a series of ever more sensitive detectors; by the end of the 1960s, he began to claim he was seeing the waves.

Weber's design was based on the idea that ripples in space-time could be sensed by the vibrations they caused in a mass of metal. He built cylinders of aluminum alloy weighing a couple of tons or so and designed to resonate—to ring like a bell at around the frequency of waves that might plausibly be emitted by a source in the heavens. Every calculation of the energy in such waves and the way they would interact with Weber's detec-

2. More of this battle will be described in the odd-numbered chapters; these provide background and commentary. The even numbered chapters tell the story.

tors implied that he did not have a hope, and when he started he did not think he had a hope either. But he went ahead anyway.

Weber insulated the cylinders from all the forces one could think of, but to see a wave it was necessary to detect changes in the length of the cylinders of the order of 10^{-15} m, the diameter of an atomic nucleus, or even less. Vibrations of this size, however, are continually present in the metal anyway, no matter how carefully it is insulated. Crucially, Weber built two of the cylindrical devices and separated them by a thousand miles or so. Then he compared the vibrations in the two cylinders. The idea was that, if there was a coincident pulse in both detectors, only something like gravitational-waves, coming from a long way away, could cause it.

Since both of the cylinders would suffer from random vibrations, there were bound to be coincident pulses every now and again just as a result of chance. But Weber used a very clever method of analysis. He used something called the "delay histogram," which is nowadays referred to as the method of "time slides" or "time shifts"—a method that is still at the heart of gravitational-wave detection forty years on and that will be at the heart of the method for the foreseeable future. Imagine the output of the detector drawn on a steadily unwinding strip of paper, as in those machines that record the changing temperature over the course of a day, but, in this case, sensing vibrations microsecond by microsecond; it will be a wiggly line with various larger pulses impressed upon it. One takes the strip from one detector and lays it alongside the strip from the other. Then one can look at the two wiggly lines and note when the large pulses are in coincidence. Those *coincidences* might be caused by a common outside disturbance such as a gravitational-wave, or they might be just a random concurrence of *noise* in the two detectors. Here comes the clever bit: one slides one of the strips along a bit and makes a second comparison of the large pulses. Since the two strips no longer correspond in time, any coincidences found can only be due to chance. By repeating this process a number of times, with a series of different time slides one can build up a good idea of how many *coincidences* are going to be there as the result of chance alone—one can build up a picture of the "background." A true signal will show itself as an excess in the number of genuinely coincident pulses above the background estimates generated from the time slides.

A time slide can also be called a "delay." The signal will appear, in the language of Weber, as a "zero-delay" excess. Nowadays scientists look not for a zero-delay "excess" but at isolated coincidences between signals from different detectors. Nevertheless, the calculation of the likelihood that these coincidences could be real rather than some random concatenation of noise is based on an estimate of the background done in a way that is close to the method that Weber pioneered.

As the 1960s turned into the 1970s Weber published a number of papers claiming he had detected the waves, while other groups tried to repeat his observations without success. By about 1975 Weber's claims had largely lost their credibility and the field moved on. Weber's design of detector continued to be the basis of most of the newer experimental work, but the more advanced experiments increased the sensitivity and decreased the background noise in the "bars" by cooling them with liquid helium. Most of the experiments were cooled to between 2 degrees and 4 degrees of absolute zero, with one or two teams trying to cool to within a few millidegrees of absolute zero. Collectively, such "cryogenic bars" were to be the dominant technology in the field until the start of the 2000s. Just two groups, one based in Frascati and sometimes known as the "Rome Group" or "the Italians," and an Australian group, kept faith with Weber's claims, promulgating results that most gravitational-wave scientists believed were false—the latter view being one which would now be almost impossible to overturn.

Nearly everyone outside the maverick supporters of Weber came to believe that Weber had either consciously or unconsciously manipulated his data in a post hoc way to make it appear that there were signals when really he was really seeing nothing but noise. This can happen easily unless great care is taken. Weber did not help his case when he made some terrible mistakes. In the early days he claimed to have a periodicity in the strength of his signals of twenty-four hours when proper consideration of the transparency of the earth to gravitational-waves suggested that the right period should have been twelve hours. Somehow, shortly after this was pointed out, the period mysteriously became twelve hours in Weber's discussions and papers, and this led some people to be concerned about the integrity of his analysis.[3] He also found a positive result that should have been ruled out because it was caused by a computer error, and, most damningly, he claimed to have found an excess of zero-delay coincident signals between his bar and that of another group when it turned out that a mistake about time standards meant that the signal streams being compared were actually about four hours apart, so that no coincidences should have been seen.

Those who had faith in Weber's experimental genius were ready to accept that these were the kind of mistakes that anyone could make, but

3. See chapter 5, note 6, below, for a retrospective reexamination of this criticism.

those who were less charitable used the events to destroy his credibility. Weber did his case further harm by the way he handled these stumbles. Instead of quickly and gracefully accepting the blame, he tended to try to turn it aside in ways that damaged his credibility. Weber's reputation fell very low, and the community tried to convince him that he should admit that he was wrong from start to finish, allowing them to give him more credit for his adventurous spirit and his many inventions and innovations, but he never gave in.

Weber died in the year 2000, insisting to the end that his results were valid and even publishing a confirmatory paper in 1996—a paper which nobody read. Weber was a colorful and determined character without whom there would almost certainly be no modern billion-dollar science of gravitational-wave detection. I have heard Joe Weber described as hero, fool, and charlatan. I sense his reputation is growing again, as it has become easier to give credit to his pioneering efforts now that he is no longer around to argue with everyone who doesn't believe his initial findings. I believe he was a true scientific hero and that his heroism was partly expressed in his refusal to admit he was wrong; believing what he did, a surrender for the sake of short-term professional recognition would not have been an "authentic" scientific act. That the published results that indicate that he detected the waves are almost certain to remain in the waste bin of physics is another matter.[4]

For a time Weber was one of the world's most famous scientists, thought to have discovered gravitational-waves with an experiment that was an astonishing *tour de force*. Many scientists now see the Weber claims as having brought shame to the physical sciences. Much of the subsequent history of gravitational-wave detection has to be understood in the light of what happened.

Long after most scientists considered Weber to have been discredited, a group based near Rome, which will be referred to frequently in this book, published or promulgated several papers claiming to see the waves. The claims were based on coincidences between a cryogenic bar in Rome and one in Geneva, on coincidences with the cryogenic bar in Australia, and on coincidences between one of their original room-temperature bars

4. But there will doubtless be attempts to revive them. An entry submitted on 2 March 2009 on the arXiv blog, entitled "Were gravitational-waves first detected in 1987?" (arxiv.org/abs/0903.0252), reports a paper by Asghar Qadir that has the potential to revive one of Weber's claims. I was to discover that this was another one of those papers which, though it has enormous potential to change things if true, was simply ignored by the gravitational-wave physics community. For a related event in respect of one of Joe Weber's papers, see *Gravity's Shadow*, chapters 11 and 21.

and Weber's room-temperature bar.[5] These claims were sometimes ignored by the rest of the gravitational-wave community and sometimes greeted with outrage.

The outrage, I believe, and have attempted to show in my more complete history of the field, can be to some extent correlated with the need to get funds to build a new and much more expensive generation of detectors. These are the interferometers which today dominate the field. An experiment like Weber's could be built for a hundred thousand dollars, whereas the U.S. Laser Interferometer Gravitational-Wave Observatory (LIGO) started out at around a couple of hundred million. If gravitational-waves could be detected for a fraction of the price, and Weber once wrote to his Congressional representative to argue just this, funding for big devices would be hard to justify. Therefore, it became a political as well as a scientific necessity to stress that the bars could not do the job that Weber and the Rome Group were claiming for them. On the basis of almost every theory of how these instruments worked, the interferometers were going to be orders of magnitude more sensitive than the bar detectors, and on the basis of almost every theory of the distribution and strength of gravitational-wave sources in the heavens, only the interferometers had any chance of seeing the waves. Furthermore, even the first generation of these more expensive devices could not be expected to see more than one or two events at best. The consensual view among astrophysicists was that the sky was black when it came to gravitational radiation of a strength that could be seen by the bars, including the cryogenic bars, and that it might emit a faint twinkle, perhaps once year, as far as the first generation of interferometers was concerned. The promised age of gravitational-wave astronomy, involving observation of many different sources with different strengths and waveforms, helping to increase astrophysical understanding, would not be here until a second or third generation of interferometers were on the air. It was only the promise of gravitational astronomy, not first discovery, which could justify the huge cost of the interferometers.

Thus the scene was set for the unfolding of a battle between the cryogenic bars and the interferometers, with the bar side led by the Rome Group. Some bar teams, such as those based in Louisiana and in Legnaro, near Padua, accepted the view of the interferometer teams and agreed to strict data analysis protocols based on a model of the sky in which signals would be rare and strong. This ruled out any chance of detecting weak

5. Weber never constructed a successful cryogenic bar, though he was given funds to embark upon the project.

signals near the noise that might otherwise have been used as a basis for tuning the detection protocols. This was the bars' last chance—it was probably the only way they could work toward an understanding of any weak signals good enough to survive the more severe statistical tests needed for a claim.[6] But the Rome Group was not prepared to accept the dismal astrophysical forecasts and exerted the experimentalist's right to look at the world without theoretical prejudice. If the Rome Group could work themselves into a position that enabled them to find some coincidences that could not be instantly accounted for by noise, and which flew in the face of theory, then they were determined to say so—in the spirit of Joe Weber. They were not willing to put all their effort into explaining away every putative signal just because it was supposed to be theoretically impossible. Thus were they to give rise to a continuing history of "failed" detection claims right into the twenty-first century, and thus did they give teeth and muscle to the history-monster stalking Arcadia's corridors.

Interferometers

Five working interferometers play a part in this story. The size of an interferometer is measured by the length of its arms. The smallest, with arms 600 m long, is the German-British GEO 600, located near Hannover, Germany. Virgo, a 3 km French-Italian device, is located near Pisa, in Tuscany. The largest are the two 4 km LIGO interferometers, known as L1 and H1, located respectively in Livingston, Louisiana, close to Baton Rouge, and on the Hanford Nuclear Reservation in Washington State. There is also a 2 km LIGO device, H2, located in the same housing as H1.

An interferometer has two arms at right angles. Beams of laser light are fired down the arms and bounced back by mirrors. The beams may bounce backward and forward a hundred times or so before the light in the two arms is recombined at the center station. If everything works out just so, the changing appearance of the recombined beam indicates changes of the lengths in the arms relative to each other — a change that could be caused by a passing gravitational-wave. It should thus be possible to see the "waveform," of a passing gravitational-wave in the changing pattern of light that results from the recombination of the beams.

6. This is roughly what happened in the case of the 2002 publication by Astone et al. (see below). In that case, however, the tuning led to their *not* finding a signal in a second period of observation, thus experimentally *disproving* their tentative claim—which seems a reasonable way to do science.

The longer the arms are, the larger the changes in arm length and the easier it is to see them, so, other things being equal, bigger interferometers are more sensitive than smaller ones. But even in the largest interferometers, the changes in arm length that have to be seen to detect the theoretically predicted waves would be around one-thousandth of the diameter of an atomic nucleus (i.e., 10^{-18} m) in a distance of 4 km. It is, therefore, something close to a miracle that they work at all, where "working" does not necessarily mean detecting gravitational-waves but being able to measure these tiny changes.

LIGO was funded in the face of bitter opposition, some from scientists who believed the devices could never be made to function. I was lucky enough to watch every stage of the building of the LIGO interferometers, and for much of that time I too did not believe they were going to work, and I was not alone even among those close to the technology. To see the first tentative indications that the trick might be pulled off, and to watch the slow increase in sensitivity right up to design specification, two or three years late though it was, has been one of the most exciting experiences of my life—perhaps more exciting that even the final detection of gravitational waves will be. But even now the big interferometers are far from perfect machines—they are still plagued by undiagnosed sources of noise which, as we will see, make their effective range somewhat less than the range as calculated from the moment-to-moment performance of their components.

Range is vitally important. Astrophysical events that might be visible on an earthbound gravitational-wave detector are unpredictable and may happen anywhere that galaxies are found. The greater the range, the more galaxies can be included in the search and the better the chance that a wave will be seen. The number of galaxies, and therefore the number of potentially exploding or colliding stars that might be seen, is proportional to the *volume* of space that can be surveyed. This volume is a sphere centered on the earth, and the number of stars and galaxies it contains is roughly proportional to the cube of the radius—the radius being the range. Thus, a small increase in range buys a proportionally much greater increase in potential detections; if the range is doubled, the number of potential events increases by eight; if the range is multiplied by ten, as is the promise for the next generation of LIGO detectors, the number of potential sources will be increased a thousandfold. When this happens the promise of gravitational-wave *astronomy* might be fulfilled.

GEO 600, because of its relatively short arms and some other problems, does not play much part in the story to be told here. Virgo, even though its arms are only 3 km long rather than the 4 km of L1 and H1, the big LIGO devices, includes clever aspects of design that should give it a better potential performance at low frequencies; this makes it a more important contributor to the detection process than its higher frequency performance alone would imply. Unfortunately, progress on the low frequency aspect of the design was slow—low frequency is always more difficult—and, in general, its development has suffered greater delays than expected and so its sensitivity has lagged further behind that of LIGO than technical limitations would imply. This will turn out to be indirectly important to the story.

LIGO's 2 km interferometer, H2, is an anomaly. A crucial element of a detection claim is the existence of coincident signals between widely separated detectors. GEO 600 is near Hannover, Virgo near Pisa, L1 is close to Baton Rouge, and H1 is on the Hanford Nuclear Reservation in Washington State. But H2 is located in the same housing as H1, so there is no separation involved. This makes coincidences between H2 and H1 much harder to interpret as a signal than coincidences between any of the other pairs of detectors. H2 nevertheless plays a part in the story.

LIGO is now known as "Initial LIGO," or "iLIGO," because one-and-a-half further generations are in progress. The half generation is Enhanced LIGO (eLIGO)—LIGO with certain components of Advanced LIGO (AdLIGO) installed early. If it does all that is intended of it, eLIGO will have twice the range of LIGO and be able to see eight times as many potential sources. eLIGO is just coming on air, though at the time of writing it has some troubles that are pushing back the start of double sensitivity and may even put its achievement in doubt. AdLIGO is to be installed in the same vacuum housings as Initial LIGO but has all new components, including better mirrors, better mirror suspensions, better seismic isolation, and a much more powerful laser. AdLIGO should be producing good data around 2015. Already I have heard people saying that eLIGO is justified by its role as a test bed for AdLIGO components or even that this is all it was intended to be in the first place. My recollection of the arguments for its construction is that the strongest pressure came from scientists who believed doubling the sensitivity would be the key to the initial detection of gravitational-waves, as Initial LIGO, for which some had held out great hope, was proving a disappointment. In the absence of pressure from some senior scientists who were sure that eLIGO would produce the desired result, it seems possible that the device would not have been built and that the initial run of LIGO would have been extended instead of disassembling the machine so soon for the new components to be installed. On the other hand, I think it is also the case that eLIGO would not have gone ahead if it had needed components that were not to be installed anyway for the use of AdLIGO, so that it could be used as an AdLIGO test-bed as well as a detector in its own right.[7] I get the sense that the scientists will feel that honor has been satisfied, or perhaps more than satisfied, if the total number of potential sources surveyed by eLIGO equals or exceeds the number that would have been observed if Initial LIGO had stayed on air until all the components were ready to be installed in AdLIGO. The calculation of how many sources have been observed involves integrating the time of observation multiplied by the cube of range. In other words, if eLIGO does achieve twice the range of LIGO, then one month of eLIGO's time is worth eight months of LIGO's (but should be worth only about six hours of AdLIGO's).

A feature of these calculations is the duty cycle. An interferometer is not always in a state to make observations when it is "switched on." First, there are periods of necessary maintenance. Then there are periods when the detector cannot achieve "science mode" because the environment is too noisy. Noise can come from what the scientists like to call anthropogenic sources—airplanes fly low over the site, piles are driven or pneumatic drills are used, trucks come close when delivering supplies, in Louisiana, trains pass, logs are felled, and, disastrously, explosive devices are fired for oil or gas prospecting—an eventuality which is going to require the shutting down of the entire detector for a month or two toward the beginning eLIGO's run. Downtime can also be caused by natural events. Seismic disturbances big enough to shake the detectors out of science mode can be caused by earthquakes, while lesser noises are caused by storms which pound the shores with big waves or winds which shake buildings or drag on the ground.

There are a number of levels of disturbance that affect an interferometer. The most severe of these is when the device goes "out of lock." The mirrors in an interferometer must be isolated from the crude shaking of the ground which, if they were fixed in place, would disturb them trillions of times more than the effect of a gravitational-wave. The mirrors are therefore slung in an exquisite cradle of pendulums and soft vertical isolation springs all surrounded by hydraulic feedback isolators to cancel out large external disturbances. "Lock" is a state where the necessary spider-

7. Jay Marx pointed out to me (private communication October 2009) that the extra budget allocated specifically for the putting together of eLIGO was only $1.4 million, so, in the context of the whole program, it seemed the right thing to do given the extra chance of making a detection.

web of feedback circuits which control the oscillation of the mirrors are in a degree of balance such that that mirrors are stationary and the laser light can bounce between them, building its strength, and be used to measure the relative lengths of the two arms with maximum accuracy. When the interferometer is in that condition, any extra impulses the interferometer's electronics has to send to the mirrors to hold them still is a measure of the signal that is being looked for. In other words, if the mirrors are perfectly balanced in a state which is undisturbed by extraneous forces, any tendency to move *may* be caused by the caress of a gravitational-wave, and the tiny restoring forces required to prevent the mirrors being affected by this caress is a measure of the wave's form and strength. When the interferometer is badly disturbed, the shaking of the mirrors becomes so great that the feedback circuits cannot deliver enough power to the actuators to hold them in the position needed for the laser light to bounce between them; this is "going out of lock."[8] But even when the device is "in lock" there are periods when the data is useless or compromised. Once more, the major source of this problem is outside disturbance: even if the mirrors can maintain lock, there may be so much going on in the feedback circuits that the effect of any gravitational-wave would be swamped. To sense the caress of a gravitational-wave the interferometer must be asleep, not tossing and turning. An interferometer, like all gravitational-wave detectors, is like the princess in "The Princess and the Pea," while a gravity wave is the pea. The pea will only disturb the princess's sleep in a meaningful way if that sleep would otherwise be deep and peaceful.

Unwanted lumps in the princess's mattress, that is, unwanted disturbances to the interferometer, are recognized by the network of "environmental monitors" checking seismic, electrical, acoustic, and every other imaginable kind of intrusion, including those caused by internal problems such as jitters in the laser, stray light affecting the feedback circuit sensors, or any departure from stability in the vacuum system. When any of these problems arise, a "data quality flag" is posted; stretches of data stigmatized by flags will, in most cases, be "vetoed." Once more, in an ideal world, the entire process would be automated: the monitors would note trouble, write the flags, and those stretches of data would be thrown away. But, again, human judgment cannot be dispensed with. How serious is a veto? Too low a bar for the posting of a flag and so much of the data would be

8. Virgo has a much better duty cycle than the LIGO detectors. This may have something to do with its clever and complex mirror suspensions which might make it less subject to the effects of external disturbances which can cause an interferometer to go out of lock.

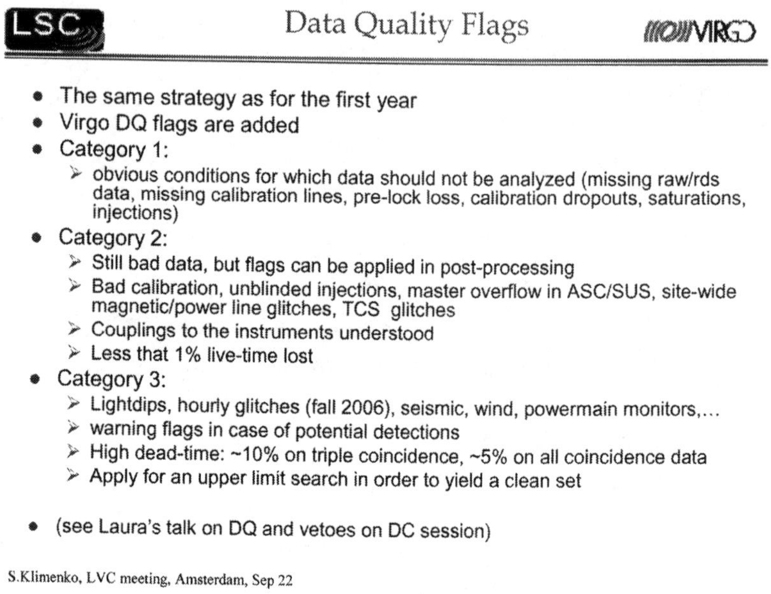

Figure 1. Three levels of veto

thrown out that the duty cycle would be reduced and reduced and that first discovery might be lost among the discards. Too high a bar and the results will be unreliable. The fact that judgment is central is institutionalized in the different way vetoes are handled for upper limits and discovery candidates. For upper limit claims, vetoes are less strict than for discovery candidates. These are the conservative courses of action. Furthermore, three categories of veto of different degrees of seriousness are set up. Figure 1 is a PowerPoint slide from a meeting which summarizes the characteristics of the three types of veto.

Of course, judgment was used to decide on the categorization—the machine could not make categories for itself. With those judgments made, it is expected that data affected by the least serious "category 3" vetoes, though it will not figure in the first analysis, will be examined later if other eventualities indicate that an interesting signal might be in danger of being ignored because of what is meant to be a "light touch" warning. This element of judgment will figure in the story to follow.

Taking all this into account, Initial LIGO had to run for about two years to gather one consolidated year's worth of data. There had been four prototype "science runs" at less than design sensitivity, so this final two-for-one-year run was known as "Science Run 5" or "S5." Enhanced LIGO's science run will be known as "S6," though there are intimations at the time of writing that there will be an S6a at well below design sensitivity and an S6b that will be nearer to what had been hoped for. Thus, when we say that if S6 reaches its design sensitivity, one month of running will be worth eight months of S5, where the "months" are integrated periods of operation in science mode. The abbreviations, H1, L1, H2, S5 and S6, and their meanings, should be remembered. They will be used repeatedly.

The range of an interferometer, at least for H1, L1, and H2, the three American interferometers, is expressed in a standardized form. A typical source of gravitational-waves would be the final moments in the life of a pair of binary neutron stars. The orbits of binary stars slowly decay, and as they get closer and closer to each other they circle faster and faster (like skaters drawing in their arms). In the last few seconds before the stars merge they will circle each other hundreds, then thousands, of times a second in a rapidly increasing crescendo which, if you could hear it, would sound like a "chirp." The standardized range is the distance at which such a system, with both components being 1.4 times the mass of our sun, and orientated in neither a specially advantageous nor a specially disadvantageous way, ought to be seen by the detector. Since whether it would be seen or not depends on the momentary state of lock, noise, and data quality, the range of any detector is continually fluctuating. From the middle to the end of S5, L1 and H1 had ranges of around 14 to 15 megaparsecs (a

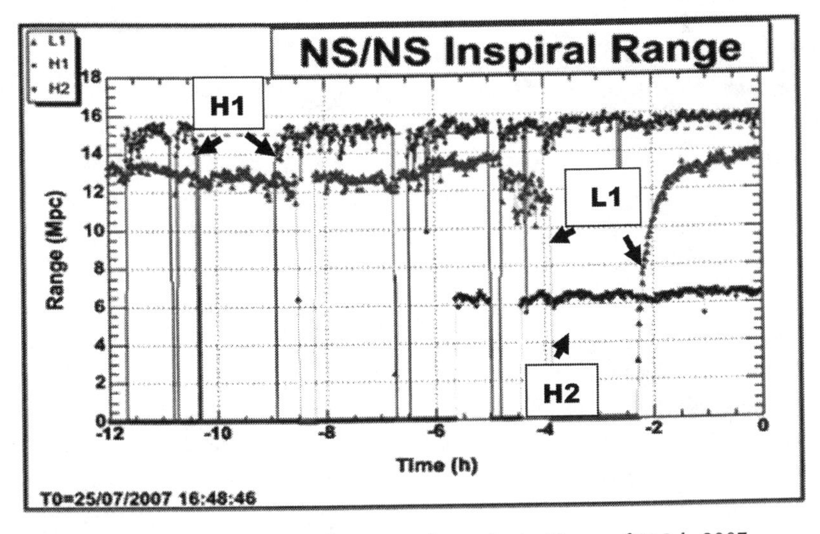

Figure 2. Ranges of the LIGO interferometers for twelve bad hours of 25 July 2007

megaparsec is a little over three million light-years) with H2's range being about half of this—as would be expected given that it is half the length of the other two. This state of affairs is represented in figure 2, which is taken from the electronic logs maintained by the laboratories.

On most days the three plots would run nearly all of the way across the graph, and the two upper ones would be almost superimposed at the 14–15 Mpc range. It is easier to disentangle the two plots on this bad day because one of the detectors has a range of only around 13 Mpc. H2, as can be seen, is not having a great day either, being down for more than half the time and having a range of only just over 6 Mpc for the rest of the day, whereas something around 7 would be more typical.

Measured by the standard of LIGO on a good day, eLIGO was, therefore, intended to have a range of about 30 megaparsecs. At the time of writing I hear scientists expressing the hope that it will reach 20 before it has to be switched off for the installation of AdLIGO components.

Data Analysis

What this introduction to gravitational-wave detection science should have made clear is that taking data from a gravitational-wave detector is not like reading the electricity meter. It should be obvious that whether some slightly unusual activity in the interferometers has the potential to count as the result of the touch of a gravitational-wave depends upon whether it is coincidental with activity on at least one other detector. If there was a third, fourth, or fifth detector with enough sensitivity for the event to be within range, then either the event must be seen in them too or there must be a good explanation for why it was missed. The explanation will depend on decisions about the scientific state of the detectors involved—whether they were in a sufficiently peaceful state to be disturbed by the pea, if such it was. As we will see, these considerations are only the start. There are subtleties upon subtleties involved in deciding whether something is sound data. These considerations reach out all the way to what the scientists were doing and thinking as they analyzed the material. As will be explained in chapter 5, this is because statistical analysis, though it looks like the purest of mathematics when on the published page, turns on unpublished details of the analysts' thoughts, activities, and statements from decades previously, and reaches forward to what they anticipate they will be doing decades hence.

A detection, then, however it is dressed up, is not a self-contained "reading." It is not nature, reflected clean and pure on the still surface of science's pool; there is no pool, only the fast flowing river of human activity on a restless earth. The first detection will be a social and historical eddy—at best a momentary stillness in the white-water of history.

The story of this book turns on a planned rehearsal of stillness-making. It concerns a deliberate test of the LIGO team's abilities, staged toward the end of S5, and known as the "blind injection challenge." Two scientists were charged with injecting zero, one, two, or three false signals into the LIGO interferometers with unknown form and magnitude according to a sequence of random numbers: the job of the team was to find them if they were there, always being aware that there might be none at all, so that any putative signal might be real.[9] The story of that adventure is the spine of the book.

9. Allan Franklin (private communication, October 2009, referring to his 2004) points out that blind injections are by no means unprecedented: "Consider the episode in which it was claimed that a 17-keV neutrino existed. One of the reasons why the Argonne experiment (Mortara, Ahmad et al. 1993) was so persuasive was that they were able to demonstrate that they would have detected such a heavy neutrino had it been present as a result of their proven success in detecting similar blindly injected events." One of the purposes of the gravitational-wave blind injections could be said to "calibrate" the whole detection procedure.

2 The Equinox Event: Early Days

For forty years and more, gravitational-wave detection physics has been full of excitement. There were the early claims of Weber, there were the disputes surrounding "the Italians," and there was the near-miracle of the big interferometers being built and made to work. Strangely—and I found some of the scientists themselves were of the same opinion—it has been a bit dull since the 2002 Rome Group's claims were crushed and the interferometer data started coming in. Now that the machines are doing what they are built to do, things have become routine. What has been going on is a huge and widely distributed effort of data analysis, none of it showing any sign of a gravitational-wave. The data hasn't gone to waste, as it has been used to set a sequence of upper limits: a whole string of papers has been published which show that the maximum flux of gravitational radiation of this sort or that sort is, as expected, less than "X." As we will see, upper limits keep the science going, and occasionally they have real astrophysical significance, but to an outsider like me, and to a proportion of the insiders, they are pretty boring. *"As scientists predicted, LIGO fails to see anything,"* does not make much of newspaper headline.

Suddenly, in the autumn of 2007, at the time of the September equinox, for outsiders like me, and for a lot of the insiders, life in gravitational-waves became exciting again. The "Equinox Event" is a coincident pulse of energy the LIGO interferometers detected on 21/22 September 2007. It rapidly became clear that it might constitute the first discovery of a gravitational-wave.

The Data Analysis Groups

At this point we need to turn aside from the main thrust of this chapter and explain a little more about how the data analysis is organized. This is necessary if the tensions and responsibilities surrounding the Equinox Event are to be understood.

Interferometric gravitational-wave detection has the potential to spot four kinds of signal. Aside from their absolute sensitivity, the biggest advantage of the interferometers over the bar detectors is that they are "broad band.'" The bars could do little more than sense when a gravitational-wave "kick" impacted on them; they would sense this as a net increase in the energy contained within the vibrations of the metal. But an interferometer should be sensitive to the shape of the signal as well as the aggregate energy it delivers. The movement of the mirrors should follow the actual pattern of the wave as it causes its distortions in space-time. Thus, for example, if the signal is caused by the final seconds of the inspiraling of a binary star system, then the very details of the "chirp" caused by the rapidly increasing rotation of the two stars around each other will be inscribed on space-time and in turn expressed in the rapidly changing separation of the mirrors (or, to be technically more exact, the rapidly changing feedback signal needed to hold the mirrors still in the face of the forces they experience). Such a signal should, then, be easy to identify because the same clear waveform should be expressed, more or less simultaneously, on the two or more widely separated detectors.

As always, real life is much more complicated because the new wave will be superimposed on the ever-present noise in the detectors and will need to be extracted by sophisticated algorithms instantiated in computer programs. The most important part of the technique is "template matching." A huge bank of many thousands of signal templates, corresponding to different inspiral scenarios, has been constructed. Thus one template will refer to the expected pattern of two 1.4 solar mass neutron stars spiraling into each other; another template will refer to a 1.4 solar mass neutron star and a 10 solar mass black hole spiraling into each other; another will be for two 20 solar mass black holes, and so on for as many combinations as one has computer power to manage. Since any specific signal might well fall between two templates and be mixed up with the noise of the detector, one can see that the technique is not as neat and tidy as it appears when first described. The duty of analyzing such signals has been given to a dedicated group known as the "Compact Binary Coalescence" or "CBC" Group, which is more widely known by its original title: the "Inspiral Group." We'll call it by this name from here on.

Any asymmetric spinning star will also emit gravitational-waves. Therefore another kind of signal that has the potential to be seen is that produced by pulsars—rapidly spinning stars that emit a beam of energy (which can be seen in its regular sweeps across the earth) and that, therefore, can be assumed to be asymmetrical to some extent. The frequency of the gravitational-wave signal produced by a pulsar will be the double the frequency of the flashes of light caused by the sweep of its beam. Therefore, if there is enough asymmetry to produce a large enough flux of gravitational-waves, it should appear as a regular pulse in the interferometer. The "Continuous Wave" group has the job of looking for this kind of signal.

A third kind of signal is that left over from the initial formation of the Universe. It is the gravitational-wave equivalent of the famous, electromagnetic, cosmic background radiation. Astrophysicists and cosmologists are particularly excited about this kind of signal because while the electromagnetic background goes back only so far toward the Big Bang, the gravitational-wave background goes back to almost the very beginning. The form of this signal is random rather than following the pattern of a definable chirp or the regular heartbeat of continuous waves. It is known therefore as the "stochastic background." It can be detected by long-term correlations in what looks like "noise" in the detectors but is really the signature of the random background waves.

The fourth kind of signal is a burst of unknown origin or form. It might be emitted by a supernova, by an inspiraling system with an awkward signature, or by something not yet understood. Searching for these signals is the job of the "Burst Group."

First Intimations of the Equinox Event

It was the Burst Group which found the Equinox Event. Over the years I have slowly come to gain more and more access to the goings on the gravitational-wave detection community, and nowadays I am allowed to listen in to the weekly telephone conferences, or "telecons," of the groups. Mostly these are highly technical and I listen to only a few of them. The first I heard of the Equinox Event was when a member of the Burst Group e-mailed me on 2 October suggesting that I pay special attention to the telecon coming up on the 3rd. He said:

> This should be an interesting telecon for you to sit in on, because there was a notable event candidate found by one of our online searches in the data from Sept. 21, and brought to the attention of the working group during last week's call. We will be talking more about it and planning to start working through our "detection checklist."

This was exciting.

Nothing is simple, and the first thing to understand is what is meant by "one of our online searches." The gravitational-wave analysts are especially sensitive to the possibility that they might be accused of biasing their data by post hoc choice of search parameters, or "tuning." Such sensitivity is good practice in all sciences, many of which try to avoid it by "blind" analysis, but, as we have seen, the history of the gravitational-wave field has made it an issue with a very high profile.

Unfortunately, however, data analysis is just like experiment in that a fair amount of trial and error is involved. The idealized model would hold that the exact recipe for data analysis could be worked out in advance before anything is actually analyzed, but the truth of the matter is that data analysis is long and subtle and things always go wrong. The problem is nicely illustrated by the subject line of an e-mail that was circulating in April 2009: "*Re: [CBC] 12TO18 Re-re-re-re-re-re-re-re-rerunning our upperlimits because of this V4 calibration issue.*" Furthermore, much can be optimized only after a series of trial runs. The solution that the gravity wave teams adopt to resolve this problem is to tune the procedure on data that is not going to be used for the main analysis. One approach is to split the data into two sections, a small one—around 10 percent of the whole selected at random—and the main section. The small section is known as the "playground," and anyone is allowed to do anything they like with it, including any amount of post hoc readjustment of parameters, until they work out the best algorithm they can for extracting potential signals from noise. Another approach is to do the tuning on data where the output from one detector has been subject to a time slide, so that anything to which the procedure is tuned cannot be a real signal. Only when all this work has been completed in one way or another, and the data analysis protocol is "frozen," is the "box opened" on the main body of data or on the data which has not been subject to a time slide. The main body is then analyzed with the frozen protocol, which must not be changed in any way. We might call this procedure a "metarule" of data analysis. The metarule is that nothing can be touched or adjusted after "playtime" is over; once the box is opened the analysis rules stay as they are.

Opening the Box

Jumping ahead for the sake of exposition, I was present when the box was opened by the Burst Group on the first calendar year of their data. This seems a good place to describe what it is like to "open a box," though this description is an indulgence, playing no part on the main theme of this chapter or of the book.

At 8 a.m. on 16 March 2008 about fifty people sat scattered in a dimly lit lecture theater on the campus of the California Institute of Technology. Everyone, as is nowadays customary at such gatherings, was alternately listening to the speaker and staring at, and perhaps typing into, the glowing notebook computers on their laps. Each notebook was linked by WiFi to the Internet. Peering over scientists' shoulders, one might see e-mail being dealt with, papers scanned or composed, or lines of computer code written or debugged. The discussion centered on whether the protocol had been finalized and whether there was anything more to do before the moment came from which they could not turn back. Whatever was not done before the box was opened could not be rectified afterwards. There ought to be great drama. Would the refined algorithms set loose on the first significant body of data from "the age of interferometry" in gravitational-wave detection physics see anything?

The moment when it came was strangely downbeat. Everyone agreed that nothing had been forgotten, and the scientists sitting in the rows of seats were given permission to open the box. The box was opened by some of them pressing a few buttons on a portable computer—looking around you could not see who was doing the work and who was not. The message was carried at the speed of light back to the banks of computers located at the main sites. A few moments later the entire database was analyzed, and the result flashed back via the satellites servicing the Internet. The scattered scientists with the portable computers reported that nothing of significance was found. An analysis that required hundreds of millions of dollars, subtle political skill, and virtuous and sometimes brutal management to enable, decades to plan, years of data collection, and months and months of agonizing over whether all the preliminary tasks have been properly done, was completed almost as soon as it was started and there was nothing to show. This is physics.

The "Airplane Event"

How seriously the distinction between playground and main data is taken is illustrated by a bizarre data analysis incident now referred to by everyone in the collaboration as "the airplane event." One only has to say "airplane event scenario" to a member of the collaboration and they immediately know what is at stake.

In mid-2004 the Burst Group were preparing an upper limits paper. Upper limits are a kind of scientific finding that was described in *Gravity's Shadow* as turning lead into gold. When gravitational-wave scientists fail to see any gravitational-waves, they turn this into a finding by saying it proves that the upper limit on the possible flux of the waves is "such and such." Since LIGO first went on air, a stream of these upper limits papers have been published. One or two of these have been of some astrophysical interest. For example, there have been papers putting upper limits of the gravitational-wave component of the energy lost in the slowing down of the rate of rotation of the Crab pulsar. Assuming all the assumptions on which gravitational-wave science is based are correct, these set bounds on the degree of asymmetry of this particular neutron star. Again, when a powerful gamma ray burst was seen in the direction of the Andromeda Galaxy, a paper was promulgated on ArXiv, the electronic preprint server, in November 2007, with the following included in the abstract:

> We analyzed the available LIGO data coincident with GRB 070201, a short duration hard spectrum gamma-ray burst whose electromagnetically determined sky position is coincident with the spiral arms of the Andromeda galaxy (M31). . . . No plausible gravitational-wave candidates were found within a 180 s long window around the time of GRB 070201.

This was taken to show that if the event had been caused by an inspiraling neutron star/neutron star system, or an inspiraling neutron star/black hole system below a certain size, it could not have been located in the Andromeda Galaxy but must have been well behind it even though it was in the same line of sight.

Such results are marginally interesting, and by all accounts, were well received at the astronomy conferences where they were presented, but most of the upper limit papers seem rather boring now that the excitement of being able to say anything at all about gravitational-wave fluxes had worn off. Most of these papers say something like "you expected there to be less than flux X of gravitational-waves of this kind: we've proved there's less than 100X." The "100X" gets steadily lower and lower as the instruments become more sensitive, but, mostly, they are still a long way from saying anything that is going to cause astrophysicists to start scratching their heads.

In the world of gravitational-wave detection, the assiduously honest analyst will be leaning over backward to make sure that nothing that is not a gravitational-wave is announced as being a gravitational-wave. In the mirror-image world of upper limits, the same analyst will lean over backward to make sure that nothing that could be a gravitational-wave is said *not to be* a gravitational-wave. This is because excluding too much gives a too low (and, therefore, a misleadingly interesting!) upper limit. It is just as important in setting upper limits that the metarule is followed and that no potential gravitational-waves are eliminated by jigging around with the analysis procedure once the box is opened. It is a violation of the metarule in the case of the upper limit paper of 2004 that gave rise to the bizarre incident.

The violation was the following: after the box had been opened, a big event was found, but then someone noticed that the loudest event was actually correlated with an airplane whose passage over the site had been recorded on a microphone. Unfortunately, the possibility of low-flying airplanes giving rise to false signals in the data stream had not been envisaged during the playtime period, so the protocol did not allow for such things to be removed. It was clear that this event could not be a candidate to be a gravitational-wave, but to remove it post hoc, "by hand," as it were, would violate the metarule for setting upper limits: it would *improve* the upper limit as a result of a post hoc procedure. But to leave it in would mean that the upper limit paper contained an artifact that made it incorrect. The debate was about whether the metarule should be violated or a false result published. This debate had been going on for about four months when, in November 2004, I witnessed one of its final paroxysms. The following are anonymized extracts from a heated discussion among about sixty people that lasted approximately half an hour and which, because it could not be resolved, ended in a vote (!) on what should be done.

S1: We've got a positive not-gravitational-wave event and I don't understand why it's still in the analysis.

S2: It is possible to remove it but it is a moral and philosophical question.

S3: In August when we discussed it, the data analysis group had a show of hands. The risk in this case is that you make the mistake that the playground is designed to keep you from making—which is that, retrospectively, you take the airplane to have caused the event but it could still be a gravity wave coincident with the airplane. Up to a week ago I was not sure but in fact I believe now that the airplane really did cause the event. [Shows plots of trace from microphone and trace from detector.] You can see that the loudest time in the microphone does not correspond to the event but clearly this stretch of data here corresponds to this stretch of data here and so its pretty clear that this stuff is being caused by the airplane, and if you look at this loud noise here—if you look at the frequency content of that—you find that it is around 85 Hertz—it is around the frequency of the airplane sound at that time. . . . I'm willing to state that I'm positive that the event was actually caused by the airplane and this event can be tied to that physical effect and that I would personally be more comfortable now than I was after the initial study about throwing that out.

S4: I am in favor of leaving the event in the upper limit calculation, and stating, of course, clearly in the paper that this is not a real event—that it has an environmental cause—but I am against changing the upper limits that we're quoting and it's because I don't know quite enough about statistics and bias to convince myself that this is a safe procedure—to go and change the algorithm by which we compute the upper limit after . . .

S5: It was always anticipated that one would check environmental signals!

S4: Yes, for vetoing a detection claim, but we did not discuss and we haven't really thought deeply about using that to change an upper limit.

S6: We did discuss it long and ago—we decided we weren't going to do that.

S4: . . . the upper limit that we quote is only meaningful when you have statistical confidence limit associated with it, and if we introduce a bias that leaves that confidence in doubt then your upper limit is meaningless. . . . We have to be conservative and take the hit on the upper limit rather than risk quoting an upper limit that's actually incorrect.

S7: If we have confidence beyond a reasonable doubt that we have an airplane, then I think we should interpret our results in that context. If someone had thought about it beforehand and just not mentioned it and then steps forward we would have no hesitation about applying it now. We would not worry about any statistical bias that would result; otherwise it is too much mathematics and not enough physical reality.

S4: The flaw is when the person steps forward and said "I knew about this before." Because that changes the algorithm. If there hadn't been an airplane that person wouldn't have stepped forward, the change wouldn't have been implemented. I really feel very uncomfortable about changing these algorithms after we've looked at the data. I think the conservative thing is to just not touch it.

S5: I think there's a difference between a posteriori adjustments like this when you're making a claim for a detection and when you're making a claim for an upper limit. I think in this case, we're not making a detection claim—we're simply saying that there's something that we believe should be excluded from the gravitational-wave search because it's contaminated in a very clear way by environmental influences, and it shouldn't be part of the upper limit. . . . I think we would look pretty ridiculous if we left it in. . . . Otherwise it's not the best we can do on the basis of the data. We're being paid to do the best job we can. I mean otherwise why spend all this money on these detectors if we're not actually going to do the best job we can with the data we have. . . .

S9: The truth is that the most conservative thing and the thing you know won't violate your statistical bounds is to leave the event in. You know that. But also the truth is that if, as a collaboration, we firmly believe that it originates with the airplane, then the bias it introduces by applying that veto over the entire analysis is minimal. The trouble is that if you did find a loud event like this and everyone didn't go back to the [plots of the traces] but you just throw it out, then you place too low an upper limit. But from what I can tell from listening to people who know the event and have studied it carefully, we should take it out because it introduces a small bias but is not going to put us into that scary regime of making too tight an upper limit.

S6: I'm convinced that this is an airplane, and if we had decided, six months ago, that we would hand-scan the events I would have no problem. The problem is that six months ago we agreed to not do that. If we remove it we're telling the gravitational-wave community that we allow ourselves the right to change our analysis after the event.

S10: We may learn from our mistakes unless we agreed earlier not to, and in this case we agreed earlier not to learn from our mistakes!

S6: We agreed to learn from our mistakes, but not to attempt to correct them because we can correct them for the next round of data analysis without doing anything retrospective.

S3: Our draft paper currently says "We observed one event with an upper limit of 0.43 per day. By the way, this event was an airplane." It's not our best job of saying what is the limit on the gravitational-wave rate.

S11: [If both limits were in] that would give the reader an opportunity to decide for himself.

Collins *(sotto voce)*: That's postmodernism.

S12: We're sure this is an airplane and we're sure the result we've produced is wrong.

S4: No—we're only sure the result we've produced is right if we leave the airplane there.

S6: It's a conservative upper limit.
S7: Let's not misuse the word "conservative" . . . It's mad!

Not long after this the meeting concluded that it would be impossible to reach a consensus and that they must vote on it. (There were many humorous remarks referring to the recent American Presidential election.) The vote went eighteen for leaving the airplane in and a much larger number (around thirty) for taking it out. The leave-it-in group was asked to concede, which they did, but not without at least some people feeling frustrated and certain that a poor decision had been made and not without one scientist insisting his name be removed from the published paper.

While everyone felt uncomfortable about taking a vote, to the extent of people looking round at me and laughingly acknowledging that "I had got what I came for," it was an embarrassingly long time later before I realized what it all meant. The procedures of science are meant to be universally compelling to all. To find that one needs to vote on an issue institutionalizes the idea that there can be legitimate disagreement between rival parties. In other words, taking a vote shows that there can be a *sociology* of science that is not rendered otiose by the universally compelling logic of science. Though we did not realize it at the time, a vote at a scientific meeting is a vote for sociology of science and we all felt it instinctively. It is not that there aren't disputes in science all the time, but putting their resolution to a vote legitimates the idea that they are irresolvable by "scientific" means; neither the force of induction from evidence or deduction from principles can bring everyone to agree.

There was one fascinating, *Alice in Wonderland*, moment during the debate when someone pointed out that, if the airplane was removed, the upper limit might be low enough to be said to conflict with "the Italian" 2002 claim—it would drive the upper limit down to the point where it ruled out the flux that the Rome Group had claimed to see. In response, a number of others opined that in that case it really would be *illegitimate* to remove the airplane because its removal could be said to be an interested after-the-fact maneuver. In that case the analysts would feel themselves very vulnerable to accusations of statistical bad faith. In other words, if the removing of this spurious piece of data had any real astrophysical significance, then it could not be removed because then it really would look like post hoc data massage; if it had no significant astrophysical significance then it could be removed!

Another fascinating feature of the debate was that two of the speakers who were most adamant in wanting the airplane removed, exposing the group to a potential statistical massage accusation, had each published papers complaining about the Rome Group's post hoc misuse of statistics. Ironically, other scientists in the room *took comfort* from the fact that even these persons, whose track record of adamant rejection of post hoc analysis had proved their statistical propriety, thought that the airplane should be removed. Strangely, no one seemed to notice the irony.

It seems to me that the right way to proceed in this case was to remove the airplane event, since any other course of action was indeed "mad."[1] The problem is that rules, as the philosopher Ludwig Wittgenstein pointed out (and this includes metarules), do not contain the rules of their own application.[2] No rule can be applied without doubts about how to apply it in unanticipated circumstances. The rules that gravitational-wave scientists impose upon themselves in their determination to avoid accusations of bias will always contain ambiguities. Scientists like to believe there is a set of statistical procedures that, once programmed into their community, will ensure the validity of their procedures and remove the need for human-like decisions. But actually, statistics are just human decisions in mathematical clothing. Every now and again, this causes the kind of trouble that the airplane event exemplifies.[3]

1. Franklin (private communication, October 2009) points out that this situation is not unprecedented. He quotes a high-energy physicist discussing the setting of upper limits as follows: "One ultimately *should* look at events in the signal regions—after all cuts have been fixed—to check whether they are due to some trivial background or instrumental problem such as the high voltage having been tripped off. If such events can be attributed to such sources, then it makes more sense to cut them and set a biased but meaningful limit rather than leave them and set an unbiased but not useful limit. . . ." (The quotation is from a report by A. J. Schwartz, entitled "Why Do a Blind Analysis?" and circulated at Princeton University in 1995.)

2. Wittgenstein 1953.

3. I like to collect examples of moments when even the most long-established rules do not cover new circumstances. One example happened during the World 20/20 cricket tournament held in the UK in June 2009—I saw it on television. In cricket, six "runs" are scored if the batsman hits a ball in the air over or onto the boundary rope or if it is caught by a fielder who touches the rope at the same time (like a home-run in baseball, but the boundary is marked by a rope on the grass and touching not clearing is the criterion). A batsman hit the ball for a potential "6" but a fielder standing just inside the rope leapt up and diverted its flight sharply upward though it was still going to land outside the rope. The fielder then ran outside the rope, leapt off the ground again to push the ball back inside the rope before it fell, and then ran in, picked up the ball, and threw it back, limiting the runs to (I believe) three. At no time was the fielder in contact with the ground or rope and the ball at the same time. I was sure it was a "6" but the umpires counted the ball as not having crossed the rope. If the umpires were right, it seems to establish a new rule allowing fielders to stand outside the field of play—defined by the rope—and, provided they leap off the ground, push potential six-hits, back into the field before they land.

The Equinox Event Breaks the Statistical Rules

The Equinox Event was discovered, as the note that alerted me indicates, by "one of our online searches." An "online search" is, again, something that breaks the rules of the strict division between playtime and data analysis proper, and, again, it is a case where it would be "mad" not to break the rule. For some reason, though, I understand, not without some argument, this institutionalized, and continuing instance, of metarule–breaking, which in scientific terms is far more dangerous than the airplane event, has been accepted—that's sociology.

An online search is a real-time study of the data which alerts the collaboration to any coincident event that stands out sufficiently under a rough data analysis to be worth further investigation. As can be seen, it comprises an analysis of main-body data from well before statistical procedures have been frozen. It subverts the effective "blinding," which the split between playtime and serious analysis time is meant to accomplish: once an online search has alerted the collaboration to the possibility of an event, only vigilance can prevent post hoc-ery. But the very idea of blinding, and the very idea of freezing the protocol after playtime, announces the fact that vigilance is not considered strong enough to counter any insidious desire to find a result.

The reason it would be "mad" not to have an online search is that gravitational-wave events may be correlated with electromagnetic events such as gamma ray bursts, x-rays, or the bursts of visible light associated with a supernova. When a putative gravitational-wave event happens that is "loud" enough to make itself felt even before the immensely complex statistical analysis that follows the "opening of the box," the gravitational-wave scientists want to be in a position to request astronomers to point their instruments in the direction from which it might have come and look for a correlated signal. This has to be done as close to the time of occurrence as possible. In any case, if the event is that outstanding, it might be the first discovery, and analysts would be "mad" if they did not want to start examining it closely as soon as possible.

Many of the dilemmas discussed in this last passage, including that of the online search, are brought out in this interview with a senior member of the collaboration:

> Collins: I hear in this analysis [of the Equinox Event] that you've been looking at this event for a while and trying different cuts on it.

Respondent: Yeah—and that's sort of the impurity of it all. . . . But what we're not doing is changing any of the thresholds, we're not changing any of the operations, any of the lines of code, we're asking questions that we consider "post-processing." However, some people will tell you that your whole procedure from beginning to end must be set up in advance, and the whole concept of post-processing, according to these folks, destroys the statistical purity.

Collins: So you'll get in the neck from some people.

Respondent: Yes—like for example Z. And he's not a fool. And his point is—I think I can represent it properly—is that in the end—of course you have to make all these wise choices about how you analyze the data—but in the end the only thing that distinguishes an event from noise is its degree of statistical improbability of having occurred from noise alone. And . . . that's a huge part of the truth, or it's the whole truth. He thinks it's the whole truth, I think it's a huge part of the truth but not the whole truth. But that's why the thing that everyone wants to show you is: "Here's the histogram of all of random events and see how far to the right this is." . . . So anything you do after you've seen it that either makes it seem more or less probable—that you can't justify as being a mechanical application of what I ought to have done before—at that point you do in fact lose the ability to say how unusual the event is.

Collins: So you said that's a large part of the truth, but what's the other part of the truth?

Respondent: There are huge ideological disputes within the LIGO Scientific Collaboration. . . . The other part is when we say we are doing follow-ups. I think Z is completely allergic to the notion that you should have something that you call a follow-up. You should define your pipeline and then you should open the [box] and write the paper, whereas—and I think this is something where S has been a great partisan—is that you do all this stuff but that's just the beginning—then you've got to see if it makes sense. And you've got to see if there's any signature in the data and look. . . .

There's a classic example in the cosmic background radiation. It's not a tragedy, like the Weber story was a tragedy, but it's sort of a well-known mistake in deciding and announcing to the world that you were . . . Someone wanted to measure the spectrum of the cosmic background radiation at the short wavelength, high frequency end of the spectrum. . . . Paul Richards, the guy who was doing it, was afraid he was going to force it to be a really beautiful black body in his data analysis. And he said, "No I'm not, I'm going to design the instrument, measure it, check everything,

> set up my data analysis pipeline and publish whatever spectrum I get." And he did—he stuck to his guns—but he also realized that the spectrum was wrong, after the fact.—It's like the airplane event!—And people make blunders, and so S's point of view is that we're inventing a new kind of measurement, we're not going to be wise in advance, but we have expertise, and we shouldn't force ourselves to abandon the ability to use our expertise. And there surely is a lot that's right with it, but once you start doing that you lose the ability to quantify how unusual the event was because you can never reproduce the chain of decisions you would have made in other cases. And so we're kind of in the soup that there is something about Z's statistical purity that is correct and yet it's inapplicable most of the time. . . .
>
> These things are delicate, and we know the classic case to either tune to make something go away or tune to make it seem bigger—different people want different things in the secret recesses of their mind.

The whole issue is expressed from a different point of view by an analyst who had been part of a team accused of overinterpreting their data. It seems in the discussion set out below that he was expressing regret at what had happened. At the same time, the conversation brings out both the unconscious desire of the analyst who, as he puts it, "falls in love" with a finding, which implies that nothing should be touched after analysis procedures have been frozen, while at the same time it brings out the need to apply experience retrospectively since not everything can be anticipated in advance. My intervention is that of a "devil's advocate."

> Q: When you have to say if something that has been detected is a real signal, then you must interface with people who remain cold. They must have experience, experimental experience also, because sometimes there are effects that you cannot imagine and only years and years of experience will teach you how many bizarre things may happen to the detector, things which are unpredictable and things which people sitting every day in front of a computer cannot even imagine. A data analyst who has a possible candidate—who studies a possible candidate—can fall in love with this data. Because the data matches some prejudice or something that has been studied already . . . perhaps the first black hole ringdown. Then you feel yourself at the center of history, at the center of the universe at that moment. And in a way there is something—you can feel that it cannot be wrong. It cannot be by chance that at this moment you have this data in front of you. So you must face—you must [try?] your

work, and your suggestions—and make a report of the result of your work to someone who has not been through this process and who hears all the details for the first time.

Collins: But surely if the data analysis group is doing its job properly and doing the statistics properly only one answer can come out?

Q: Data analysis work is not so simple. It is not pushing, first, the red button and then the yellow one then the black and something comes out. There is some role of the human being in the process.

The First Equinox Telecon

My notes from the telecon, when I first heard the Equinox Event (EE) discussed, intimate that I was a little disappointed. The whole thing was very low key. Routine business was dealt with and the first mention of the EE came more than twenty minutes into the discussion. Only after more than an hour did someone say, "[It] looks like a very good candidate."

I suspect that one reason that the telecon was a low-key affair was that the scientists wanted to be professional—they knew that fair data analysis requires vigilance and that too much excitement might compromise it. So things were done in an orderly fashion, allowing the data to "speak for itself," if that was what it was going to do.

A second reason was that everyone knew that even if this did turn out to be a reportable event, there was a good chance that it had been deliberately inserted into the data stream as part of the blind injection challenge; there was no point in getting too excited about something that might turn out to be an injected artifact.

The third reason for my disappointment, as a listener-in on an early discussion of what might just be the first discovery of a gravitational-wave, was that the banal nature of the event was being revealed. It simply was not an "event" in the sense that the word is normally used. It was not a sudden explosion, or the appearance of a comet, or a weird freak of the weather, or a moon landing, or a new speed record. It was a slightly unusual concatenation of numbers in a torrent of numbers. Without a computer continually monitoring the stream of numbers, there would be nothing there. It was all statistics. My notes taken at the time express it thus:

> I think the weirdness is that there is nothing to "see," just a load of carefully worked out statistical procedures (I guess this is how it is in high energy [physics] too). There is something unsatisfying about the "construction" of a signal from a lot of numbers and procedures. It is something to do with

> being faced right up to the fragility of the inference—it is not just a philosophical nicety but a visceral thing. You feel the fragility—it is just numbers.

Still, they were interesting numbers, and as weekly telecon followed weekly telecon their meaning became steadily more defined.

Two types of reality-transformation were applied to the numbers. First, there was an ongoing statistical examination. How did the numbers look under the different analysis pipelines that the Burst Group operated? And had the Inspiral Group seen anything at the same time? Second, there was the "Detection Checklist" to be worked through.

Sigma Values and the Language of False Alarm Rates

As we have seen, in exploratory physics signals are weak and can only be recognized statistically. As explained, a "discovery" consists of some combination of events, which makes its presence felt as numbers manipulated by a computer program, that is *so unlikely* to have arisen as a result of chance that they have to be counted as a signal—a systematic intervention by nature. The most widely used convention for expressing the unlikelihood of something being due to chance is in terms of a number of "sigmas" or "standard deviations." The theory is that there is smooth distribution of noise in the output of the apparatus so that, if plotted on a graph with size of noise along the bottom axis and number of noises of that size on the vertical axis, the result would be bell-shaped. That is to say, there would be a lot of small noises which would give a high peak around the central zero point of the bottom axis, slightly fewer a little way to the positive and negative sides of the zero point, and fewer and fewer noises as one moves away in a negative and positive direction. At the "tails" of the distribution—the bottom of the bell—there are only a very few big noises well away from the center point. If the noise is "well behaved"—which is to say it follows the right mathematical function—its distance from the center point can be expressed in terms of standard deviations. If the curve has a "Gaussian" profile, then roughly 68 percent of the noises are found within one standard deviation of the zero point, 95 percent within 2 sigma, 99.7 percent within 3 sigma, 99.99 percent within 4 sigma, 99.9999 percent within 5 sigma and so on. One can say, then, that if some signal lies beyond a distance of 2 sigma from the center there are only 5 chances in 100 that it could have been due to noise; if it lies outside 3 sigma, there are only 3 chances in a 1,000 that it could have been due to noise, and so forth; a

5 sigma result would be equivalent to one chance in 1,000,000 that the event would have been caused by smoothly distributed noise. Different sciences have different standards as to what counts as a sufficiently large sigma value to make something count is being so unlikely to be due to noise that it should count as a discovery. Exactly what this standard should be plays a large part in this story.

In gravitational-wave detection there is also another way of expressing the unlikelihood that a signal could really be noise. It is also sometimes said that "this effect could arise by chance only once in so many years." Why is this, and how do the two methods of expression relate to each other?

In the case of the LIGO interferometers, the noise plot has too many large noises to fit the smooth mathematical model. These "glitches" (see below) give the distribution a pathologically "long tail"; there are too many big noises out at the extremes of the distribution. Therefore the calculation that gives rise to a statement in terms of sigma cannot be done because the model does not fit.

What is done instead is that the "false alarm rate" is directly measured over the coincidence data produced from the time slides. Because the set of data is limited to what can be generated by using the time-slide technique on the output of the detectors, rather than being the infinite set implied by the existence of a smooth mathematical relationship, there is less certainty about how right it is; it becomes hard to know how much confidence one should have in one's statement of confidence. Still, a likelihood of any one event being due to chance can be induced from whatever data is generated by the time slides. In this case the result was that a coincidence of this amplitude and degree of coherence would turn up by chance roughly once every twenty-six years.

Now, given that the Equinox Event was a three detector coincidence, and the three detectors were all up and running in "science mode" for a total of 0.6 of a year during the second year of the S5 run, which was the basis of the analysis, it means that a single time slide generates 0.6 of a year's worth of spurious coincidences.[4] In fact, one thousand time slides were carried out—each with a different delay. This produced six hundred year's worth of spurious coincidences. Around twenty spurious coincidences that looked as convincing as the Equinox Event were found in the

4. I am not 100 percent certain of the exact choices that were made, but they make no difference to the principle and very little difference to the numbers.

six hundred years, and that means that such a spurious coincidence will show up about once every twenty-six years.

The one chance in twenty-six years can be converted to say that there was 0.6/26 = 0.023, or 2.3 percent likelihood of such an event turning up by chance over the course of the run itself. This probability can then be translated into the sigma level that would be associated with it had it actually emerged from the calculation based on a smooth mathematical model. This language is familiar to physicists and can help them decide whether they "should" take it seriously or not. The 2.3 percent would, under these circumstances, be associated with a roughly 2.5 sigma result. Physicists can, then, refer to the Equinox Event as a 2.5 sigma event (even though the figure may not be accurate) when they need a rough indication of how the result compares with other results in the same field or the confidence typically invoked in other fields. A 5 sigma result, which is the standard for contemporary high-energy physics, would be equivalent in the S5 run of an event that would occur no more than 0.6 times in a million years.

In passing it should be mentioned that the Inspiral Group had seen something in the data that was not incompatible with the Burst Group's Event, but it was not enough to reinforce their claim and would not even have been noticed if the Burst Group had not seen it first.

The Detection Checklist

Of central importance to any potential announcement is the application of the Detection Checklist. This is a long list of hoops that a candidate gravitational-wave has to jump through if it is to survive. Each analysis group develops its own checklist. The Burst Group's version as of October 2007, with its seventy-three hoops, is shown in condensed form in appendix 1. An italicized remark indicates that a task had been completed by the date in question. A more complete version of the list would show who was assigned future tasks, who had completed finished tasks, and what were their sources of evidence, with hyperlinks as appropriate. Though this checklist is six pages long, it is worth reading through quickly to get a sense of what is required to make a credible detection claim. The groups do not allow themselves to make facile claims, and the degree of caution revealed by the checklist process seems appropriate.

Within the Burst Group, interest in the event increased in succeeding telecons as it continued to jump through the checklist hoops without any obvious failures. But, that said, the interest was not very great and not everyone felt it. I think I was more excited than most of the scientists. I decided I would try to test just how much interest there was by looking at the way the news was spreading. First, however, I had to make sure I wasn't about to give away any secrets.

There is a degree of rivalry between the analysis groups; each of the four groups would like to be the first to see a signal. On the other hand, rivalry is muted because, mostly, the groups are looking for different things. In the case of the Burst Group and the Inspiral Group there is an overlap.

The Inspiral Group is looking for the typical "chirp" signature of the final seconds of the inspiral of a binary star system. It does this by matching thousands of templates against the data stream. Matching templates in this way should give a big advantage in detecting a candidate signal, because true signals that match the template will stand up above the noise. The way the process works is that, when there is a correlation between signal and template, the template "lights up," as it were—suggesting that "something that matches this template has been detected."[5] But, as already intimated, in the real world nothing is so straightforward. Sometimes the detector noise will conspire to highlight a template falsely. The Inspiral Group applies a second test, the chi-square test, to see if the data really does match the template well. If the chi-square test is applied too loosely, noise fluctuations will still sneak through and highlight templates falsely. This was a worry expressed in an e-mail circulated by one commentator before the techniques were fully refined:

> I was struck by the fact that almost all of the loudest events that the Follow Up team examined were loud events that one could tell BY EYE were not binary coalescence signals. Why is it that the CBC pipeline keeps such events through to the very last step of the pipeline? Isn't it possible for a matched filter search to tell whether a putative signal actually looks like one of its templates? . . .
>
> How is it that these junky . . . signals pass the various signal-based vetoes? Why are the thresholds set so loose that this can happen? . . .
>
> Is it that the group is [so] eager not to lose a single possible detection no matter from what weird corner of parameter space. . . ? Isn't there a

5. My house overlooks Cardiff Bay, and, for amusement, I used to log all the new ships that entered the docks, trying to read the names with the aid of powerful binoculars. The angles or the weather was often such that reading a name was a matter of "extracting signal from noise," just as with gravitational-wave detection. It was striking that if the names were written in English—that is, if I had a set of templates to match them against—it was far easier to read them (to initially decipher and then confirm what they were) than if they were written in Russian or Greek script—where I had no template.

> danger that a genuine but weak signal will be lost in the flood of junky loud signals? . . . [Or is it that in some cases] the group has abandoned a matched filter search without admitting it? . . .
>
> We'll want our readers to hear the "ring of truth" in our detection claim. Inability to reject flagrant junk signals will call any detection claim into question.

On the other hand, as the third paragraph of the e-mail implies, if the chi-square test is applied too tightly, and if the templates are not accurate predictions of the true waveforms—and remember there are only a limited number of templates to be applied to an almost infinite number of possibilities—it will also suppress the signals. For some binary star systems, the predictions (templates) are not sufficiently accurate. There are, then, many subtleties in "tuning" the Inspiral search.

The Burst Group, however, is just looking for a coincidence between signals of no particular form, and it could be that they will spot an inspiral which the Inspiral Group will miss for some of the reasons described above. This could be embarrassing, and at least one bit of gossip that later came to me implied that the Inspiral Group felt somewhat shamed by the fact that the Burst Group had seen something that might be an inspiral and they had not. Given this potential rivalry, I had to make sure that the Burst Group did not mind my discussing "their signal" with members of the Inspiral Group. At the same time, from a different source I heard that the Inspiral Group had their own different detection candidate. I made e-mail inquiries of a leading member of the Burst Group about whether I was free to talk to the different groups about the other's work and obtained the following reply, which allowed me to go ahead with my inquiries:

> Although we in the burst group would be pleased to "see it first," we keep no secrets within the LSC-Virgo sphere—at least several members of the inspiral group already know about it. In fact, one of our follow-up checks is to see whether the same event candidate is found by other searches, and in the end whatever statement we can make about the event candidate should reflect having looked at it in as many ways as we can.

By this time I was keeping a running log of everything that was happening—notes on telecons, copies of e-mails, interview transcripts, and so forth. Here is an extract from my running log made in early October 2007:

I then telephone "A" (burst group) and "B" (inspiral group), to talk about these events. Both telephone calls are recorded.

Neither of them is very excited. Neither of them knows much about the other group's event (B knows nothing at all). A does not much care about either event—he says of the Oct 06 (Inspiral) event that they get something like this every couple of months. (B says it is twice a month.) An SNR [Signal to Noise Ratio] of around 6 [which the Inspiral Group have for their event] is barely different to noise.

A says he is not very excited about the Sep 21 [Equinox] event because it is right at 100 Hz which is where all the glitches are anyway—he obviously think it is a glitch.

B is surprised not to have heard of the event since he had lunch with A yesterday, but I explain that A does not think much of it.

In sum, though I am excited, that's my problem. These guys are too busy with their teaching or whatever to think it worthwhile to investigate even the event pertaining to their own group, never mind that in the other's group. But, of course the 21–09 event will get to the Inspiral Group in due course as a result of normal working down the detection checklist.

In sum, neither of these has caused enough excitement to make anyone change their normal routine—not even to send a group e-mail around.

Note that I have taken an active part in telling at least one member of the inspiral group that the burst group has something. . . .

The point about this episode of fieldwork is that it tells me that my excitement is not shared. It is not shared because in the case of the October event it's just "same old same old." No one believes these things are GWs—though any one of them could be—in fact there is a very good chance that the first real GW will be close to the noise.

In the case of the burst group, at least for A, it is something similar. The thing is in that glitchy area. . . .

So there is this problem, which one can sense in the very lethargy (I should say, lack of much special activity—the item was in the middle of the burst group agenda, even!) that is going to make it hard to actually agree that a GW has been seen. Because hundreds have been nearly seen. How is the epistemological break going to be engendered?

And it is going to be an anti-climax if it is just statistics, or failure to find an alternative explanation. BUT THIS VERY LACK OF EXCITEMENT IS A KIND OF TRIBUTE TO SCIENCE—50 YEARS TO SEE NOTHING EXCITING—MARVELLOUS AS A TYPE OF ACTIVITY. WHAT AN INDICATOR OF THE PURITY OF DEDICATION. [Capitals in original notebook entry.]

Though this extract is from notes taken very early on in the process, it captures the whole "mindset' " dilemma in a nutshell. Few people were particularly excited about the Equinox Event (the Burst Group's potential event) because it looked just like lots of things that have happened before and it like much of the noise that is found in the detector anyway in the region of 100 Hz (cycles per second), which is where it had been found (a point that will turn out to be of great importance in the argument that will come later). The science of gravitational-wave detection, as I note in the capitalized passage, has become the science of showing that potential signals are not signals—a monk-like activity which places integrity and dedication to duty far above reward—something admirable in itself. So unexcited were people that they were not even telling others in the wider collaboration about it—not even over lunch! I found myself telling some of the other scientists about it and having to explain that the fact that I am the one who is acting as the conduit does not imply anything sinister.

And yet, there is a good chance that the first gravitational-wave to be detected will be a marginal event—something that does look like a noise or a glitch. There is a good chance that the first event will be like this, if people are willing to see such an event at all, simply because the distribution of the sources in the sky means that there are likely to be many more at the outer edge of the range of the detectors than much closer in. "Much closer in" captures a much smaller volume of space than "right out at the edge." The number of sources that can be seen at different distances increases as the cube of the distance, so there are one-thousand times as many sources at a range of 15 Mpc—which, in Initial LIGO, is at the edge of detectability for a standard source—than there are at 1.5 Mpc, where things should be easily detectable. In the case of one of these distant sources, only the long drawn-out application of refined statistics and the checklist will extract it from the background. It is going to be unexciting—just statistical manipulations. As I note sometime later in my running log:

> So that is why it is all so depressing: a real event is just the outcome of messing around with lot of statistics and saying "this is unusual." No atom bombs explode, no power is generated, nothing new at all can be done!!!!

How, then, is the collaboration going to shift people from this instinct of dismissing potential signals, which can itself be felt to be a kind of holy duty, expressing the utmost purity of intention, to seeing a potential event as a real event that deserves all the enhancement it can get outside of post hoc statistical massage? That is the mindset problem.

Of course, the fact that people are going through the checklist, and one of two of them are finding the Equinox Event increasingly interesting, shows that the mindset problem is not necessarily fatal. If data analysis was a purely algorithmical process without any input from human judgment, it *couldn't* be fatal. But there is always judgment involved, as the airplane event and lots more to be discussed illustrate. So the mindset problem could be fatal when it comes to those marginal decisions that do involve judgment.

As it happens, as events unfolded the Equinox Event began to gain salience. There is something you can *do* with such a potential event—discuss it. As sociologists know, discussing things makes them real. The next meeting of the LSC-Virgo collaboration would be in Hannover, and the Equinox Event was going to be discussed. The buildup is indicated by the following e-mail alerting the "Detection Committee," the committee charged with overseeing the announcement of any event and deciding whether to take it to the collaboration as something that should move forward to publication:

> To: the LSC Detection Committee: . . . October 11, 2007:
> A candidate event has been detected by the burst group. None of the search groups are ready to present a case to the Detection Committee but there is knowledge throughout LSC that something is afoot. The burst group is considering giving a brief presentation of their activities regarding this event at the forthcoming Hannover LSC meeting without giving conclusions. It would be a good idea for the Detection Committee to meet to discuss strategy and to formulate the questions that we believe will need to addressed. Suggest that we have a telephone meeting. . . .

When the meeting took place there were immensely long discussions over procedure and exactly how the event was to be discussed. It was concluded that, to keep the excitement damped down, it should be spoken of as just a run-of-the-mill "loudest event" of the sort that is bound to appear in every analysis run. In my notes I express myself astonished that the scientists should be so cautious—the announcement was going only to the collaboration, itself an organization with very restricted membership, my presence in which is an anomaly—though, perhaps, the organization was large enough for leaks to be a concern. As it happens, however, the name "Equinox Event" was soon to become the standard reference, making any attempt to treat it as just an ordinary loud event futile.

Just before the Hannover meeting there was another Burst Group telecon in which the likelihood of the event being due to correlated noise was estimated using three different procedures, producing results of 1 every 100 days, 1 every 2 years, and 1 every 50 years. Extra significance can come from analyzing the data in ways that take more of its properties, such as the energy profile, into account when working out how like each other are the signals in separate detectors. Techniques can also be applied to eliminate glitches with known causes, thus making "the event" stand out higher above the background. The latter method is where the danger of "tuning to the signal" might arise.

The Hannover Meeting

The high point of the Hannover meeting was going to be the discussion of the Equinox Event. But the Inspiral Group seemed to steal a bit of the Burst Group's thunder with an announcement of their own. Apparently, the Inspiral Group, quite independently of anyone else, had decided to run their own blind injections on themselves. They had done this, as they were at pains to stress, without causing any potential trouble to anyone else.

It is the case that the detectors are subject to a continual stream of artificial "signal" injections, known as "hardware injections," which are used to monitor the sensitivity of the devices and the analytic procedures. These false injections cause no trouble because they are not secret—the very point is that the groups know what has been injected when, so that they can check to see that their pipelines were sensitive enough to detect the injections had they been real. The Inspiral Group had simply elected to keep a couple of these regular injections secret from themselves, and they were able to announce at the Hannover meeting that they had found something in the data. I must say I found it puzzling that the group had told the Hannover meeting about this before opening their own private envelope, so that they could also tell the meeting, there and then, whether they had found something real or not. But others said that the Inspiral Group exercise was a good one and the analysts were entitled to their fun. Others said that the exercise was not as useful as the "proper" blind injections because the regular hardware injections were perfectly formed, so it was obvious they were a false not a real signal, which would always be dirty in one way or another. There was no reason to make the calibrations dirty, whereas the true blind injections would be of dirty signals so as to present a more realistic challenge. Anyway, the Inspiral Group's self-imposed blind injections acted as "hors d'oeuvre" to the Burst Group's discussion.

The Burst Group went through the checklist. They explained that though the Equinox Event may not stand out very much in terms of just being a coincidence, when a test for the coherence of the waveform in the detectors was applied, it gained far more significance. A coherence test is always going to be a reasonable thing to do on any coincidence and was planned before the box was opened. The Inspiral Group also reported that they had something of low significance that matched the Burst Group finding but that it did not show up strongly in an inspiral search. The Burst Group spokesperson concluded by explaining that nothing they had seen was "ready to go out the door" and that they were merely reporting work in progress. Nevertheless, they said that, if nothing was found to argue that this was an artifact, they would be expecting to take the result to the Detection Committee.

3 Resistance to Discovery

The driving idea behind the blind injection challenge was to do some social and psychological engineering—to get the community out of its negative mindset. Knowing that there might be blind injections in the data, the scientists had to be ready to see something, not just reject everything. Before going on to explain this in more detail, let me make clear the truth-status of the claim that *I* have just made about the idea behind the challenge. This rationale also covers the status of the claims made in the introduction, such as: "the strongest pressure for its [eLIGO's] completion came from scientists who believed doubling the sensitivity would be the key to the initial detection of gravitational-waves."

The status of these claims is that they are what I heard people saying as I "hung around" the community, acting as a quasi-member of it and acquiring "interactional expertise."[1] Now, a claim of this sort does not prove that these were the motivations that informed everyone. I was hearing snatches of conversation and explanation that fitted into my continually developing understanding of how the community in which I was embedded was working. I made no attempt to do a representative survey to find out how

1. The concept if interactional expertise is developed in Collins and Evans 2007. It is expertise gained through deep immersion in a technical discourse in the absence of practical ability. Experimental tests indicate that it enables sound technical judgments; it is the crucial expertise used by managers when they take on new technical projects (Collins and Sanders, 2007).

many people thought "this" and how many people thought "that." I didn't do this primarily because I do not think surveys are as revealing as hanging around, conversing, and interacting as much as possible, while trying, as far as possible, to become like a member of the group one is studying. I have nothing against surveys as an additional source of information and have tried to use them in this way from time to time, but I do not do it often because they are unreliable and can be counterproductive.[2] They are unreliable because few people answer them, and they can be counterproductive because they distance one from the community. Your friends learn about you from acquaintanceship, not from getting you to fill in questionnaires about your eating, drinking, and reading habits—at least not beyond the introductory round of Internet-dating. Finally, I am after something more than the aggregate of individuals' motivations. I am looking for the "spirit" of the discussions; this is expressed by what can and cannot be legitimately stated in public. I am looking for what has been elsewhere called "the development of formative intentions" or legitimate "vocabularies of motive" rather than trying to work out what is in a set of different individuals' heads. Working out what is in individuals' heads is the extremely difficult and demonstrably fallible job of law courts and the like. I am not equipped to do it. My approach, of treating the kind of things people say to me as indicating what is say-able at that time, rather than looking for their "true" motives (as though we even know our own motives with certainty), I call the Anti-Forensic Principle. The Anti-Forensic Principle indicates that sociology is concerned with the nature and "logic" of cultures—in this case developing local cultures—not the guilt, innocence, or motive of any specific person. One acquires an understanding of the nature and logic of a culture by chatting in coffee rooms and corridors, not by doing surveys.[3]

It is the case, however, that not long after the blind injection challenge was put in place, various scientists offered explanations of why it was a good thing that were different or additional to the "changing the mindset" explanation that I have just put forward. Some scientists indicated that these other reasons were what motivated the blind injection in the first place. I can't be sure that these reasons were not the initial motivating factor for at least some these people. It does not matter: a good proportion

2. An example is presented below.

3. The investigator's interactional expertise (Collins and Evans 2007) provides the warrant for claims and questions about the science expressed here in the first person or simply stated without further justification. For "formative intentions" see Collins and Kusch 1998. For "vocabularies of motive" see C. Wright Mills 1940. For the first invocation of the Anti-Forensic Principle, see *Gravity's Shadow* (412).

of the community thought it was being done to change the mindset of the community, and this made a lot of sense because it was felt that the mindset was in need of changing. That the challenge served other purposes too, even if these were not the driving force, is also beyond doubt; these other purposes will be explained in due course, but first the "mindset problem" will be examined.

How does a mindset become established and get transmitted to new generations of a community? In large part it is through relating "myths" or telling "war stories." The term "myth," used in this context, does not mean a false account. Rather it refers to the first part of the definition as found in my *Chamber's Dictionary:* "an ancient traditional story of gods or heroes, especially one offering an explanation of some fact or phenomenon; a story with a veiled meaning." Here there are no gods, nor even heroes, but only antiheroes, and the meaning is not veiled. The subjects are Joe Weber and/or "the Italians," and the stories of their wrongdoing are circulated and recalled with words such as "to analyze data like that would be to risk doing what Joe Weber did," or "to claim that would be to make statements as irresponsible as those of "the Italians." You can hear those phrases repeated in the corridors of interferometer meetings whenever some judgment is being made about data analysis. Recalling those stories provides a guideline for new scientists on how they are to act and reinforces the views of the scientists who already know how to act. To act in any other way would be to violate a taboo and act like a scientific degenerate.

Joe Weber and Statistics

It is now widely believed, though never decisively proved, that Weber managed to get the results he got by manipulating his statistics retrospectively.[4] Weber had a tendency to "tune to a signal." As I argue in *Gravity's Shadow*, if he did do this, Weber would have been doing the thing that served him well when he commanded a submarine chaser in the Second World War. When you are trying to find a submarine, you turn the dials of the detection apparatus backward and forward as you tune in to the telltale echo. And it is vital to find the submarine if it is there. If it is there and you fail to find it, some of your compatriots will likely die. If you "find it" and it is not there, nothing worse will happen than some waste of time and possibly ammunition. In submarine chasing it is much better to choose a strategy

4. There were many other ways in which Weber could have gone wrong outside of poor statistics; it is just that the statistical explanation has become standard.

that will result in false positives than false negatives, and the skillful commander will find the submarine even if it is at the cost of quite a few false alarms.

But tuning to a signal is dangerous in physics, or, at least, it is dangerous in exploratory physics. Suppose a standard has been set such that a set a numbers reflecting some kind of potential discovery will only be counted as a discovery if it would arise, as a result of chance, only one time in one thousand searches. The trouble with tuning to the signal is that each adjustment of the tuning dial represents a new search. Thus, if you twiddle the dial 100 times before you find the 1 in 1,000 you have done a hundred searches. An event that is only likely to turn up one time in one thousand in one search is likely to turn up one time in ten in a collection of a hundred searches. Thus, if you report only one of those searches, the event will look like a discovery, but it does not meet the standard of a discovery at all.

If the meaning of the last paragraph is not immediately obvious, it is worth going back to it until it is because a great deal of what follows turns on its logic. A major claim of this book, remember, is that, however much a discovery is presented as a unitary reading, it is really just an eddy—a small area of seemingly still water—in the onrushing stream of history. One can see, if one follows the logic of the last paragraph, that a paper that claims a 1 in 1,000 value discovery—something which to follow the metaphor, is an undisturbed patch of water in the stream—might well be swirling around at a speed appropriate for a one in ten event. It all depends on what happened immediately upstream of the eddy, such as whether the water was disturbed by lots of twiddling of dials. Notice that how much dial twiddling took place upstream is not generally mentioned in the scientific paper which reports the result—the paper simply reports the result of the final calculation of unlikelihood, and that is why it can be so misleading. To understand the true meaning of a published statistical claim you have to be a cross between a historian and a detective. And, as we will see later, you have to be able to read minds too.

The fact that statistical reports, though presented as timeless readings, should actually be read as historical accounts is one of the reasons it is impossible to know for sure if Joe Weber's "results" really were the outcome of what is known as "statistical massage." To know that would be to know exactly what Joe Weber did with his dials, and we don't. The fact that it is widely believed, to the point where it has become a myth, or regularly told war story, belonging to the field of gravitational-wave detection, is, however, a fact with real consequences. It makes the whole community extremely sensitive to what can go wrong. It is certainly a powerful contributor to the mindset and provides one of the resources that scientists can turn to in their debates should they want to count an apparent signal as insufficiently convincing to count as a discovery.

"The Italians"

Not all Italians are "Italians." When someone in the LIGO community refers to "the Italians," he or she has in mind a specific group who ran cryogenic bars and were based in the Italian high-energy physics laboratory widely known by the name of the small town, just outside Rome, in which it is situated: Frascati. As some of the Frascati physicists also held jobs in Rome's universities, they are also known as "The Rome Group." There are about half-a-dozen "Italians." Nowadays, as with Joe Weber, "the Italians" are used for myth-relating and mindset-setting.

Over the last half-decade all the detector groups worldwide have gathered together to share their data with one another. In the early days, each new detector group believed it had achieved the technological breakthrough that would finally achieve the elusive detection, but in each case years of frustrating work proved that the promised levels of sensitivity were not going to be reached any time soon. LIGO was the only group to reach, or nearly reach, their design sensitivity in some approximation to their projected timetable, and even it, in spite of having by far the biggest team, reached the promised landmark about three years late. No other interferometer team has yet come as close to its design sensitivity as LIGO.

LIGO's policy, under the leadership of one-time, and now once more, high-energy physicist, Barry Barish, was to bring the international groups together. The logic of gravitational-wave detection means that in the long term data-sharing is inevitable because only by "triangulating" the signals seen on detectors spread across the planet can the direction of a source be located. Each individual detector is almost completely insensitive to the direction from which a wave comes, and it takes four widely separated detectors to pinpoint the source, working from the different times that the signal—assumed to be traveling at the speed of light—impacts on each one. Barish was used to working in international collaborations such as CERN, so, bringing the groups together was a natural thing for him to do.

The Virgo detector, located near Pisa, is the only one which can begin to compete with LIGO, and the initial relationship between the groups was one of rivalry. Virgo has a better suspension which, in principle, should allow it to detect signals of a lower frequency. The spiraling together of a binary system composed of large black holes should emit waves at low frequency for some considerable time; such waves would be invisible to a higher frequency detector, but would make a distinctive waveform easier to spot.[5] There are also many low-frequency pulsars that could not be seen by LIGO and for which the signal would build its signature over time. This gave the Italian-French team some chance of making an independent detection even though they had only one detector of 3 km arm length as opposed to two at 4 km, so that they could not back up a claim by seeing coincident signals between two machines.

Too much rivalry could be fatal for all the groups: suppose LIGO saw an event but Virgo did not? Virgo might insist that this implied that there was something wrong with the LIGO claim. I do not know whether this was part of Barish's, or any one else's, motivation, but in a case like this, to borrow the words of Lyndon B. Johnson, it is better to have everyone inside the tent pissing out than someone outside the tent pissing in.

An agreement between the main groups became easier when it turned out that the especially good low frequency performance of Virgo was not going to be realized until some time after the intended higher frequency performance had been achieved. Indeed, for a long time LIGO had the better low frequency performance. This, with the dawning of the realization that no one was going to make the first discovery very soon, took the heat out of the rivalry, and by around the end of 2006, Virgo and the LIGO Scientific Collaboration (LSC), which already included GEO 600, agreed to collaborate and hold joint meetings. The collaborating group was named LSC-Virgo. The clumsy name indicates that an institutional merger may not be a complete merger in every sense. As we will see, residual suspicions play a part in the story.[6]

The point is that there are many Italians in LSC-Virgo who have the same attitudes toward "the Italians" as do the members of the LIGO group. Indeed, as compatriots, they may be more embarrassed by what "the Italians" did. The problems of "structural balance" in these relationships are complicated by the facts that one of "the Italians," Eugenio Coccia, has become head of the Gran Sasso Laboratory, one of Italy's leading physics institutions, and that a good number of "the Italians" have now joined the

5. It has become less clear that massive black-hole binary systems near to the end of their lives will have had time to form during the lifetime of the universe, and, to date, no such binary systems have been detected by any means.

6. To use the language of *Gravity's Shadow* (665–67), while the two groups exhibit "system integration" sufficient to give rise to "technical integration," "moral integration" is incomplete.

LSC-Virgo collaboration. The tensions are illustrated by a rather good joke made by one of my friends in LIGO who, echoing a line from *Dr. Strangelove*, remarked to me when he heard that several of "the Italians" were to be present at future meetings of the closely guarded LSC, "but they'll see the Big Board!"[7]

So what did "the Italians" do to become quasi-mythical exemplars of bad behavior? They continued to claim to have seen gravitational-waves of a similar energy to those Joe Weber had claimed to see long after Joe Weber was discredited. Worse, in two cases they announced results where the signal was in coincidence with Joe Weber's signals. There were four sets of announcements altogether involving the Rome Group of which two were particularly notorious, with the last being the one most often mentioned today as "a warning to children."

The first occurrence, which attracted almost no attention, was a paper published in 1982 showing correlations between Joe Weber's room-temperature bar and a cryogenic bar run by the Rome Group but located in Geneva. This paper claimed to find a "zero-delay excess" of events associated with a 3.6 standard deviation (or 3.6 sigma) level of significance. Remember, a "zero-delay excess" means that the two widely separated bars vibrate strongly at the same time more often than they spuriously appear to vibrate at the same time when the two data streams are offset in time. The offset comparisons, or comparisons with delays, or comparisons with time slides—which are three ways of saying the same thing—show what should be expected as a result of coincidences between noise alone—the "background." A finding that there are significantly more genuinely simultaneous coincidences suggests that something external caused that zero-delay excess.

Each time slide will produce a slightly different result for the number of chance coincidences. From these results one can create a histogram that shows how many times different numbers of spurious coincidences show up in the complete set of time slides. One will find that the histogram is peaked at some average level of spurious coincidences, while the further one departs from that average number the fewer slides will there be that exhibit that number of coincidences. As you go right out to the tails of the histogram, which are a long way from the average, there is very little

7. In *Dr. Strangelove*, the "Big Board" is the graphic representation of the location of American warplanes and their attack plans found in the top-secret "War Room." An American general makes the remark when he learns that a Russian is to be invited in to learn the plans as a last chance of averting an accidental nuclear war.

chance of finding slides with those very many or very few chance coincidences. A 3.6 sigma result means that one would expect such a result to be generated by chance coincidences caused by noise alone only about one time in ten thousand—some way toward the tail of the histogram. Attaining a 3.6 sigma result would provide a fantastic level of confidence in psychology or sociology, where 2 sigma, or five chances in one hundred, is regularly accepted, but it would not count as much of a result in contemporary high-energy physics, which prefers 5 sigma, which is about one chance in a million. We will return to the choice of confidence level; for the moment it is enough to note that in 1982 "the Italians," along with Joe Weber, published a result that was strong enough to be irritating if anyone had taken any notice of it, but they didn't.

People took much more notice of a 1989 paper, which was again a collaborative effort between the Rome Group and Joe Weber. In this paper they claimed that their respective room-temperature bars, which were the only detectors "on air" at the time, had detected gravitational-waves emitted by a supernova—the famous Supernova 1987A—which was visible from earth and which emitted a flux of neutrinos that was also detected. In that paper they calculated that the energy needed to enable their detectors to see the waves would have needed the total consumption of two thousand solar masses. Although the consumption of this much energy in a supernova was incompatible with either known physics or known astrophysics, they published anyway, exerting the right of experimenters to see theoretically impossible things. Their announced result also caused a fuss because at this time there was an active campaign to fund LIGO; if the result was valid, then the assumptions on which the huge and expensive LIGO were based were wrong. The result reached the science news media but was soon ruled out of court by the majority of the physics community.

The next incident of "bad behavior" was the 1996 announcement, in the form of conference presentations, that the Rome Group and the group in Perth, Australia, led by David Blair, had found suggestive coincidences between their bars. This result never got as far as the journals, but it was discussed at several conferences—because of the determination of Blair, and to the extreme annoyance of the nascent interferometer community. Guido Pizzella, the then leader of the Rome Group, told me that he felt too bruised by the reaction to the Supernova 1987A claims to expose himself to further scorn. In 1996 Blair told me, however:

> I think it's not right . . . because of the fuss that there's been in the past, to deny the data. . . . [W]e were not going to be bullied by people who have

their own agenda. We believed that what we had seen was reasonable and interesting, and that you should tell the story as the story goes—as it unfolds.

Blair made the conference presentations.

Once more the energy calculation showed that, if Perth and Rome were seeing something, the assumptions upon which hundreds of millions of dollars were being spent were flawed. The members of the interferometer community were furious; they felt that every "serious physicist" should accept that to detect gravitational-waves, especially in the large numbers required to have a significant "zero-delay excess" in the bars, one had to wait for the interferometers, which were going to be many orders of magnitude more sensitive. From the point of view of the interferometer community, the only purpose that could be served by these announcements was to cause trouble, to remind people of the days of Joe Weber's shameful claims, and to project an even more flaky image of gravitational-wave physics to the already skeptical scientific community.

The final incident was when "the Italians" published a paper in November 2002 claiming that, in 2001, they had seen coincidences between two of their own cryogenic bars, one located in Frascati and one located in Geneva. At this time of growing integration in the international community of gravitational-wave physics, the publication came as a shock because the paper had been submitted before being seen by the community, which would likely have tried to suppress it.[8] "The Italians" knew they would find no friendly reviewers among the LIGO-centered groups and so went straight to the peer-review process of the journal *Classical and Quantum Gravity*, which published what, on the face of it, looked like an interesting result. Its special interest lay in the fact that the zero-delay excess coincidences were clustered within certain hours during the day and that the particular hours moved around the clock during the course of the year in such a way as to suggest that the source of the energy was related to the galaxy and not the solar system.

This claim echoed one made by Weber for his own results in the early years. Unfortunately, in the case of Weber the effect disappeared after a while and it dropped from view. The power of this kind of claim lies in the fact that, while it is easy to imagine some spurious source of coincident

8. The flames were fanned when *New Scientist* (9 November 2002) picked up the controversy and, finding that some physicists were unwilling to offer comments to their reporters, wrote an editorial with the tag line "Hushing up scientific controversies is never a smart move."

disturbance correlated with the changing solar day—much creaks and groans as the earth turns its faces to and away from the sun, with different things being heated and cooled, tidal forces exerting their effects, and so on—it is very hard to think of anything that is correlated with the sidereal clock. The sidereal clock shifts around in respect to the solar clock as the year passes. As the earth goes round the sun in the course of a year, the way it is orientated in respect of the center of the galaxy at any one time of day goes through a complete cycle. For example, if London faces the center of the galaxy at noon on a certain date, it will be sideways on to the galaxy three months later, will face away from the galaxy three months after that, be sideways on again but the other way round three months after that, and be back to full face again three months after that. If the clustering of signals seen by the pair of bars moves round the clock in the same way, it suggests that something to do with the galaxy, not the sun, is causing the clustering, and it is hard to think of what it might be if it is not gravitational-waves.

The counterargument, presented as a verbal assault, was that the statistics of this clustering did not stand up to examination. Once more, statistical analysis, it was said, had been converted into wishful thinking. The true statistical significance of the clustering, when the calculations were done properly, was only one standard deviation, and this amounted to nothing. Whether this was exactly true or not depended on a fine subtlety of statistical thinking which will be discussed in chapter 5. Here, however, it is the reaction to the claims that is being explored.

In December 2002 there was a meeting in Kyoto of the Gravitational-Wave International Committee (GWIC), which coordinates the various worldwide gravitational-wave detection projects, including cryogenic bar detector and interferometer projects. The Rome Group publication was discussed at the Kyoto meeting. The following excerpts from the discussion give the flavor of the debate. The chair, Barry Barish, who was also the director of the LIGO project, introduced the item as follows:

> As we start having results in this community we need to figure out our own guidelines and standards. For those who come from outside of this community, it's fair to say that this community doesn't have a very good reputation—that the past history of having results that haven't stood up hasn't served the community very well, and I think it's important to have credibility and to do things right.
>
> I come from a community in particle physics where there's been a lot of good results, and exciting results, but also wrong results, and that commu-

nity had developed throughout the years certain ways of presenting results, and vetting results, that may or may not be the model for here but is at least one we should discuss.

There's a couple of things that stimulate this discussion for me. One is we're just getting to the point in LIGO where we're going to have our own results . . . and I'll show you a picture of our present plan but it's kind of developing—of how we go through keeping the results private, how we present them, and in what ways we bring them to publication. And there's a recent publication by, er, er, your group [indicating one of "the Italians" who was present], that raises certain questions, I think, and I want to raise those so we bring them for discussion.

First let me say some ideas and show you one picture from us, which may not be the right ideas but at least I can tell you what the reasons are. They come from some experience in particle physics. In particle physics there's a lot of ideas that when they first come out people object to, and some history that's made when things finally come to publication. There's a certain vetting process that's been healthy. And I propose we need some of that in our own community before we present things. . . .

And in a community that hasn't been fully credible in its past history, it may be very important—you have to build a credibility if people are going to believe results that come out of this field. . . .

As we have results, we really need a way that we all try to follow—not exactly—the same rule, that we try to follow as we go forward to present these results so that we present the best kinds of paper which contain the best information and not debate them after they're in press—I think that's the worst problem is if you debate them after they're in press. Inside our community it makes even things that can be very right—they can be right or wrong—but it makes the whole thing controversial—unnecessarily.

The project director of LIGO, Gary Sanders, remarked:

Let's say there's a paper comes out and the whole community comes back and says we don't believe it. Let's say that happens. Where would you rather have that in? After the preprint comes out before publication, where the author of the publication has a chance to hear this and think about it, or you submit the publication and hear about it later. Now it's on the record and now it is, as [. . .] described, somewhat divisive in the field. It divides the field into those who are willing perhaps [to] take more risks, to put it in the terms, and those who would like this field to develop as a conversation.

To this the following response (lightly edited for English) was made by a spokesperson for the Rome Group:

> So, first of all let me say something. The publicity—I really did not want the publicity. The journalist at *New Scientist* called me because, she said, of the paper in *CQG*. The paper was written in a way in which we were very careful in the title and the abstract and the final conclusion. It's not a claim—it's not a claim—it's just results. Even the analysis is very simple—I mean just reporting coincidences versus the time. And it describes the procedure of data selection. It is very simple and, let me say, it is very poor. The paper on the statistical analysis will be presented here by Pia Astone at this conference. And now that GWIC has proposed some guidelines we can follow these steps in future.
>
> May I say that in the absence of guidelines from GWIC we followed the guidelines of our group. And the guidelines of our group have always been, like in many other communities, to send the paper to a journal, without any restrictions on referees, expect the referees reports, and then modify the paper or not—depending on the outcome. And then, once the paper is accepted, circulate the paper in the community.
>
> I think this is a procedure of many people. People put a paper on the web only when it is accepted by journals. Not all the people, but many people in the experimental field do that. So this is just to follow the story.

Barish responded:

> But not so much in fields that get picked up by *New Scientist*. That's why particle physics doesn't do it that way. Because the fields that tend to get picked up and made very public have a different—you live in a fishbowl—it's a different kind of environment. So it's true, but I don't think you can identify [those who do what you said as] a high profile community. We're a high profile community.

The Rome Group spokesperson then added:

> Let me say something else. Of course the feeling was that this was not a normal paper because we had the feeling that we had hit something that may be important. So we discussed it in the group—how to proceed. First of all we made a lot of controls . . . and then if we should first send preprints or send the paper directly to the journal. The majority of the people were of the opinion that we should send direct to the journal and receive the

reports of the referees as a part of the procedure in which the referee plays the role of a sample of the community, let us say. . . .

I have to say that the referee reports were very good, so, if the paper is correct, why not diffuse this information? We make it known: other groups now know that they should look at a time when they have a good orientation to the galactic disk, and we show exactly the time and the date of the events to give the community information. If there are flares, if there are X-rays, if there are gammas at those times, how else can you reach everybody? Of course you can put something of the web but you can also publish. . . .

He remarked later:

The reason why I personally as a group leader accepted that we should publish the paper was because it was published not as a claim or as a evidence, but it was reported as a study of a coincidence actually taken by the detectors at their best performances and only with this in mind—that it was not a claim, just an *indication* [my emphasis]. Also to declare how we will do the next analysis. So we told before what were the parameters, we made the choice, the selection, the procedure, and then using that procedure we found something that is not a claim but which gives some—let's say, can fix the procedure for looking to the next one and for contributing to the discussion in the community giving the feeling that doing [things] in a certain way one sees something that is unexpected but which can be very important. . . .

I understand perfectly the sensitivity of the field, and for that reason we published the paper as a study of coincidence not as a claim and we didn't look for journalists at all. This was something that happened in spite of our attitude not because of it.

A representative of the non-"Italian" Italian groups summed up some of their worries as follows:

I think we are discussing something different. This is that this is a high profile field because there is a lot of public expenditure in it. The reason why there is a lot of public expenditure is that this is a forefront field of physics and we all agree that the reputation of the field is very important. . . .

The other Italians, not in "the Italian" group, thus shared the concerns of the LIGO group.

In 2003 a statistical analysis carried out by Sam Finn, which argued that the statistical significance of the galaxy-related clustering of signals was really only one sigma and therefore counted for nothing, was published by *Classical and Quantum Gravity* in the form of a response to the Rome Group paper. "The Italians" replied to this charge. Their response, however, was not published in the journal, and it found its final resting place in the electronic preprint archive. That, as far as the bulk of the gravitational-wave detection community was concerned, was the end of the 2001 coincidences and the 2002 publication. I believe that whether "the Italians" were right or wrong in respect of their findings, the statistical argument was too revealing to be forgotten. It will be looked at in detail in chapter 5.

Discoveries as Binary Events

The charge of weak statistics was not the only one thrown at "the Italians." Another accusation could be heard, but only in the corridors rather than the public sessions. Nevertheless, corridor discussions are equally "formative" of the ethos of the field. What was said was that "the Italians" were acting sneakily in the way they defended themselves. The Rome Group spokesperson insisted that the paper "was published not as a claim or as a evidence but was reported as a study of the coincidence actually taken by the detectors at their best performances and only with this in mind—that it was not a claim, just an *indication*." To the Rome Group this seemed a perfectly reasonable way to defend their actions, but the large proportion of the gravitational-wave community thought it dishonorable. In 1995 a senior American scientist had explained this to me in reference to some of the earlier "Italian" results:

> It was mostly at these general relativity meetings. . . . They would give their presentations in such a way that they would lead you, they would show you this data, and they would show you the events, and they would show you some statistics they'd done, but never enough of it so that you could really get your arms around it. And they left you with this tantalizing notion that they could go either way. They either could make a claim that they had detected something if they had wanted to, if they had gone the next step in their presentation, or they could back off and say, "Well, yeah, maybe the statistics isn't good enough." And they left you at that critical juncture. . . . What would happen is that you had to draw the inference. Now that gave

> them the freedom at any point later on, or maybe even . . . to say, "Well, if we choose to say this, we have detected it, or if we choose to interpret that way, we haven't detected it, because the statistics isn't good enough." It was this . . . ambiguity—OK? That got to me—OK?

Scientists worry that a research claim that is open to different interpretations could too easily be used to the researcher's advantage. A publication of this kind that turns out to anticipate a more decisive claim confirming the positive interpretation could grab the Nobel Prize even though the work had not been finished; if the positive interpretation is not confirmed, however, there is no penalty because no mistake has been made—the initial publication was too equivocal.

In the introduction I argued that discoveries are eddies in the onrushing stream of history. But most physicists do not see it that way. Physicists see a discovery as something that either did or did not happen: the discovery switch can be either "on" of "off" but not in between; the symbol is either a "1" or a "0." And the Nobel Prize is the icon of the digital nature of the discoverer's world. You can win or share a Nobel Prize or not win or share a Nobel Prize, but you cannot sort-of-win or sort-of-share a Nobel Prize: to act as though you *might* have done something that deserves a Nobel is to threaten to wreck the whole "discovery" edifice which has so long supported physical science.

And that is why in 2009, in a hotel restaurant in Arcadia, I could listen to a senior gravitational-wave scientist telling me that he did not want anything he published to

> read like that Rome paper from whatever year it was—2001 or whatever—where what was infuriating about that was its coy playing with "we don't really have good evidence but we want you to think maybe." And that paper haunts us. And people draw different lessons from it, but what haunts me, and some other people, is not wanting to be accused of trying to straddle the line of winking at people when we aren't actually able to stake our honor on it. . . . trying to stay on the sober side of coyness and winking.

That is how the "myth" of "the Italians" continues to play its role and construct the ethos of what it is to find something new about the natural world. There is to be no winking. The sentiment will return when we approach the end of the story.

Blind Injections and Their Problems

It was at the Hannover meeting that I first learned, to my surprise—and I discovered that I was not the only one who was taken by surprise—that one of the purposes of blind injections has been retrospectively discovered to be fending off press speculation about unusual activity/excitement in the collaboration and to discount leaks. Even if the collaboration becomes noticeably heated in their analysis of an event, no one can know if it is a real event or a blind injection until "the envelope" is opened. The "no one" includes "deep throats," unless they go in for the most mundane kind of cheating by looking at the channels which record the injections—which everyone has vowed not to look at—and journalists (and me) who wouldn't know where to look. Only two people in the collaboration are responsible for the injections, and, crude cheating aside, only they know what their random numbers procedure prompted them to insert and record in the envelope. Even cheating could be avoided by enciphering the channel into which the fake inputs are inserted, but the scientists are worried about going in for this degree of complication in case something stops them from decoding the channel again when the time comes. In any case, even if someone did cheat, it would not be that some incorrect physical fact would be announced to the world; it would just be that the intended social engineering did not work properly.

The social engineering—and this has to be assumed to be the main purpose of the blind injection challenge—was intended to change the mindset of the scientists from nondetection to detection. They had to find the blind injections to demonstrate their ability to find a gravitational-wave. It sounds like a great idea, but it turned out to be a lot more complicated than anyone, and that includes me, who can legitimately claim to have some expertise in understanding "social things," had realized it was going to be.

The biggest danger was that the blind injections would have exactly the opposite effect to what was intended. Extracting a signal from noise is a lot of work. If scientists thought that all this work was being wasted on an artifact, they might be still less willing to do it! One scientist put it this way: "All your enthusiasm gets sucked away. . . . It's messing with your head . . . in a complicated way." This was quite a widespread opinion, and another scientist said it had a very real consequence for him:

> I concluded early on that the [Equinox] event was an injection and have not been willing to devote as much urgency in investigating it, as I would

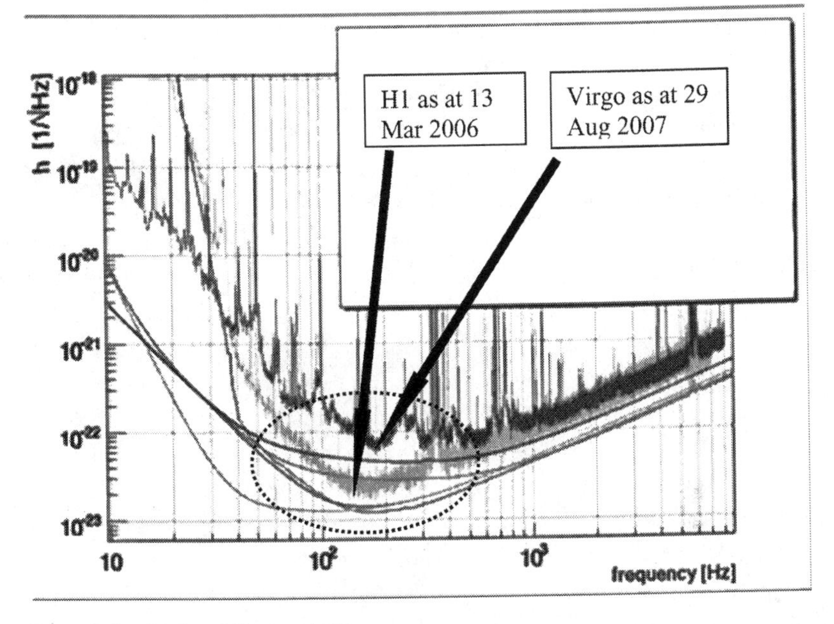

Figure 3. Sensitivity of Virgo and H1

> have otherwise. I'm just not willing to drop everything else to hunt down a deliberate false alarm (although I'm glad others are).

The other big problem was that quite a bit of energy went into trying to work out, not whether the event could represent a real signal, but whether it was in fact a blind injection. This was well outside the spirit of the exercise, but if one could work out that it was in fact a blind injection, then it might be more tempting to adopt the approach of the scientist quoted above.

As it happens there were clues that could be used to indicate that the event was more rather than less likely to be a blind injection. First, the whole exercise had not been put into place until quite late in the final two-year/one-year run known as S5, and the event was in the right time period; this made the odds about 3:1 for an injection. Secondly, the collaboration between LIGO and Virgo was relatively new, and it was not possible to inject a corresponding signal into Virgo. Therefore the event had to be injected in the frequency waveband where Virgo was relatively insensitive compared to LIGO, so that it could be seen by LIGO but not by Virgo. Figure 3 shows the sensitivities of Virgo and H1. Sensitivity is indicated (inversely) on the vertical axis (the smaller the strain that can be seen the better) with frequency on the horizontal. The dotted oval indicates the area where LIGO is sensitive enough to see a small event and Virgo is not. As can be seen, this waveband is around 100 Hz or 100 cycles per second.

Restricting the blind injection to within this waveband meant that the absence of a signal in Virgo was not a dead give-away. As it was, for those inclined to think that way, the fact that the event occurred in that low frequency band, though it was not a clear indicator, did substantially increase the odds that the event was an injection. There was a feeling among many of the scientists that it would be just too much of a coincidence if the first gravity wave to be detected just happened to fit into the restricted parameter space demanded by the blind injection exercise.

Another interpretation of the utility of the blind injection exercise, and it is one that would never have occurred to me, is that it would calm the analysts down in case some promising candidate was discovered, and that this would produce more considered and therefore more reliable analyses. That viewpoint is reflected in the conversation which concludes this chapter.

Another problem, also reflected in that conversation, is whether the sequence of events that follow a blind injection can really be the same as that which would be followed in the case of a real event. In other words, where does the process end and at what point is the envelope opened? The longer the envelope stays closed, the more time is wasted going through spurious exercises; the sooner it is opened, the less like a real event analysis is the fake analysis. As it was, the idea was that the envelope would stay closed long enough for the groups to write draft papers as though for publication. It turned out, however, that they wrote only the abstract because working only that out was, as we will see, uncomfortably time-consuming and labor-intensive. More worrying is the fact that collaborations were set up with groups of astronomers looking at the electromagnetic spectrum—x-rays, radio waves, gamma-ray bursts, and so forth. The idea, as mentioned earlier, is that these groups should be alerted so that they can look for bursts in their own wavebands that correlate with gravity waves and that appear to come from the right general direction. But can one waste these other groups' time on what might be an injection? This issue does not seem to have been settled.

The advantages and disadvantages of the blind injection idea are summarized in table 1.

In October 2007 I carried out a small opinion poll to find out what scientists in the Burst Group thought about the blind injection. Not untypically, only five replies came back, but the variation is interestingly large.

Table 1. Advantages and disadvantages of blind injections

UPSIDE	DOWNSIDE
Encourage analysts to work hard because there should be something to find	Discourage analysts from working hard because could be wasting time on an injection rather than an event
The doubters might get egg on their faces if something they say is not worth pursuing turns out to be an injection	It takes the excitement out of the analysis—people just shrug instead of being animated
Discourage analysts from working too hard or too obsessively so work is not hurried but careful and deliberate	Waste valuable analysis and thinking time on artifacts; cause spurious and unnecessary excitement
Help keep a lid on potential findings because they might always be an injection—the outside world can be told this	You can't really go all the way because it wastes more and more time as you go further (e.g., to the electromagnetic groups)
Enables detection procedures to be rehearsed before the real thing	*Don't look now but a corresponding downside will be discussed on p 135.*

Table 2. A little survey—October 2007—on what the Equinox Event might be

It is an injection	It is a signal though it may not justify a discovery claim	It is correlated noise
45	45	10
85	5	10
40	30	30
40	20	40
75	10	15

The replies are shown in table 2. Scientists were asked to report their feelings about the nature of the Equinox Event by assigning probabilities, summing to 100, for each of three possibilities.

Hannover Coffee Discussion

At the Hannover meeting I found myself discussing the event with four of the scientists over coffee. Here is how the conversation went:

Collins: I'm interested in your best guesses about what's going to happen to this burst event.

A: It will be declared as an injection.

Collins: Yeah, but if it's not declared as an injection?

B: The trouble is it looks like a lot of other things so that's going to make it difficult. But if it's highly statistically significant, we can't throw it out so we'll have to write a paper. . . . We might not want a detection paper, so we'll have to write a paper that says "we saw something, but we can't say for sure what it is."

A: My feeling is that it'll be an extremely interesting candidate, but I think we will not be publishing results based on few gravitational-wave events. If it is only one event, we have to see more of those.

B: We have to publish something, otherwise we would have no burst publication from S5 and that would be the only other option and that would be also . . .

C: Otherwise you're doing an airplane to it and you're throwing it out without any reason to throw it out.

B: Exactly.

D: So I think what will happen is that this decision will be made before we decide whether it's a blind injection or not. If I was a gambling man I would also wind up saying it will probably wind up being a blind injection, but it will be interesting to have to make that decision before we know.

C: I don't think we can know what bias knowing the possibility of a blind injection could be in there has on that decision.

D: Oh it's a huge bias.

B: But the other thing we can't do is go ask the rest of the astrophysical community if they saw something around this time, before we open—well I guess we could do it but it seems crazy to do that—before we open the envelope.

A: No. I think we could do the following thing—"Look, we have got here a program which says that we will do blind injections, and therefore this candidate might be a real event or it might be a blind injection—we don't know that but we would like to find out whether you had any events." . . . That will make them not so excited, but if you do this only once a year. . . .

B: I think it's coloring peoples' actions to have, in fact, blind injections. If there were no blind injections, we would be behaving differently, and maybe the blind injections helps us to behave a little more honestly in some ways.

C: And we're calmer because there are blind injections.

B: But maybe we're not taking it seriously enough—because there are blind injections.

. . .

C: But don't forget, the blind injections do another thing, which is—a big worry—let's say we can't find an explanation—let's say we didn't have the blind injections—we can't find an explanation for this—we don't want to stand behind it as a publication of a gravitational-wave—you know, what do we do? Do we throw it out? Do we not publish on S5? The blind injections are going to keep us thinking it's OK to come to a conclusion that we think there's something there in the data because we think there could be something there in the data. You know, if the blind injection wasn't there would it be biased the other way—there would be a lot more hysteria, a lot more pressure to throw it out because we're worried, you know would it get airplaned.

A: You know, we have this problem, when do we excite the LSC and when do we excite the entire world? I think to excite the LSC [Detection Committee] a one in ten-year false alarm rate is good enough for me. . . . And for the external world, its one in a hundred.

C: The original number I heard for a primary search . . . was one in one thousand days.

4 The Equinox Event: The Middle Period

As October turned into November 2007, the talk was of how fast to move the analysis forward. The first question was whether the event should be written up and passed on to review committees before the box had been opened on the main analysis. After a lot of discussion it was decided that this could not be done—the full analysis had to be completed first, and this would take time.

The argument was that only after the full analysis had been done could one be sure how unlikely the Equinox Event really was. Only after the full analysis had been done would one really understand the background. If the background were "stationary"—that is, even across the whole stretch of data—one could do as many time slides as one liked on a short stretch and develop as much understanding of the background as one liked. But if the background is nonstationary—that is, some stretches are more heavily populated with glitches than others—then a longer stretch of data is needed to handle the background with more confidence. This means that the effective sensitivity of the detector becomes a function of how long it is switched on. Suppose one had switched on the machine for only two months and seen such an event. Would one have to say: "We cannot count that as an event because the machine was not on for long enough to reassure us that the event was unusual"? Or, to look at it another way, should S5 had been extended so that the extra time on air effectively rendered the machine, not only more likely to see something that might occur, but also innately more sensitive because one could speak more authoritatively about the unlikelihood of anything that did turn up. So it turns out that sensitivity for even a single event is a function of time on air, and this is not something that seems to have been written into the design of the apparatus. Up until this point the logic always seemed to be that likelihood of seeing an unusual event, but not sensitivity, was a function of time on air. Another very good reason for staying on air longer is that seeing a second event would markedly change the confidence in the observations.

In any case, it was hoped that the full analysis would be completed by January or February 2008 and that, therefore, the envelope could be opened in the collaboration meeting of March 2008. Actually, it was not to be opened until a year later—a full eighteen months after the equinox of 2007—and even then, for some, it was opened prematurely.

December 2007

There was another meeting in the second week of December in Boston, where things seemed to be dragging more and more. Below are some extracts from my notes made at the time. In reading these notes it should borne in mind that the sociologist's interests are not always the same as the scientists' interests, though a good few of the scientists share some of the sentiments expressed below. A clean and unmistakable first discovery would be a sociological anticlimax just to the extent that it would be a glorious scientific triumph; the sociologist would have much less to talk about. Likewise, insofar as upper limits are scientific successes, they are of little interest sociologically except that they are perceived by scientists as being a success. Results that the scientists have to argue about are much better from a sociological point of view, for they display the process of decision-making. Here, then, is an account of what was happening around this time from the sociological perspective:

> I sit through paper after paper that finds something initially interesting but winds up showing that it can't be separated from noise. . . . All this "can't be this, can't be that, must be less than this, must be less than that,"—it is all beginning to get exhausting and numbing. Why is there no trace of gravitational-waves *anywhere*. Oh for the days of Joe Weber and the Italian claims.
>
> [Someone remarked] that even if, after they open the envelope, they find the Equinox Event was not an injection, they still have to check to make sure that no one injected anything maliciously and to do this check properly will take a very long time. One could go on like this forever—there is always

another check to do and then another you can dream up; the structure of checking is just like the problem of Tantalus—there is always some new check you can dream up. And this is where things are going. *The apparatus is becoming less and less sensitive*. The equinox event satisfies all the criteria set out before the analysis but now it has been decided that it is not enough. A new criterion has been invented—that the individual contributors to a coincidence must not look too much like typical bits of noise. This means the apparatus has another blank spot in its detection footprint. The scientists say it can see "this far" but now it is working they say "it can't see this far after all." . . .

There is no excitement about this event any longer as there was in Hannover. . . . Is it that the injections have taken the fire out of every ones' bellies? Is it the weight of history pulling everyone down? In a casual remark to [. . .] I point out that I know of no episode in the history of science where the scientists were commended for handling history properly—science is discovery, not nondiscovery. Is it just fear of saying something unclear—because the equinox event is quintessentially liminal and cannot be handled? People are scared of having to say what they will have to say—"they are not sure." But as even [. . .] said, the most likely event is a liminal event. And this means if you eliminate liminal events you did not tell the truth about the sensitivity of the apparatus.

The Start of the Middle

What was also going on in this middle time was further refinement of the analysis. Various things were happening. More time slides were being carried out, so that the background became better defined. Work on additional pipelines was being completed, and one of these took into account, not only the fact of the coincidence, but also the coherence of the signal on the separated detectors—did it make sense that the signals seen on the detectors could have come from a single point in the sky representing a single source? The answer from this pipeline was "yes," and that implied a major increase in the likelihood of the event being real, or the unlikelihood that it was a chance coincidence.

The following figures can be read as a visual illustration of the kind of statistical analysis that was going on. These figures were developed after the position and frequency of the putative signal had been extracted by the computer; knowing these features of the signal, the appropriate filters could be applied to the raw data to make it visible in the data stream. The figures, then, do not represent the process of extracting the signal statistically but may help in getting a sense of what kinds of judgments were implicit in the statistical analysis.

Across the whole frequency range, noise swamps the raw output from the interferometers. Low frequency noise completely dominates any putative signal, and when examined the raw data looks like a smooth curve. It is only after these very loud noises have been removed that any structure begins to emerge. Fortunately, the sources of the large noises are sufficiently well understood for them to be filtered out without risk of distorting any gravitational-wave signal that is there; the whole art of interferometer design is to allow noise at frequencies well away from gravitational-wave signals but to minimize noises that occur within the gravitational radiation waveband that the machine is exploring.

Figure 4 shows the result of the application of stages 2, 3, and 4 of a filtering process, informed by what had already been done in the statistical analysis, to the run of data from an interferometer which includes the Equinox Event. The first trace shows what is left after the very loud low frequency noise is removed. The result is a trace dominated by loud high frequency noise. Removing that leaves the middle trace. Here the shape of the Equinox Event begins to emerge, but it is still confounded by other noises that now do fall within the waveband of interest. Fortunately, some of these are also well understood—for example, the noises caused by vibration in the mirror-suspension wires are well understood and well defined—so they can be eliminated by narrow filters. Examples of such imperfections can be seen in figure 3. They are the tall spikes at certain narrow frequencies. After the last filters eliminating these spikes are applied, the signal stands out fairly clearly, as can be seen in the rightmost trace.

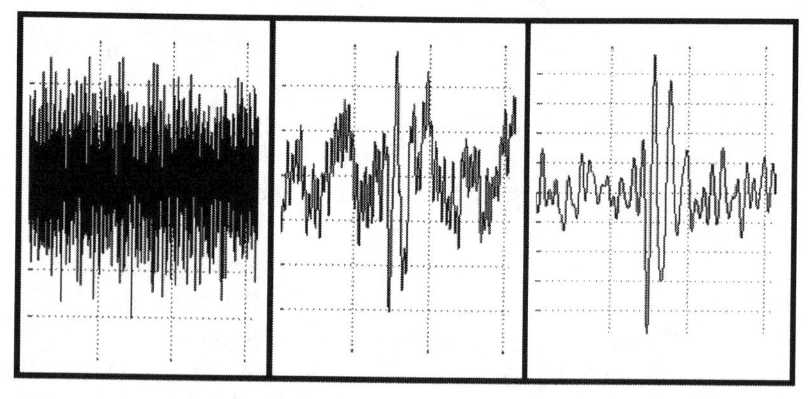

Figure 4. Stages 2, 3 and 4 of a filtering process that reveals the initially hidden Equinox Event signal. (Traces by Jessica McIver.)

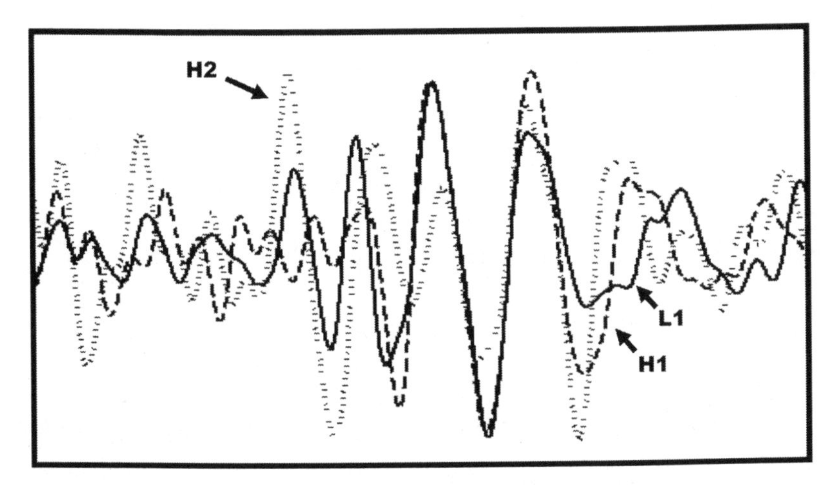

Figure 5. Equinox Event: H1 (dashes), L1 (solid), H2 (dots). Jessica McIver did the scientific work to produce the plots shown in figure 5—designed to be viewed in color—for the LSC-Virgo collaboration. She was kind enough to make new versions suited for black-and-white reproduction especially for this book.

Figure 5 shows enlarged versions of the filtered Equinox signal with the outputs of H1, L1, and H2 overlaid. The statistical analysis which is being discussed in the telecons reflects, in effect, the extent to which lines like these fit over each other, though the lines are, as it were, virtual—they are something like what is implicit in the statistics. The confidence expressed is a measure of the rarity of such close-fitting high amplitude outputs when it comes to overlaying the noise on noise as revealed by the time slides exercise.

To put the matter another way, one has to look at the extent to which coincidentally high amplitude signals from the three detectors, which we can think of as being represented by something like figure 5, overlap. One then has to ask whether this signifies the effect of some common external cause on the three detectors. The only way to know is to see how often the signals might overlap in this way if there were no external cause. We know there is no external cause when time-offset outputs are being compared. The answer obtained from the time slides is that something like this comes up about twenty times in six hundred years, or roughly once every twenty-five years. Given no other sources of information, that is what a discovery is! It is the decision that this event is so unlikely to have happened by chance that it must represent something "real." Hence it is easy to understand much of what will be found in the remaining chapters. We will examine the role of what is culturally acceptable in the scientific community as a publication and what counts as an acceptable discovery as opposed to an outrageous claim; each of these can change. We will look at a carefully worked out flowchart for the process of decision-making. And we will attend to arguments about the exact wording of the way any putative finding or nonfinding is to be announced to the world.

More work was also being done on vetoing out noise so that the background would go down. If the source of a noise was understood, it ceased to be merely a random noise and became instead a disturbance that could be eliminated from the analysis, reducing the number of chance coincidences. (This is a dangerous game to carry out post hoc because one might be tempted—subconsciously—to remove noises that do not look like the "event" while not removing those that do look like the event.)

By the telecon of 19 December 2007, confidence was growing that the Equinox Event would jump the hurdles set by the group and be taken on to another stage. The following extract from that telecon gives a good sense of the state of play and the difficulty of turning a blip in the statistics into the binary "yes" or "no" of a "discovery."

> A: [Does taking this] to the Detection Committee mean that the Burst Group thinks this is a gravitational-wave? Is that the take that other people have? . . .
>
> B: A, you're probably right, but the definition of what is a detection candidate is subjective, and I don't think it can be unanimous. That's why I think we should be presenting it, not necessarily as our detection candidate—we can present it as one event in ten years. We should present it in mathematical terms rather than put in words what this is about. And we should ask the Detection Committee to view it like that. . . . We should present exactly what is the significance of this event and we should ask them to view it with that in mind. . . . For someone a detection candidate might be at the level of one every hundred years, for others, one every ten years, and I don't think it is our goal right now to agree exactly what that means within the group. As long as we agree that outstanding events should be brought to the Detection Committee for further consideration by the collaborations. . . . [We have heard] when in doubt, we should bring things to the attention of the collaborations.
>
> C: [But] we cannot just present something to the Detection Committee, saying . . . we give this for your review and if you like it we will publish it and if you don't like it we will not publish it. It's not the way to go. The Group should stand behind it and have a clear opinion about it. . . . The Group should defend its opinion.

B: But if you remember [the] flowchart, at the end of the day there is going to be a vote of the Council, in order to decide whether we publish or not. . . . We can have an internal vote based on what we have in hand right now, and if this is the only event we see in the S5 [and Virgo] run, and if that statistical statement on the significance of the event is correct, is that crossing everyone's threshold for publication or not?

D [This is a paraphrase]: We must have a consensus, but the consensus might be to make a weak statement about it. . . .

B: . . . I will be delighted if the Burst Group really has consensus on what exactly this event is and whether it should be published or not. But I would not be surprised if there are opinions all over the spectrum from "this is background we simply don't understand" to "this is probably interesting but we cannot publish" to "go on and publish." So I think we should be prepared within the group and within the collaboration, if it goes down this road, to have opinions that are going to fall over the spectrum. And I think it is fair given that most of the work we have in hand right now does not have this event as maybe the gold-plated event that one would like to see for the first detection.

Glitches

At a meeting in December 2007, one of the Burst Group members had remarked that their confidence in the event was growing. But another member responded: "But our confidence in our confidence is not very high." What he was getting at was something else that was pulling in the other direction. Right at the beginning of October, when I had telephoned around to gauge the level of interest in the event, I had found that one of the Burst Group members had said he was not very excited because the event was right at 100 Hz, which is where all the glitches were found. In Hannover I had told another respondent that, as a sociologist, I was depressed by these glitches, since it seemed to make it unlikely that the event would ever amount to anything. The more it amounted to, the happier I would be, because the more sociology I could get out of it—not to mention my "he's finally gone native" sheer excitement about the possibility that a gravitational-wave had been discovered. This respondent had tried to reassure me, however:

Respondent: Well, some days it depresses me, and sometimes I say "well, that's why we have statistics." Because you get to ask yourself the question—this event happens so often in H1 and so often in L1 and how often do they happen at the same time with roughly the same degree of match as what we see here? So it is in fact possible in principle to overcome that. But it has to be something about the characteristic of the event; it is extremely unusual even if its building blocks are usual. Like, for example, the individual events are seldom this strong—we know they were simultaneously strong in both. And they happened so close in time that the wiggles really overlap. . . . But we've known the data had glitches, yet we look for things that shape. We've always said we could find . . .

Collins: In other words, the shape of the glitches is a reasonable shape for a signal as well.

Respondent: Yes they are—they are! Absolutely. We're not being lunatics, we're just doing something that's dangerous. . . . And that is the depressing thing that I learned this summer [during a project in which I was trying to separate glitches from event candidates. Glitches look like signals.] . . . [E]very single glitch in this detector has only one [signal like] pattern [by and large].

Now, however, the "it's too much like a glitch to count as anything" argument was growing in momentum at the same time as the statistical significance was growing. On 11 January the same respondent e-mailed me with the following remarks on the matter, referring back to a telecon of a few days earlier:

> H1 shows a number of other glitches within 10 sec of the equinox event. . . . There was no agreement on the ultimate question of whether background rates are thus mis-estimated. . . . People are going to do more thinking about this, and there will be more discussion in the future about the extent to which one should say that H1 was stationary or else if it was misbehaving at the time of the event. (I'm in the "misbehaving" camp now.)

The detectors all have systems for vetoing data which are not of good quality. Where the data is bad, a "data quality flag" is set, and the data are thrown away or used only with great caution. No data quality flags were showing in the areas of data contributing to the Equinox Event, but it was now being argued that this was a problem of the procedures. Close examination of H1 showed a lot of glitches near the time of the event. These had not been noted because the time intervals within which vetoes were set were too coarse to pick out this area as any different from other areas. More refined examination was showing that the contribution made by H1 to the Equinox Event looked just like a number of other glitches that were happening in H1 around the same time. If the contribution of H1 was really a glitch, and it was looking more and more as though it was, then what was being seen was not a coincidence between random noises, nor a demonstration of the effect of a signal, but a correlation between a noise in L1 and an artifact, that was understood and expected, in H1—in other words, something of no interest at all. Let us call this "the glitch hypothesis."

The First Positive Account

In spite of these doubts, by August a junior member of the team, whom I will call Antelope, had written up the results in a report which spoke of a "Burst Candidate." The existence of this report became widely known and gave rise to a heated telecon on August 20. I found out about it from e-mails and minutes. What follows below are extracts from the minutes of a telecon, with a brief introduction to it by a respondent. They give some sense of how the scientists think about a possible detection:

> There was a heated telecon when it became known that Antelope had started writing up a Detection Report on what he is calling "Burst Candidate 070922a" which takes a rather enthusiastic approach to the E.E. People wanted to know why the Burst Group was sitting on this possible detection, instead of immediately forwarding it to the Detection Committee. The main points of the discussion as recorded in the minutes were as follows:
>
> A: need to update 1 per 50 yr number based on Year 2 analysis, when we open box in few weeks . . .
> B: there will be a total of 1000 time lags in several sets.
> D: don't you know already that this is of least 1 per 10 years?
> B: you know from many ways it is even stronger
> E: even 1 per 5 years is interesting
> B: we don't have good statistical measure just from FAR [False Alarm Rate], look at probability distribution—this is less than two sigma. . . . In most experiments, don't look at 2 sigma results.
> D: Need to move forward with process, this is strong enough
> A: We might decide that at 2 sigma we can't claim a detection. . . .
> G: The system has failed. It has taken too long to finish this. Statistical perfection isn't the best goal. . . .
> D: Send this now to Detection Committee?
> H: This would be great

I: Don't think this should be a claim presented to a jury
E: Will never be unanimous
J: There are other purposes to bringing this forward. For ex: Does this influence commissioning and running?
H: We know there is a problem with glitches that look like this event.
I: agree that this a big problem
D, J: we need to know whether it is an event or a glitch
H: This is hard to do, given the population of glitches
F: Are we procrastinating?
E: No, Burst Group isn't procrastinating. . . .
G: We need to find a better way to engage the Collaboration.

The Amsterdam Meeting

In September 2008 the collaboration came together again in Amsterdam. The group was trying to work its way to a consensus that would be acceptable to the wider collaboration on how the Equinox Event was to be

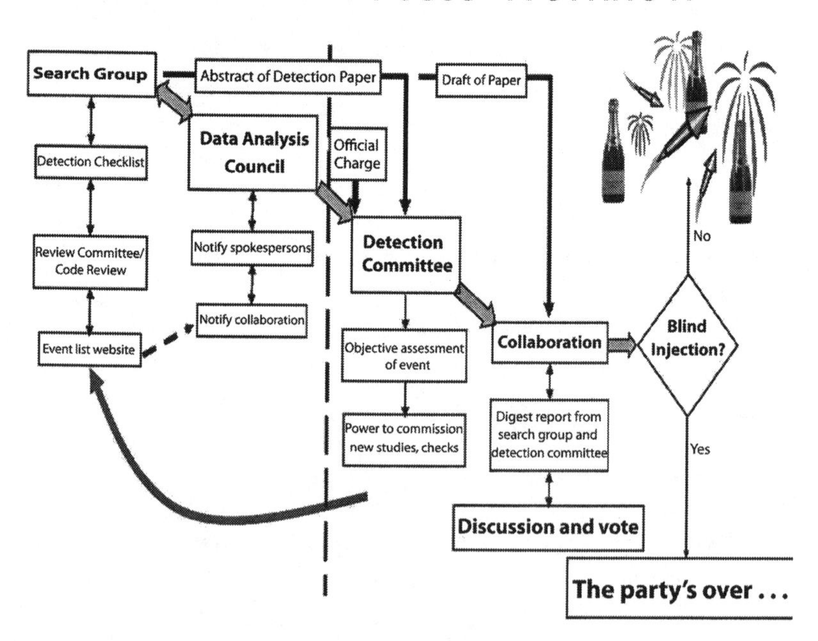

Figure 6. Organizational chart for making a discovery in LSC-Virgo. This version of the chart was first presented at the Hannover meeting.

handled and spoken about. One thing I have discovered about physicists, or at least this group of physicists, is that they love to try to solve problems by inventing organizational structures. I have often been surprised that, when I have asked a question of a senior member of the collaboration about what the members are thinking about this or that conundrum of analysis or judgment, the reply refers to the committees or bureaucratic units they are putting together to deal with it. It is as though a properly designed organization can serve the same purpose as a properly designed experiment—to produce a correct answer. At the Amsterdam meeting the collaboration was reminded of the institutional process that would lead to a discovery by means of the organizational chart shown in figure 6.

In case readers have any lingering doubts about the social nature of scientific discovery, this PowerPoint slide should dispel them. It is a matter of social organization. Furthermore, as can be seen, the "proximate cause" of the first discovery of a gravitational-wave will be, not a physical event, but a vote. As was explained in the context of the "airplane event," a vote is a sure sign that something sociological is going on.

The Abstract

There were two talks about the Equinox Event in Amsterdam. The first was a description by the group of what they thought they would say about it in the proto-publication that was being drafted. This description would take the form of an "abstract" to be presented to the meeting. Days were spent arguing about what this abstract should contain. I collected several drafts as the final version evolved. An early version contained the following wording:

> The estimated rate of the background events with the same strength or stronger is once in 26 years. . . . From examining publicly available data, no electro-magnetic counterpart has been found. The observed sky location is not consistent with the galactic center or the Virgo cluster. . . . Because of the moderate significance of the event, and its close resemblance in morphology to the expected background, we claim no detection.

A subsequent version read:

> A single event above these pre-determined thresholds resulted. It is of moderate significance, XYZsigma, and of resemblance in frequency and

Figure 7. Members of the Burst Group draft the abstract to be presented to the Amsterdam LSC-Virgo collaboration meeting.

> morphology to background events. On this basis, the event is not considered as a genuine gravitational-wave burst candidate.

The next and final version, largely developed during the lunchtime discussion shown in figure 7, but with some subsequent modifications, said only:

> The analysis yielded three events, which were above the search thresholds based on measured background rates. One of these events passed all veto conditions which we had established in advance. We examined the events in detail and we do not consider any of them to be genuine GW candidates.

Figure 8 shows the moment in the presentation when that sentiment was expressed as a PowerPoint slide. The last slide in this talk (figure 9) is a set of bullet points summarizing what had been said.

The Equinox Event was not going to expire that quietly, however. One senior member of the collaboration clearly found the negative conclusion galling and asked how they could be so sure it was not a gravitational-wave candidate.

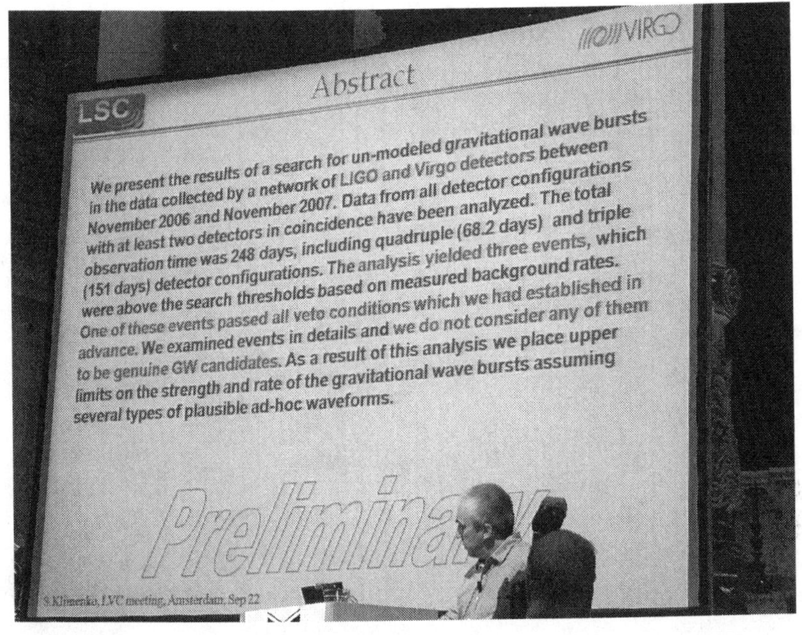

Figure 8. In Amsterdam, the Equinox Event is announced not to be a gravitational wave.

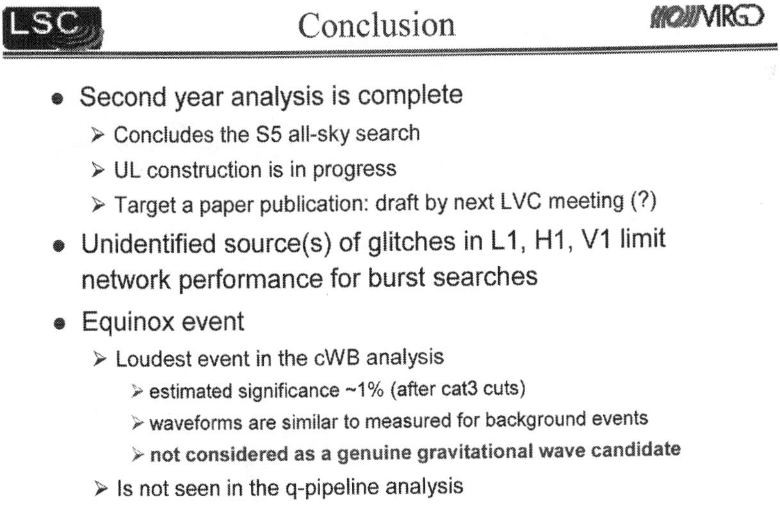

Conclusion

- Second year analysis is complete
  - ➢ Concludes the S5 all-sky search
  - ➢ UL construction is in progress
  - ➢ Target a paper publication: draft by next LVC meeting (?)
- Unidentified source(s) of glitches in L1, H1, V1 limit network performance for burst searches
- Equinox event
  - ➢ Loudest event in the cWB analysis
    - ➢ estimated significance ~1% (after cat3 cuts)
    - ➢ waveforms are similar to measured for background events
    - ➢ **not considered as a genuine gravitational wave candidate**
  - ➢ Is not seen in the q-pipeline analysis

Figure 9. Equinox Event riddled with bullets

A second senior member made a point against the glitch hypothesis. The second bullet point in the last set of three expresses the glitch hypothesis: that, since the event looked like a glitch and occurred in a part of the spectrum full of glitches, at least one of its components was likely to be a glitch. He said:

> I don't understand why the second bullet is a significant consideration: That tells me only that whatever astrophysical event this may have been was unresolved by our instrument with 100Hz bandwidth. . . .
>
> Our instrument has a finite bandwidth. If you can't resolve the event, everything will look like the frequency response of the detector.

This turned out to be a key argument against the glitch hypothesis. The detectors have what is called a "response function." Though they are said to be broadband instruments that can follow the intricate ups-and-downs of a gravity wave as it passes through them and draw them on its output, once more nothing is so simple, and the output is a function both of the waveform and of the resistances and affordances of the elements of the complex machine itself.[1] The second senior member was saying that the machine cannot always "resolve"—that is, exactly follow—the waveform of events because its responses are not equally fine in different wavebands, or areas of the spectrum. Therefore, he said, what hits the instrument might very well look like a glitch, so the fact that this event looks like a glitch does not show it was noise rather than a gravitational-wave even if it looks like noise rather than a gravitational-wave.

Here is the tension between statistics and craftsmanship that will not go away. On the one hand, the premise upon which calculations of the sensitivity of the detectors have always been made is that it is all a matter of statistics. One the other hand, the upholders of the glitch hypothesis feel themselves empowered to say that the meaning of statistics can be adjusted as a result of a craftsmanlike examination of the behavior of the device at the time the data were recorded. As one of my respondents put it in an e-mail:

> This is what you call "double counting" [see below], and what [has been condemned as] "seeking reasons not to believe," but I think that it is a highly responsible use of "experimentalist's craft judgment," which must . . . be applied especially for the very attention-getting case of a first detection.

1. "Affordance" is a philosophical term: e.g., a door handle is designed to "afford" turning.

One danger is that the outcome of such an examination cannot be legitimately used to enhance the statistics—that would be post hoc data massage—but it can be legitimately used to dilute the statistics because that is a conservative move. A conservative mindset, then, has opportunity to exercise itself.

Antelope Again

I have to confess that I do not understand the organizing principle behind the agenda for these meetings, but it is nice for this account that, after all this, Antelope's paper told a different story. His summarizing, bulleted, PowerPoint slide is figure 10.

As can be seen, the first entry under "The Interesting" reports an independent analysis by a scientist we will call "Bison" and his group. This had come in too late for detailed evaluation. The Bison group's approach produced a likelihood of only once every three hundred years for an Equinox-like event to be produced by noise alone. If the Bison group's approach were taken seriously, the previous presentation's report of the Equinox Event's death would certainly have been premature. The majority of Burst Group analysts, however, felt that the Bison group's results were unreliable

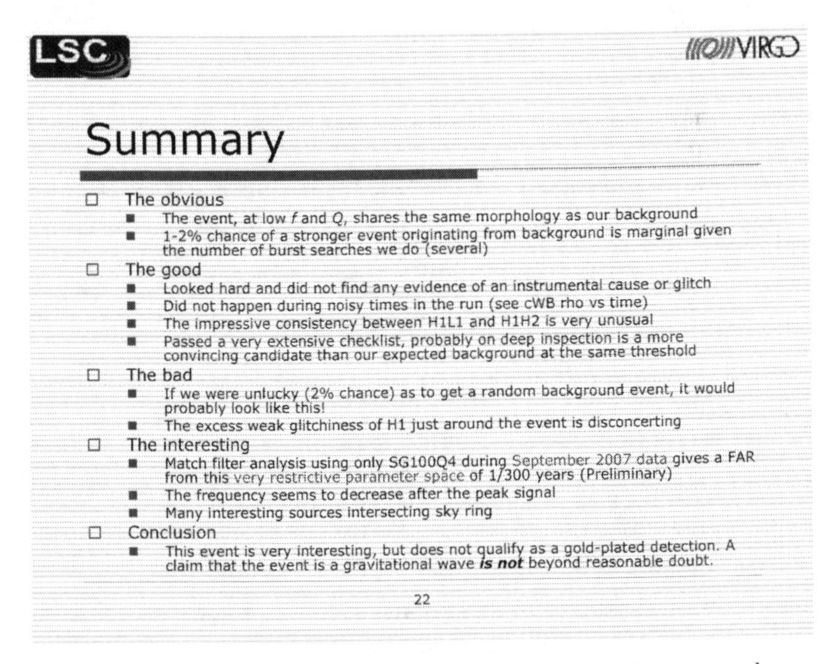

Figure 10. Rumors of the Equinox Event's death may have been greatly exaggerated

in consequence of the "trials factor" (see chapter 5)—the Bison group had tried too many ways of analyzing the data before reaching a conclusion.

The final bullet also runs against the tone of the previous presentation, which was represented as approaching group consensus. Instead of:

> [The Equinox Event] is not considered as a genuine gravitational-wave candidate.

We have:

> A claim that the event is a gravitational-wave *is not* beyond reasonable doubt.

Note how much effort went into the drafting of the exact wording of the abstract presented by the Burst Group and note how much difference the slightly varied wording in Antelope's version makes. This is a theme that will recur. On the one hand, we have physical science—that quintessential home of the supposedly exactly calculable; on the other, we have hours of argument over the exact wording to be used in describing a passage of scientific activity.

Political Interests

The gravitational-wave collaboration that comprises the LSC and Virgo has now gathered into itself nearly every person in the world who has any expertise in gravitational-wave detection physics, and if science can only remain healthy so long as what Robert Merton called "organized skepticism" is encouraged, then that skepticism had better come from inside the collaboration. Barry Barish, who led LIGO to the point of reaching its design sensitivity, frequently made the comment that the only skilled critics of the work being done were to be found inside the collaboration and that, therefore, the organization had to work out its own internal means of subjecting its work to critical scrutiny. The discovery flowchart (figure 6) institutionalizes this principle. But the collaboration can also be seen as a series of healthily competing analysis groups, each ready to try to dismiss each other's findings.

Thus there seemed to be some competition between members of the Burst Group and members of the Inspiral Group. Members of the Inspiral Group seemed more prone to be dismissive of the Burst Group's detection candidate than were members of the Burst Group itself. A senior theoretician who contributed much to the methods of the Inspiral Group made a statement—and it is a statement that he repeated at most of the meetings I was go to—to the effect that the Burst Group on its own could never make a detection. The underlying argument was that a signal profile from the Burst Group could fit no known template—if one did, it would be another group's business—and that unless such a formless signal turned out to correlate with some other event visible in the electromagnetic spectrum, it could never be separated from noise. He said of the Equinox Event and by implication of other similar events that might be found by the Burst Group:

> We don't know where it is; we don't know what it is; and we're not even really sure that we saw something. It's hard to publish that. That's the difference between these searches and the other ones. Right?

Another leading member of the Inspiral Group referred to the conclusion of Antelope's paper and said of it:

> I would never put that statement in the paper because we simply do not have enough information to say that, so we simply couldn't make that claim. The fact is that the truth really is something that we probably will never know [unless it's a blind injection]. We'll never have access to events that are at this level of SNR, so the claim that the event is a gravitational-wave "is not beyond reasonable doubt"—I think the claim itself is [meaningless]. . . . Whether the event is a gravitational-wave or not we may not be able to answer.

Once more, this remark, if taken seriously, would make it very difficult for the Burst Group ever to claim they have found anything unless it were so loud or so well correlated with other events that the group's refined statistical procedures became almost irrelevant.

Another disagreement over the value of the Equinox Event was to emerge between LSC and Virgo. The LSC wanted to use the blind injection challenge in general, and the Equinox Event in particular, to rehearse the entire detection process. The consensus seemed to be that any event that was likely to occur by chance in the region of less than once very ten years should proceed through the detection flow chart right up the Detection Committee—the last stage before it was taken back to the collaboration for a vote. It should be completely clear, by the way, that not a single person I ever spoke to, nor me, for what it is worth, believed that the Equinox Event ever had a chance of gaining the imprimatur from the Detection Committee of "worthy of publication as a discovery claim." The following passage of debate indicates the tenor of discussion at the Amsterdam meeting:

> A: . . . We should actually be able to state how many sigma the candidate is. Gosh, if we can't state that we're in real trouble. [Interjection—we can state it for this one—it is 2.5 sigma—laughter.] You tell me it's 2.5 sigma, [...] says he doesn't believe it, other people here say we should use trials factors [see chapter 5], OK! I would like to feel we could arrive at a consensus about what it is and the collaboration stands behind the . . . [inaudible] sigma, and right now I don't hear that.

From the floor a question is put about what would count as a detection claim.

> A: I don't have an answer to that, and it goes back to what I said yesterday. What we did in the Inspiral Group and we asked the question—"What's interesting?" And its, kind of, one per one hundred years. But that's the low end of interest. And we aren't sure if it's one per thousand, or one per ten thousand years, that is really where we start to feel comfortable. We don't know the answer to that. So, honestly, as far as a criterion for stating, "we have a detection," my impression is that the Inspiral Group feels that we have a two orders of magnitude uncertainty in the false alarm probability.
>
> B: . . . [P]eople do this analysis in high energy physics, and they find 3 sigma all the time and they usually disappear when they do the analysis again. So 3 sigma is not significant at all. 4 becomes interesting, and 5, "this is serious."
>
> C: You can't define sigma and the reason is the long tail. [Statistical analysis is based on models which follow smooth patterns. Here the pattern is not smooth because of the glitches—which show up as an anomalously long tail.] . . . If the long tail persists and then you've got to argue about whether . . . you are willing to say "this long tail kills us" and "we just don't know how to get rid of the long tail," then you've got to deliver it. On the other hand, if you are willing to go out on a limb a little bit, and say "hey, it's a long tail, it's very improbable, one or two percent is, you know, one in a hundred years is definitely of interest, because you've got the tail"—you would never say that if you didn't have the tail—then yes, it's a different problem.
>
> D: I cannot imagine us making a first detection just based on some number—you know, one per one hundred years or something—may well go

above that—"that's it, we've made a detection." If it's one per ninety-nine years then negligible. So . . . we'll have some reasonable number like one per one hundred years, and then when it comes close we'll follow up and we're going to claim the first detection because we also saw that event in a telescope or in a neutrino detector as well, or something like that.

The Equinox Event, as can be gauged from this discussion, just did not have the "gravity" to be the golden event that would convince the world that a first detection had been made. It was all a matter of nuance—Antelope's account versus the accounts that were more hostile to it being discussed as a potential, though regrettably weak, real event. Nevertheless, the Event did seem to meet the LSC's criteria for going before the Detection Committee for further examination. In spite of this, members of the Virgo group were adamant that it should not go even that far. LSC people, on the other hand, wanted to exercise as much of the detection procedure as was possible and were less than happy that the procedure had been cut short.

Why did the Virgo group take this negative attitude? Once more I will invoke the Anti-Forensic Principle (see the start of chapter 3): I don't know! What I do know is that two competing accounts were in circulation. One was that Virgo did not want to see the Equinox Event treated too seriously because Virgo was largely a bystander (except insofar as they shared in the data analysis). Virgo was not included in the blind injection challenge because the protocols were not ready. (Remember, the Equinox Event fell into a waveband where Virgo was much less sensitive than LIGO and thus could not be seen by Virgo; this was something that encouraged analysts to guess that it was a blind injection and put less effort into analyzing it.) This account, then, said that Virgo did not want to see significant gravitational-wave–related activity going on unless they were involved.

The other account, given to me by a senior Virgo spokesperson (S), was that Virgo was more sensitive to the way these things can get out of control than the LSC because they had seen it happen in their own backyard with the Rome Group. They understood the Detection Committee to be there to give the imprimatur for publication, and they wanted to cut off any such possibility from the outset. After agreeing that Virgo had indeed "vetoed" the promotion of the Equinox Event to the Detection Committee level, he said:

S: I think that much of this has to do with the past history of several crucial Virgo members who were operating bars and so they had previous experience of possible announcements that then went into nothing. And so this makes us extremely cautious because what has happened is the way this Detection Committee was formed, which is in disagreement with what [I later heard was the purpose of the Committee and this was that the Committee] was really the ultimate judgment before publication. And so people felt that the Equinox Event was not at the level to go to that step—this is the reason. I think there has been a different perception of the role of the Detection Committee and the other point is this past history of people.

Collins: So you think that the American version of the Detection Committee is really more something that will rehearse and is quite likely to throw things out?

S: Yes—I talked a lot with [. . .] and he was saying that he was seeing the Detection Committee as something that would try to investigate further and even, in a sense, sweep away the fears from the analysis group by asking other questions. Also one thing that was one of the worries of [. . .] was that the analysis groups don't have too much of a grip on the detector itself. They analyze data, and they have less experience of really having the detector working and so knowing that it's a delicate instrument. And so I think this was one of the reasons for the Detection Committee as [. . .] was putting it. But then, it was presented in a drawing, where there was a flow diagram, as the ultimate judgment, and so . . . and we were afraid of collective madness.

C: Some Americans think there is another problem—which is that Virgo does not want anything to be a detection until Virgo is sensitive enough to have a positive share in it.

S: I think this is not right. . . . Absolutely not.

C: Honestly, hasn't anyone from your group ever pressed this on you?—Don't let's let anything go forward until we've got Virgo running?

S: No—no, absolutely not.

C: So if it turned out that this Equinox Event had been bigger, you would have been happy to see it go forward?

S: Yes.

S then pointed out, quite correctly, that there has to be a possibility that one detector team will see an event and other will not because, depending on the characteristics of the source and because of the different orientations in three-dimensional space of the LIGO and Virgo detectors, the signals seen in one may not be seen in the other.

On the other hand, if one wanted to make a case for the more political interpretation of Virgo's resistance to the highlighting of the Equinox Event, one could point out that they also resisted, insofar as they could, any positive mention of it in the draft abstract for the publication to be discussed below, where, at best, it would have been said to be a candidate that had insufficient statistical significance to be counted as a discovery, so that no "collective madness" could have ensued. Of course, Virgo, like the LSC, is a big group, and not everyone in it may have had the same motivations.[2]

Whatever their motives, the Virgo group stopped the Equinox Event going forward to the Detection Committee. Many LSC members believed that, as a result, a valuable opportunity for rehearsing procedures, which, if stringent enough, would give the analysis groups more confidence to forward insecure claims, had been lost.

In terms of the metaphor offered at the outset, the net outcome of all these debates was that Equinox Event was never allowed to become a lasting eddy in the stream of history—nothing within which either new physics or new social life could reorganize itself in an enduring way. The Equinox Event had stilled the stream for a moment, but now it was lost once more in the turbulent water of the social and physical world.

2. On the other hand, in later private correspondence (October 2009), S said that he saw the problem being not so much with insiders as with outsiders taking any result, however poor the statistics, and trying to find some event in the heavens that correlated with it. This could give rise to a lot of speculation and "noise" in the press that had nothing to do with real gravitational-waves.

5 The Hidden Histories of Statistical Tests

In today's frontier astronomy, all one "sees" are numbers. The numbers represent the things that would once have been seen in classical physics (though these may subsequently have been counted and represented as numbers). The great advantage of looking at numbers rather than things is that it makes it possible to impute the existence of things that would be too faint to spot in any more direct way. Thus, if you want to see some faint object in the heavens, you have to allow its emissions—light, radio waves, x-rays, neutrinos, or whatever—to have an effect on you. If you use your eyes as the receiver, you need a lot of emissions. But if you use an array of fancy electronics that can count the impact of individual photons over a long time, you can calculate whether there are a few more photons coming from this point in the sky rather than the background, and that can tell you that there is something there, emitting photons, but at a much lower level than would be required to make much of an impact on your retina. So the development of cleverer and cleverer ways of gathering fewer and fewer photons to make an ephemeral "spot" means that the frontier of observation is always moving out and taking fainter and fainter objects into our observational purview.

The cost of this progress is having to move straight to statistical assessments of the significance of numbers without the prior intervention of the eye.[1] Of course, there is no such thing as a

1. Sometimes misleadingly represented in the newspapers with false-color computer images that look like photographs.

"direct" observation, but sometimes the degree of indirectness seems to stretch the very idea of "observation" to the breaking point.

If it weren't that seeing is done today with numbers, there would be no gravitational-wave detection science, because gravitational-waves are far too faint to "see" in any other way. The science is a matter of the most intricate calculations meant to extract meaning from tiny electrical currents generated within each giant interferometer, as these currents struggle to compensate for—and thereby hold the whole apparatus in equilibrium—mirror-movements thousands of times smaller than the nucleus of an atom that are brought about by minuscule changes of the average phase of photons circulating within the interferometer.

Going back to photons emitted from some point in space, the question is whether there are "a few more photons than there should be" coming from "that" point as opposed to the background, or that might be caused by light scattered off dust in the atmosphere, or whatever. Inevitably, the calculation involves statistics: "There are a few more photons coming from that point—but could it be just a matter of chance scattering that has created a bright spot, or does it mean there is really some object doing some genuine extra emitting?" In the case of gravitational-waves: "Is this movement of the mirrors, represented by those numbers, something special, or is it just the kind of random movement that is going to happen every now and again in any case?"

As has been seen, the answer comes in the form of "unlikelihood." An "observation" consists of a statement: "It is so unlikely that this concentration of photons/set of coincident movements of mirrors could have arisen by chance that it must represent something real."

In the published papers and the announcements that win their authors Nobel Prizes, that unlikelihood is represented by another number. A number is the kind of thing we think of as "objective." That is to say, numbers appear to arise as a result of well-defined states of affairs in the world. It is possible to argue about whether this is a big pile of apples or a small pile of apples, but if there are fifty-six apples, then "that's it." That is why decision-makers love numbers—the apparent "that's it" quality of numbers seems to relieve decision-makers of the responsibility of judgment.

Frequentist and Bayesian Statistics

There is a long-running argument among those who use statistics in the physical sciences between the Bayesians and the Frequentists. Indeed, the argument is of such long standing, and gives rise to such passions, that people on either side sometimes jokingly refer to their preferred preference as their statistical "religion." It is worth looking at the religions for a couple of reasons. First, examination of the argument between the religions is one way to begin to show that the interpretation of statistics is always subjective, however objective it looks. Second, the Bayesian approach, as I shall argue later in the chapter, can be used to justify a publication strategy involving weak claims. This chapter starts with the first reason and ends with the last, other elements of statistical subjectivity being explored in the middle part.

The crucial difference between the two statistical religions seems to be that the Bayesians believe that the unlikelihood statement that is at the center of all statistics-based claims must take into account what you already believe about the world—the "prior probability" of the claim being true. The Frequentists believe that prior probabilities are too subjective to feature in a statistical calculation which results in a number and that the number, therefore, should reflect only what likelihood you calculate, not whether that calculation is credible or not.

Now it is obvious that prior credibility plays a part in the assessment of unlikelihood for both Bayesians and Frequentists. If the Frequentists were looking for a star and their calculations suggested that the telescope had spotted a fire-spitting dragon in the sky, they would be unlikely to report it. In an e-mail, one of my respondents, a committed Bayesian, put it this way in respect of the search for gravitational-waves:

> I think the criteria for the first detection are mostly sociological. It's the level of evidence and group credibility we need to convince people we have seen something entirely outside their experience (a GW source) rather than something entirely within their experience (detector noise, aeroplanes etc). What level that is depends mostly on the attitude of those we want to convince, and their prior predisposition to interpreting the data as gravitational, and less on the data themselves. Once we've done that, GWs are magically within their experience. Life gets easier, and we are free to act like normal astrophysicists (i.e., speculate wildly and mess up a lot without being chastised for it!).

Prior expectations play a role in far more mundane ways too. For example, the claim that there is a coincidence between two signals on two widely separated detectors and that there are only a limited number of coincidences in the background depends on a prior model of a gravitational-wave in which the wave comes from a limited area of the sky and travels at the speed of light. Given that this is the velocity, one knows that the two components of a "coincidence" between two detectors separated by two thousand miles cannot occur more than about 1/100th of a second apart. That means that when one is trying to work out the background of coincidences that might be caused by noise alone, one can ignore all pairs of incidents that are separated in time by more than 1/100th of a second—those aren't "coincidences." If gravitational-waves traveled at the speed of sound, one would have to take into account all "coincidences" where the two events were separated by three hours or less, and one would be able to rule out background coincidences only where the time of impact of the two components was more than three hours apart. In this case, gravitational-wave detection as we know it would be impossible. And yet it has never been experimentally or observationally "proved" that gravitational-waves travel at the speed of light—that is part of the point of the very science that is being described on these pages. In still more detail, the same circularity applies to any attempt to detect gravitational-waves that depends on a template or even, as in the case of the continuous sources and the stochastic background, on a rough model of the source. The effective prior models of the Frequentists have also been at the heart of the criticism of all of Weber's and the Rome Group's positive claims.

The Bayesians simply say that all these prior expectations should be an explicit part of the calculation before you complete it, and they should be represented by a number. The Frequentists prefer to disown unlikely results "after the event." Mostly, but not always, Frequentists and Bayesians come to the same conclusions at the termination of the calculation, but the "not always" can be important, and is important in this story.

The religious war continues because the Bayesians believe the Frequentists are throwing away or misrepresenting valuable information and/or concealing their use of it, or using it as a post hoc decision-making mechanism. On the other hand, the Frequentists point to the fact that it is very hard to put a number to the "prior probability" of something being right; to put the prior information into number form is try to disguise "subjective" guesses as "objective" information. The Bayesians say that it may be a guess, but at least everyone can see what the guess is, and they can criticize and make their own guesses if they want. The Bayesians say that everyone agrees that miracles (unexpected results) require extra evidence if they are to be proved and that their approach makes this requirement part of the explicit procedure, whereas the Frequentists just agree that something is implausible with a nod and wink and then move on.

The Subjectivity of Frequentist Statistics

Joe Weber's practice of "tuning to a signal," discussed in chapter 3, has already shown how Frequentist statistics can be subjective. Every twist of the knob counts as a separate look, or "cut," of the data. If the unlikelihood as reported in the publication is 1 in 10,000 that the result was due to chance, then, if there have been two cuts, it is really 1 in 5,000; if four, 1 in 2,500; if ten, 1 in 1,000; and if there have been one hundred cuts, it is 1 in 100. If there have been many cuts and the paper makes no mention of them, then what looks like a surprising and significant result can be simply a consequence of what is known as "statistical massage" and of no physical interest at all. If all this is done deliberately, then it is cheating. If it is done inadvertently, then it is just bad statistical-craft practice. Immediately one can see that to understand a statistical result, and that includes a Frequentist result, one has to know what was happening at the bench and at the computer keyboard from long before the paper was finally written up. One has to be, in other words, a perfect historian if one wants true objectivity. One has to be a perfect judge of character if one thinks one should take this but not that scientist's word for what they did, and, even if one's assessment of their trustworthiness is flawless, one still has to rely on their own understandings and their memories of what they did. This is part of what is meant by "hidden histories" in the title of this chapter.

But that is not the end of the history. What also matters is what happened outside of the laboratory, as can be illustrated by the case of parapsychology experiments. It is a embarrassing fact that there are an awful lot of experiments that seem to show that, for example, a person trying to use psychic powers to guess the images on cards that are being looked at by someone in a remote location who is trying to "transmit" the image tends to guess right slightly more often than they should if the result were pure chance. The statistical significance of these experiments is generally better than those thought publishable by psychology journals, and the design of many of the experiments seems sound even after the most painstaking examination by determined critics.[2] The resolution offered by some of the more honest but determined critics is that even if the unlikelihood of the result being due to chance is 1 in 1,000, for every positive experiment

2. One should ignore those critics and skeptics who find ever more ingenious ways to explain how a protocol *could have been* violated; no science can stand up to that approach.

that has been reported, 999 have been carried out which obtained null results (or negative results), and they have simply not been reported—they have been consigned to the "file drawer." If this is the case, the one positive result is entirely unremarkable. Only positive results are reported because only positive results are interesting, but they are nullified by the "file drawer problem."[3] For the file drawer to be a genuine problem does not imply that anyone is cheating. Rather, the point is that to know the objective meaning of a positive result one has to be a perfect historian, this time not only of the individual scientists' past experimental life but of the past experimental lives of all the other scientists doing similar work. Only when all this is taken into account is there a chance that the right number will be used in the judgment of the degree of unlikelihood that the result was due to chance.

The file drawer problem affects physics too. High-energy physics is said to have moved, in the 1970s, from a 3 sigma level of significance to higher levels as a result of such considerations. Allan Franklin writes of this as follows:

> Thus, [in the 1960s and 1970s] the observation of such [a 3 sigma] effect was evidence for the existence of a new particle for any one experiment. But, in fact, the data implicitly refer to a sample space containing a much larger number of experiments. . . . So, quite correctly, [Arthur] Rosenfeld argued that one should not consider only a single experiment and its graphs, but all such experiments done in a year. This made the probability of observing a 3 [sigma] effect considerably larger. Changing the criterion to 4 [sigma] lowered that probability considerably. (Franklin 1990, 113)

According to Franklin (private communication), the 4 sigma criterion was the norm by the late 1970s.

But Franklin (and we must assume Rosenfeld) still don't get to the heart of the matter. Why pick a year as the boundary of the sample space? A year is entirely arbitrary. The sample space is all experiments of that type that have ever been done. And that, of course, leaves open the question of what is an experiment "of that type"? Furthermore, why was 3 sigma

3. Incidentally, the parapsychologists argue that so many positive experiments have been done with such high significance that even if everyone had been doing negative telepathy experiments since the start of civilization and leaving them in the file drawer, it still would not counteract the positive conclusions.

counted as satisfactory in the first place and why was 4 sigma counted as satisfactory afterwards?[4]

Nowadays, and we will return to discuss the matter further, high-energy physics takes 5 sigma to be the publishable level. I asked Jay Marx, the current director of the LIGO project, and himself an ex-high-energy physicist, why this was.

> Collins: How did 5 sigma come to be established in the field that you come from?
>
> Marx: Years ago difficult experiments were done to study the weak interaction. Some of those experiments with published high significance—3 sigma and greater—later turned out to be wrong, while experiments that had a 5 sigma effect mostly turned out to be right. The result was a common wisdom—or mythology—that one should not be confident in a result unless it was a 5 sigma effect. I was taught that as a student. It is because difficult experiments can be subject to unknown systematic errors. When you have a quoted confidence level, it assumes you know all your systematic errors, which may not always be true. A 5 sigma confidence level seemed to give one enough confidence because it gives a wide enough berth to cover the unknowns.

4. For an early discussion of how certain statistical result can mean different things to different people see Pinch 1980. There continue to be statistical disputes and ambiguities in the most technical elements of contemporary science. Here is an e-mail circulated by one of the more accomplished statisticians in the LIGO collaboration (the issue continued to be debated for some time):

> I talked with 'X,' one of the BaBar statistics gurus, who is a confirmed Frequentist, but I think in the end we agreed that it doesn't matter whether you are a Frequentist or a Bayesian (as far as I'm concerned, if we're writing down and integrating over probability distributions for the distance to M31, we're Bayesian).
>
> But the main point is this: When setting 90% CL *upper limits* on something (like an efficiency, for which we can define a probability distribution), the ONE-SIDED interval [e90%, 100%] (or [0, D90%], resulting in the 1.28 number for Gaussians) is what people in high energy physics assume you to mean. When I described the approach of taking the 2-sided interval and choosing the worse number (resulting in the 1.6 number for Gaussians) he replied (I paraphrase): that's conservative, wrong, crazy. He never heard of doing that for a X% CL upper limit, and didn't understand why anyone would want to do that (and I couldn't enunciate an argument because I don't understand it, even after reading Patrick's note saying that approach "is the most natural.")
>
> Of course, I'm not saying that just because people in HEP do things one way, we should too; but HEP has been setting 90% CL upper limits since before I was born (and before anyone knew the difference between Frequentists and Bayesians).

Collins: So the actual number of 5 just grew up as a tradition in the field as a result of experience?

Marx: Right: We're saying we don't believe it unless the significance is extremely high because there may be more uncertainties than are reflected in a published error.

The problem of mistakes is, probably, still more severe in the social sciences, which typically take a 2 sigma level of significance as the publishable standard. Two sigma implies that, other things being equal, a result might be wrong five times in one hundred. There is no rationale that I know of for the different significance levels in different sciences except what can be accomplished in practice. There seems, therefore, every reason to suppose that a large number of social science results are wrong, especially as social scientists seem generally unaware of the problems and are not particularly careful about using many cuts (or tunings) until statistical significance is achieved, nor do they think about or declare the process by which published results are selected from the body of un-publishable analyses.[5]

The Trials Factor

Even when everyone is self-consciously aware of the dangers of the processes just discussed, it does not mean that the problem has gone away. When they are self-consciously worrying about it, physicists refer to the problem as "the trials factor," and it is a real concern for LIGO data analysis.

The organization of the work has a bearing on the trials factor. In chapter 2 the four groups that are responsible for looking for different kinds of signal were described. The Equinox Event, it will be recalled, was spotted by the Burst Group. A member of one of these groups, whose anonymity had better be preserved, good humouredly described the different characters of the competing groups in terms of categories from the TV series, *Star Trek*. This was in July 2007, before the Equinox Event had appeared. The Inspiral Group, he said, are like the Borg. According to Wikipedia, "The Borg is a species without individuality where every member is a part of 'the collective' in an attempt to achieve perfection. They assimilate species and their technology when it suits them." That is, they are very efficient and hard work-

5. A biological scientist with whom I entered into casual discussion told me that in her field 2 sigma is also the norm and that they expect about 50 percent of published results to be wrong.

ing, strongly and authoritatively led, and continually expand their activities. This style of organization, I would suggest, is appropriate for a group that must organize a search through a massive series of wave-form templates.

The Burst Group is in contrast chaotic—my respondent likened them to the Ferengi, which Wikipedia describes as follows: "They and their culture are characterized by a mercantile obsession with profit and trade and their constant efforts to swindle people into bad deals." The point is that they are much less well organized, without strong leadership; each member of the group insists on doing things his or her own way: There is a "bazaar" of competing different methods all running in parallel. The inefficiency but creative freedom of the bootlegger is typical of their approach, perhaps made necessary by the quintessentially unknown properties of the bursts they have to look for.

The Burst Group, then, has split itself into competing factions, each doing their analysis in their own way. Incidentally, the term "pipeline" has come into use for a method of data analysis that culminates in a result associated with a statement of statistical confidence. The Burst Group, then, has within it a number of groups each using a different pipeline, all of which they have developed in competition with one another. This is excellent from the point of view of data analysis creativity and cross-checking, but it creates a trials factor problem. When analysis groups proliferate, it looks as though the significance of the results of any one group may have to be divided by the number of different pipelines all looking for the same thing in different ways.

Here is how one member of the community put it, referring not just to what was going on in the Burst Group but to what was going on across the whole data analysis collaboration:

> I just want to say that, if you look across the collaboration as a whole, we have four different search groups, each of which is running a handful of different analyses, so there's something like ten different analyses running—more than ten analyses running across the collaboration—so if you want to say something like we want to have below a 1 percent mistake rate in our collaboration, that means, right away, that you need false alarm rates of something like one in every ten-thousand years if you want to be sure. It's a factor of ten because there's at least ten different analyses running across the collaboration.

But this is swampy ground. If it is the case that the significance of a positive Burst Group result is affected by a negative stochastic background result—the result of a group looking for an entirely different phenomenon except that gravitational-waves are involved—why stop there? Why isn't a positive Burst Group result affected by experiments being done in a completely different branch of physics? Or what if someone, somewhere, is stealing LIGO burst data and doing other kinds of analyses on it, all producing negative results, yet unknown to the collaborators? Does this mean the original results are vitiated? This kind of problem was already hinted at when the Rosenfeld argument for raising the standards of high-energy physics from 3 to 4 sigma was discussed—the boundaries of the appropriate sample space are vague.

Can I sabotage a positive result by doing some quick negative runs on the same data? One scientist put the point very neatly:

> That's the other problem, there's a bunch of [different] numbers associated with this. . . . Somebody . . . wrote a pipeline . . . that has never been looked at or studied, which says once in three hundred years. I could write a pipeline tomorrow, I'll bet, which could not see the Equinox Event, thereby degrading its significance. And all I have to do is write a really bad pipeline. In fact I can write eight bad pipelines that are bad in different ways so they're all uncorrelated, and when I do that all eight will miss that event, that will downgrade its significance immediately. So if I wanted to I could kill that event because of the trials factor.

As this respondent points out, the trials factor problem occurs only if the methods are "uncorrelated" or "independent." Must one, then, actively discourage other people from analyzing the data in independent ways in case they reduce the statistical significance? And, as the scientist who made the remark immediately above also pointed out to me, reinforcing some of the other examples, whether a second result detracts from a first result or adds to it may depend on exactly how it is done. In this case, one of the pipelines actually did produce a positive result while a second actually did produce a negative result; the implication drawn by some was that the significance of the positive result should be divided by two. But, as he pointed out, if, in the case of the "negative" result, the threshold for what counts as "positive" had been lowered only very slightly, it too would have turned out positive, not halving the significance at all. Something weird, or at least indeterminate, is going on here. As my respondent put it: "[There can be situations] where the whole trials factor argument is meaningless—it's just stupid."

The discussion of the Bison group's 1-in-300 year claim illustrates the trials-factor problem in day-to-day practice. It was said the Bison group had tried twelve different ways of matching coincidences, of which only one had delivered anything remarkable. The 1 in 300 should be divided by 12, giving 1 in 25, which was in the same ballpark as the existing group consensus. Bison's defense was along the lines that the twelve trials were not independent, so there was no dilution. This defense, as it happens, made no headway with the other physicists but indicates the extent to which these questions are debatable.

To summarize: to take the trials factor into account, one must first decide what counts as *the same experiment*, and only "the same" experiments count in the calculation. Experiments that are different do not count. Then one has to decide on the boundaries of the social and temporal "space" which will be searched for experiments of the same kind. Then having collected the set of experiments of the same kind, one has to decide which trials within each experiment were independent and which were correlated, and here is it only the "independent" ones that count. Even if these qualities—"sameness," "in the right space," and "independence"—could be defined, they would still come in gradations, not in "yes's" and "no's."

And even if one did know, logically, as it were, the exact right way to define all these imponderables, how would one gather the facts of just how many trials had been conducted in one's immediate location and elsewhere? Once more, as well as solving all these quasi-philosophical problems, it seems one needs perfect knowledge of activities of all the actors in the world who are potentially involved if one is to resolve the number that correctly represents the confidence one should have in a conclusion.

What Did You Have in Mind?

If only it were so simple! But there is yet another layer of difficulty: the meaning of a statistical result depends on what was in the minds of the research team. Suppose I ask for your birthday and you say "July 25." I say "Amazing! That was the date I had in mind, and the odds against that are 365 to 1." Well, if I did already have that date in mind, the odds are indeed 365 to 1. On the other hand, if I did not really have it mind, there is nothing of any interest going on.

To see how this works out in gravitational-wave physics, we can go back to "the Italians" and their 2002 paper. It will be recalled that a central claim of the bitterly received 2002 paper was that the two bars registered an excess of coincidences at a regular part of the sidereal day. In other words, over the course of twenty-four hours there was one hour that registered a peak of activity and, perhaps, another, twelve hours later, that registered a very small peak. The second peak would be expected given that the earth is transparent to gravitational-waves, so that the orientation of any one bar detector in respect of the galaxy effectively repeats every twelve hours, but it could have been worrying that it was not more marked.[6]

As mentioned, a crucial criticism of this finding was made by an American analyst, Sam Finn. He said that he had worked out that if you take a purely random distribution of events, in which the number of events is equal to the total number of events reported by the Rome Group, and divide them up into twenty-four bins in two different ways, there is a one-in-four chance that the two resulting distributions will differ by the amount in the Rome data. The question being asked here was whether the data justified the claim that the peak was correlated with the earth's relationship to the galaxy—the sidereal day—rather than with the sun—the solar day. Finn's point was that the fact that there was a peak when the data were analyzed according to the sidereal day and not when the data were analyzed according to the solar day was statistically unsurprising and represented virtually no gain in information: if you modeled the procedure with random numbers, you could get a difference in event rate of this size one time in four.

Furthermore, Finn argued, even the original claim that there was a zero-delay excess in the first place, irrespective of the clustering, was statistically insecure. He calculated that, in any random distribution of Rome-like data over twenty-four bins, the statistical unlikelihood of one bin standing out to the extent that had been presented was equivalent to only a little more than one standard deviation—again about one chance in four.

In their paper, the Rome Group had reported that the likelihood that the clustering that they had found would be due to chance was only 1.35 percent. How could there be such a difference between their account and Finn's?

The explanation of the difference is simple but revealing. The Rome Group's level of significance was justified if the analysts had set out to look for a peak associated with a *particular hour* during the day rather than

6. The business of twelve versus twenty-four periodicity came up in the early days of Weber's work (see *Gravity's Shadow* and chapter 1, above). It now seems that the analysis of "the Italians" shows that a twenty-four hour periodicity is acceptable, even though Joe Weber's finding a twenty-four hour periodicity was taken by many, and subsequently by Weber himself, as being an impossible result. In retrospect, Weber's initial finding might have been acceptable, but the fact that it seemed to change in the face of criticism was not!

a peak in any unspecified hour. Finn's calculation that a likelihood of 27.8 percent was for *any hour* turning out to have such a peak of coincident events.

Now one can see the problem for understanding the meaning of such a statistical claim. If the Rome Group had done their analysis and only later noticed that the hour containing the peak happened to be one where the detectors were most sensitive to the galaxy as a source of signals, then their procedure would fit Finn's metaphor—set down in his published reply—of firing the arrow first and drawing the bull's-eye later. This would be a clear case of post hoc statistical massage, the equivalent of my saying I had your birthday in mind before you mentioned it even though I hadn't. But if "the Italians" had set out to look only for peaks in the hour when the detector "faced" the galaxy square on, then their 1.35 percent chance calculation was correct—the bulls-eye would have been drawn prior to the flight of the arrow—as when someone really has the guessed the date of someone's birthday before they tell it.

Someone, let us call him "X," told me that he believed "the Italians" had not chosen that hour in advance. X reported to me a conversation he had had with a member of the Rome Group in which it had been confessed that they would have reported the result irrespective of the hour in which the peak had occurred. And there is a rationale for this position. As it was, there was some problem about whether there was one peak or two in the course of twenty-four hours. If the galactic center was the source, then two peaks would have been expected. But the one peak could be explained by the particular orientation of the two detectors if the source was not the center but the whole galactic disk and the disk was put forward by the Rome Group as the source. Yet the galactic disk is not the first source one might think of because of the concentration of stars at the galaxy's center. Furthermore, the Rome Group had, at different times, speculated about other sources, such as a "halo" of dark matter surrounding the galaxy, which would justify alternative directions as potential sources.

On the other hand, the Rome Group defended itself against accusations of any such post hoc data selection in a paper circulated on the electronic preprint server:

> The Galaxy is certainly the privileged place of the sources attainable by present GW detectors and we think that the experiment described . . . should be considered as based on the "a priori" hypothesis of signals originating in the Galaxy. This was clearly indicated in [a] previous published paper: . . . "No extragalactic GW signals should be detected with the present detectors.

> Therefore we shall focus our attention on possible sources located in the Galaxy." (Astone et al. 2003)

On other occasions, however, such as in their published defense of the 2002 claims (see below), they seem less concerned with evidence about the timing of statements about prior expectations, suggesting that it is legitimate to consider any models that seem sensible, whether developed before or after the analysis, so long as they are not *chosen* to increase the saliency of the results. This claim is justified by a Bayesian approach, which will be discussed shortly. But as far as the Frequentist analysis is concerned, one can see that it points straight at the motive behind the choice—which is the *internal state* of the analyst when the choice was made.

The sociological point is this: We would be ill advised to try to work out what the Rome Group had in mind—the discovery of persons' internal states is a perilous enterprise, and for the sociological purpose at hand there is no need to try to discover them or even speculate about them.[7] Nevertheless, it remains that for the purpose of Frequentist statistical analysis it is vital to know the internal states, and this is made evident by the amount of time and energy that the physicists have expended in trying to establish what they were. X, who was an accomplished statistician, treated the spoken report of what the Rome Group would have done had the clustering hour been different as highly germane to what was being claimed. And, Bayesian analysis aside, in concrete terms, it amounts to the difference between a 1 in 4 likelihood of the 2002 results being due to chance—a result almost certainly not worth pursuing—and a likelihood of 1 in 75—something that one might well want to follow up.

The way this dilemma is normally resolved is that analysts are expected to state what they are looking for in advance, as Astone et al. claim that they did. For now, let us sum up what has been established. In spite of the vaunted objectivity of Frequentist statistics, to know the true meaning of a statement of probability in a published paper, one must know the history of the reporting teams activities, and one must know the history of everyone else's "similar" experimental and analytic activities. One must define a time period, and a set of physical locations, over which one is going to count activities as "similar," and one must work out what "similar" means and what an "independent trial" of a similar experiment means. Then one must know what the analyst had in mind, which is the sort of thing that is

7. Again, this is the Anti-Forensic Principle.

normally defined only with a considerable degree of uncertainty, and often in the face of determined disagreement, after a prolonged legal trial.

The Bayesian Approach

What I have tried to establish in the last section is that Frequentist statistics are full of subjectivities and irresolvable uncertainties. This counters one of the arguments used against the Bayesians—that Frequentist statistics are objective while Bayesian statistics are subjective. Actually, both are subjective, and both make it hard for the uninitiated to recognize that subjectivity by presenting the conclusions to their analyses in the form of outcomes of calculations. The Bayesians proclaim that putting a number on one's prior beliefs forces one to state them clearly and that this is a virtue.

The case for the Bayesian approach is most easily made, once more, by thinking about the approach of "the Italians." In a long and technically dense paper published in *Classical and Quantum Gravity* (*CQG*) in 2003, Astone, D'Agostini, and Antonio offer a Bayesian defense of the 2002 paper by Astone et al. They argue that the meaning of the 2002 paper depends on one's prior expectation—that is, on the number of gravitational-wave events per year that are thought reasonable. If that number is very high, then the paper shows that the expectation is wrong. If the number is very low, then the paper provides no new information. But if the prior expectation is in the region of what was apparently found, then the findings give strongest support to the Galactic center as their source (interestingly, under this analysis, not the Galactic disk). They do not claim to have proved anything beyond this—that is, they say, not they have definitely found gravitational-waves, but only that their findings add something in the way of information in the case that their prior assumption about the expected rate, marginally backed up by the actual findings, is correct. They claim that this is how science works—by adding one small piece of information to another. And it seems reasonable to say that since even the Frequentists have to use prior expectations, or models, if they are to do any science at all, changing the pattern of priors even in such a small and provisional way cannot be a bad thing.[8]

8. Astone, D'Agostini, and Antonio's long Bayesian paper, I should add, had no impact whatsoever. Google Scholar shows only one self-citation. Here, then, I am engaged in the rather peculiar exercise for a sociologist of treating seriously a paper which the scientists themselves treated as "invisible." I believe my license to do this lies in the fact that I am not using it to support the 2002 paper—which is what the scientists were interested in—but to add something to a much more general point about the gradualist nature of scientific discovery.

It has to be said that the position of "the Italians" has not been completely consistent through these twists and turns. At one time they claimed that it had been clearly stated a priori that the galaxy was the preferred source, but even then there have been switches between galactic center and galactic disk. It could also be argued that a confirmed Bayesian would have kept quiet about the whole thing, given the very low prior probability of seeing anything much with detectors that were so insensitive according to prevailing theory and also given the context of consensual astrophysical opinion about the nature and prevalence of potential sources. It is very non-Bayesian to talk of "the experimentalist's right to look at the world without theoretical prejudice." (It should be borne in mind that I said that—it is my phrase. I did not hear any of "the Italians" actually say it, though the sentiment is clear enough in the quotation from David Blair [see chapter 1.3] and in much of the discussion of these events in *Gravity's Shadow*.) So it might be said that "the Italians" were Bayesian when it suited them and not Bayesian when it did not. But, once again, I am going to invoke the Anti-Forensic Principle: I neither know nor care about any of this because it is not my job. In any case, though it should never be erected as a guiding principle, in practice it may be good for scientific progress to take one position in one context and another position in another context; human beings do it all the time. Here, however, I am interested only in the logic of any consistent position one could extract from the various arguments.

We can now ask whether it is good or bad for science to publish papers such as Astone, D'Agostini, and Antonio's 2003 paper in *CQG*, which make only small claims, and which are only valid if certain initial, optimistic, assumptions are correct. It is obvious that you are not going to want them published if you are championing a much newer and more expensive technology that is based on more pessimistic prior assumptions about the flux of gravitational-waves such as to render the old technology obsolete. But let us set this aside and try to consider "the view from nowhere."

Here is a problem for the Frequentists: if you demand that a paper should never put forward what one of my American physicist colleagues dismissively referred to as "indicazioni," but only firm discovery claims, then it is hard to establish prior hypotheses. We have seen that, even in Frequentist terms, what "the Italians" had in mind crucially affects the meaning of the results—either their clustering was subject to a likelihood of being random of 1 chance in 4 or 1 chance in 75, depending on whether the hour of interest was chosen before or after the results emerged. But the only firm and clear way to establish a prior hypothesis is to broadcast it through publication.

The alternative is a catch-22. If you have a provisional claim, you should not publish it, but the only way to make your next claim less provisional is by publishing a provisional claim. Only by publishing their claim could "the Italians" put themselves in a position to avoid, after their next round of data gathering, the charge of painting the "bull's-eye" after the fact. And, as it happens, their next round of data gathering does not seem to have supported the now clearly stated 2002 result, so even the LIGO-based Frequentists should be delighted. The 2002 publication made it impossible for "the Italians" to shift their ground, should they have wanted to shift it, in order to support a new interpretation of the aggregate data as new results came in. As a result of the fact that new data did not support the clearly stated 2002 claims, we now know that the 2002 paper contained no new information at all—it has disappeared in as clean a way as it is possible for something to go away in science. And what harm has been done? On the other hand, if additional data had reinforced the 2002 claim, it would have had a cumulative importance far beyond that of a brand new announcement of the aggregated data set: a "bold conjecture," as Popper would say, would have been made, wide open to falsification, that would have enhanced the "scientific-ness" of data that confirmed rather than falsified it.

The four grounds for not publishing are, first, embarrassment about the history of the field and its many unsubstantiated claims, which scientists feel make the subject a laughingstock among the scientific community—this is not a "scientific" reason. Second, the interest of the interferometer groups in having the only viable technology—again this is not a ground that can be legitimately announced in public. Third, the view that science like this should be involved only in the binary process of discovery/nondiscovery, or Nobel/no-Nobel, in which case "indications" claims are sneaky, back-door attempts to take credit from others who better deserve it—which has to do with reputations and rewards, not knowledge. Fourthly, a preference to present science as a producer of certainty, in which case disputes and disagreements should be kept in-house—something which I will argue in the envoi is not the best way to do science.

The view that disputes should be kept in-house I have described elsewhere as a preference for "evidential individualism," in which each individual or laboratory is privately responsible for the entire chain of scientific discovery from provisional indication to published observation claim.[9]

9. *Gravity's Shadow*, chapter 22.

Its counterpart is "evidential collectivism," where the wider community takes responsibility for the scientific truth through public debate of openly published provisional claims. That evidential individualism played a part in the rejection of the Rome paper by the community is easily seen in this selection of quotations from the 2002 GWIC meeting at which the matter was debated:

> In particle physics it's traditional that the particle physics community . . . For 99 percent of physics results it goes through a process of vetting inside the particle physics community before it goes to the press and before it goes to an archival journal.
>
> We [should] . . . agree on a kind of set of guidelines of how we proceed in this community and how we present things . . . because we are going to start having results and as we have results we really need a way that we all try to follow—not exactly—the same rule, that we try to follow as we go forward to present these results so that we present the best kinds of paper which contain the best information and not debate them after they're in press—I think that's the worst problem is if you debate them after they're in press. Inside our community it makes even things that can be very right—they can be right or wrong—but it makes the whole thing controversial—unnecessarily.
>
> The question . . . is where should the controversy be? Should the controversy be after a paper is published in a journal and should it take place in the press, or should the controversy take place between the release of pre-prints for more general comments?
>
> The question is when you go public with your data—is that when you publish it or is there a step in-between where you expose it to the community, which is us.

What ought to be published? Should indications of possibilities be put into the public arena or should everything be kept in house until certainty is reached? We'll come back to the question.

This chapter, as well as opening up the question of whether uncertain results should be published, has revealed that even though statistics makes calculations, it is really representing judgments. Statistical tests have histories which affect their meaning but which are never completely knowable. A statistical test is like a used car. It may sparkle, but how reliable it is depends on the number of owners and how they drove it—and you can never know for sure.

6 The Equinox Event: The Denouement

The book began with my sitting in Los Angeles airport waiting for my flight home after the meeting in Arcadia at which "the envelope" was opened. We are now back in Arcadia, at the beginning of that meeting, and I will shift to the present tense.

Though not everyone is agreed that all the proper work is finished, the consensus is that it is time to reveal what was what. The time, from first noticing a promising event candidate to finishing the work on its analysis, has been eighteen months. This seems too long if gravitational-wave detection is ever to be a science that can rank with the other kinds of astronomy. One needs a result that can be compared pretty quickly with other astronomers' sightings if the promise is to be fulfilled. As it is, the opening of the envelope has been promised over and over again only for the moment to be put off, as people realized that still more prior analysis needed to be done. Remember, the problem is that no one wants to do retrospective analysis, and that as soon as the contents of the envelope is known, any further analysis is going to look "wise after the event." So, within reason, people have wanted all the analysis that could be done, to be done, before the envelope is opened. As it is, there have been hold-ups in completing some pipelines and in finishing the full analysis needed before "boxes" could be opened. Though there are people who think necessary prior work is still incomplete, the sheer time that has elapsed is making their view seem unreasonable. It might have been otherwise if the Inspiral Group had hurried their analysis along as fast as they intend to in the future.

The Abstract

The Burst Group has now written their draft abstract. This is the abstract that will be submitted for publication if it turns out that the Equinox Event is not a blind injection. It has to be written because it reports the upper limit results for S5, and, if the Equinox Event is not an injection, then it must figure in the upper limit calculation. The rule is that the papers have to written with conclusions reached, and firmly and unambiguously stated, before the envelope is opened; there will be no backsliding. The Inspiral Group has found nothing of any significance. The Burst Group has to deal with the Equinox Event. There are three possibilities for how it might turn out—injection, correlated noise, or an event which is below the threshold for reporting as a discovery. The concept of "injection" does not belong within the conceptual universe of the paper, so one of the other two choices has to be made. I understand there have been huge debates over what should be in the abstract, the contents being prefigured by the argument at Amsterdam about Antelope's concluding remark. The full abstract of the proto-paper as finally drafted will be found in appendix 2. The crucial sentences referring to the Equinox Event read as follows:

> One event in one of the analyses survives all selection cuts, with a strength that is marginally significant compared to the distribution of background events, and is subjected to additional investigations. Given the significance and its resemblance in frequency and waveform to background events, we do not identify this event as a gravitational-wave signal.

Those sentences might, instead, have looked something like the following:

> One event in one of the analyses survives all selection cuts, with a strength that is marginally significant compared to the distribution of background events, and is subjected to additional investigations. This event cannot be ruled out as being a gravitational-wave signal, but it resembles background events in frequency and waveform and its significance is too low to justify a positive identification.

This seems a small difference, but it is not; that is why the exact phrasing was argued over for so long. Again, it is worth noting in passing how the work of science is done: there is calculation, but, in the last resort, there are words. The Burst Group has expressed itself in such a way that the most ready interpretation of the event is that it is noise. They have chosen not to say that it could be an event that is too weak to count as a discovery. At least some members of the collaboration believe one of the crucial components is a glitch. Should they have wanted to express themselves in another way, they have been pressed not to do so by strong negative opinions such as those expressed by members of Virgo.[1]

What if it is a blind injection and not noise? If a blind injection, then it was intended to look like a gravitational-wave signal, and the fact that it also looks like noise just goes to show that gravitational-waves sometimes look like noise. In that case, has the Burst Group made the wrong choice? Given the weakness of the event, making the wrong choice would not be a serious mistake in terms of physics but it might be a mistake in terms of mindset. The members of the Burst Group will not have allowed themselves to be excited about the possibility that they have seen a weak gravitational-wave; they will not be filled with regret that their proto-event did not come up to scratch. Caution will have trumped hope.

Opening the Envelope

The room is full and pregnant with anticipation. Jay Marx, the director of LIGO, is to open the envelope (figure 11). He stands at the podium, PowerPoint at the ready. He makes a nicely judged and well-received wisecrack that relieves the tension a little: "I've given lots of talks to this group, and this is the first time I've not seen two hundred people checking their e-mail."[2] Marx then shows a jokey slide of a very old envelope, but people are becoming impatient and this does not work so well.

He then puts up his first proper slide. There is a moment's silence as the group takes in its meaning (figure 12).

1. It was members of Virgo who also insisted that the actual statistical confidence associated with the Equinox Event should not be mentioned in the abstract.

2. It is the habit of physicists (one which I have acquired), to work on their networked laptops throughout all such events, whoever is speaking. It does not necessarily seem rude because they could be making notes, looking, online, at the PowerPoint slides being presented, or checking calculations made by the speaker. But mostly they are writing programs, working on their own talks, or answering e-mails.

Figure 11. Jay Marx prepares to open the envelope

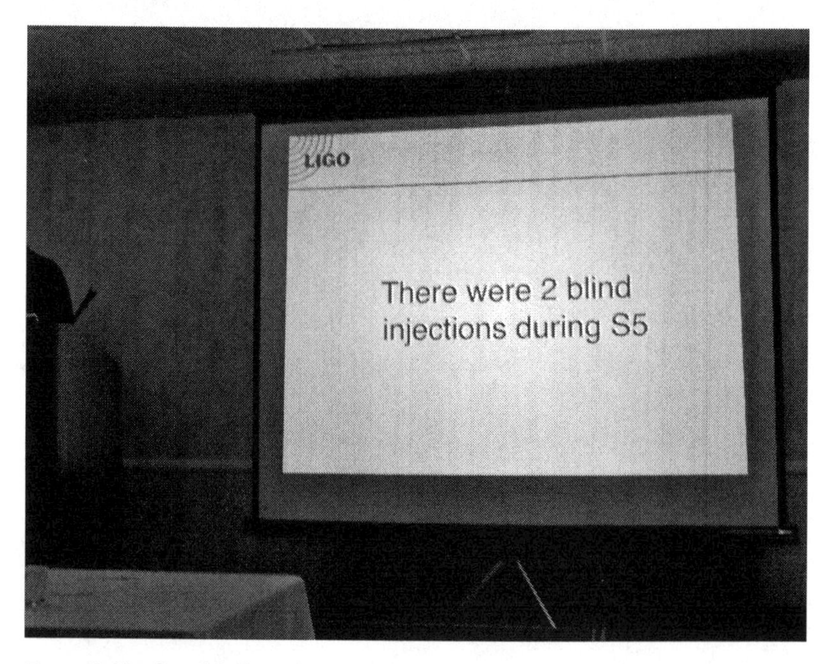

Figure 12. The first slide from the envelope

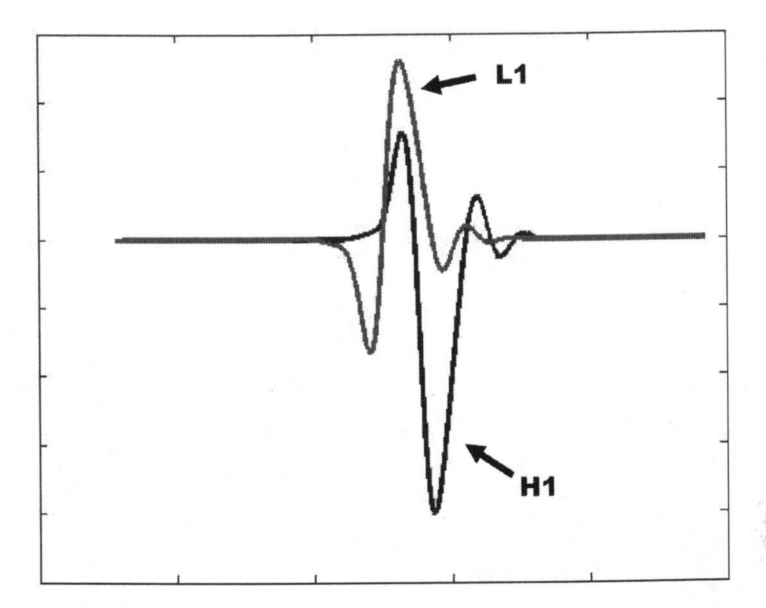

Figure 13. The injection

There were two blind injections in S5! All the discussion had been about the Equinox Event, but there were two blind injections, not one. At best, one of them had been completely missed.

And it had: there was an injection on 13 September of a loud and clear inspiral that did not look like a glitch in any way and that had not been seen at all. The second injection was the Equinox Event, and the Burst Group had nearly exactly identified its characteristics before they decided it was too much like noise. The injection team's input into H1 and L1 is shown in figure 13.[3]

Marx says: "We observed something significant [inaudible] for one of them. But we didn't have the chutzpah to say anything."

There followed a strange few minutes as the audience came to terms with what they had heard. A lot of people, including me, had expected this envelope opening to be an anticlimactic event. It was felt that it would anticlimactic if the Equinox Event was not a blind injection because it would just be the noise that everyone thought it was, and it was felt that it would be pretty uninteresting if it was a blind injection because "so what?" But

3. Comparison with figure 4 shows the match between injected and extracted signal and illustrates the ever-present noise in the detector and the influence of its response-function on what is seen.

Figure 14. Aftermath of the Envelope Opening—about one third of the room

actually, the fact the Equinox Event was a blind injection did not turn out to be uninteresting at all when the moment came—that it had not been treated with more respect suddenly felt like a matter of real concern. This is not to mention that there was a second injection that had not been seen. The audience reaction was a mixture of the jokes and laughter that relieve tension, and that Marx had licensed in his introduction, interspersed with moments of quiet stillness purporting something serious. My note, made at the time, and referring to the Equinox Event, reads: "The room is really quite silent. People are wondering why they didn't go for it."

The next half hour was fascinating as people came to terms with what they had heard and began to stake out positions. We will concentrate on the Equinox Event, as the other injection—the missed inspiral—was an altogether simpler matter, though one about which regret was expressed during the discussion by a senior member of the Burst Group:

> This is vetoed by a category 3 veto because there was high micro-seismic . . . but it is a category 3 veto and because we want to analyze data—because we want to detect things in that data—and we have not produced that histogram. These injections might have been gravity wave signals and before we

> opened the box we forgot, I forgot to ask for those histograms, which I often do. . . . We always said we will look in all the data we analyze and we didn't look at all the data we analyze.

Within a few moments a member of the Inspiral Group had found the inspiral event by analyzing the appropriate stretch of data on a laptop and had shown how the analysis algorithm correctly highlighted the right template with the right masses of stars, distances, and so forth. It had been missed because, as the quotation indicates, it had occurred in a stretch of data vetoed by data quality flags. The vetoes, however, were category 3—the light touch vetoes that invite a further reexamination. That reexamination would have happened in due course, but they had not done it prior to the opening of the envelope. Both an in-between procedure called "looking at the histogram" and another which involves looking for coincidences between H1 and H2 would have indicated that the vetoed data needed more attention. Both were meant to have been carried out alongside the main analysis, but they were not done under the pressure of completing the main analysis in time for opening the envelope. The event would have been detected eventually because those stretches of lightly vetoed data would have been reexamined. The incident illustrates the logistic choices to be made where vetoes are involved. The lesson learned would bring about a reordering of priorities. There is nothing more to say about it.

The Equinox Event is more complicated because the outcome could still be made out to be a success. If the event was not clear enough to count as a detection, then the LIGO community could be said to have demonstrated virtue by refusing to give the coincidence credence beyond that which they would give to any other chance concatenation of noise. The first reactions were indicative of the argument to come. The senior theorist who often invokes the trials factor problem spoke up early and forcefully to make his usual point. He said that it was impossible to claim a detection for an unmodeled event unless there was a correlation with something else in the electromagnetic spectrum or the like. This received a robust response from the floor: "We don't need a model. If we see something that is statistically significant, we should have the guts to say so."

Surprisingly, a very senior scientist known for his caution, especially when it comes to confusing signals with artifacts in the devices, said:

> Maybe [I would] argue with [the senior theorist] about this. The only reason for us hesitating was the discovery that there were events [glitches] that looked very much like this in the data. That's the only reason. But the fact

> is that none of them—and if you look at the rates for them at the most optimistic level, you would have gotten rates that were still smaller than what we've now got. . . . People asked the question, would you publish this event, and then I would have argued. That was a completely different discussion. The discussion would then have been: should we wait until the middle of S6 and we see more events? . . . On the other hand I think that the thing that made us hesitate is that many of the events that we look at as background looked similar to this. I think that was very emotional and, OK, I feel guilty.

Still more surprisingly an analyst who had done much to destroy the credibility of the Rome Group's 2002 paper remarked:

> I want to make a comment at a somewhat higher level here. We are, right now, and in the foreseeable future, working at the edge of detectability. That is, we may get very lucky and have a very loud source that we can believe in unambiguously, and, you know, some people will go out and buy their tickets to Stockholm and the like even before the paper is published. But for the most part we are almost certainly going to be in a case where we are not going to have the kind of confidence that some of us would perhaps like to have, and I think we need to get used to the idea that we may have to, as a group, say we have seen something and put ourselves on the line over it. And that is not necessarily a bad thing. But I think we do have to get ourselves into the mind frame that we could be wrong. One of the advantages, however, in astrophysics is no one will ever know. [Laughter.][4]

This remark was echoed by another audience member:

> I was going to say something that sort of parallel to what [. . .] said. Right now we have a conservative mindset, that is, that we don't want to publish that we saw something when we didn't. And I think there's nothing wrong with that, but I think we have to decide are we willing to live with the possibility of not seeing something that is really there in order to be conservative or not. So it's a sort of cost-benefit analysis. . . .

In these last three comments I thought I heard the coo of "Italian" pigeons coming home to roost.

4. It would be right to treat the remark about astrophysics as the joke the audience took it to be.

A senior member of the collaboration, expressing a Bayesian sentiment, said:

> [O]ur conservatism will diminish. When we are looking at our fifty-seventh detection, we will be less conservative. However, we are always going to be at edge of sensitivity for lots of events. So the question of digging signal out of noise is always going to be with us—not for every event but for lots of our events.

Another audience member regretted how things had turned out:

> I am very afraid of what this means. . . . I think . . . the Burst Group burst searches should have been considered more seriously. Maybe not be ready to go out and tell [the world, but] we should certainly have considered it as a detection and we didn't. I think the psychology of many people has [been] "it looks like a blind injection so let's not worry too much and we'll open the envelope." I was hoping it was not a blind injection. I really thought it was real. I didn't think we could claim it [as a discovery], but I didn't think it looked like the background at all. [And some exchanges later:] I don't think it was taken seriously by all of us. *We didn't talk about it in the corridors. . . .* [my emphasis]

Another senior member of the collaboration said: "[The last] point is a very, very serious one. We have not yet gotten ourselves into the mind state where we can detect something."

The final comment, or near-final comment, from the floor is one we will return to. Someone said: "Are we too detection oriented? In other words, why do we need to make a detection in S5 or in S6? What is the rush? . . . What is the scientific basis for rush?"

The Immediate Aftermath

The next couple of days were intense. Some people became very upset. A highly respected member of the collaboration (HRM) was said to have said during a meeting that the blind injection challenge showed that the collaboration was a failure and would never detect gravitational-waves. HRM is sure, however, that all he intended to say was that the "exercise showed that the collaboration was afraid to make a claim of detection based solely on GW data AT THAT POINT. . . . [The collaboration will make a detection but] was farther from that point than I would have hoped at that time" (private correspondence, September 2009).

The febrile quality of the discussions is shown by the fact that those who had been responsible for the downgrading of the Equinox Event felt they were being attacked and were very unhappy about it. I interviewed the highly respected member (HRM) during the meeting. He said to me:

> I'm actually quite disappointed. I think more than anything else it shows that we haven't as a community really gotten beyond, I think, an unreasonable fear of being wrong about a first detection. Unfortunately I think everyone's perceptions are colored by Joe Weber's experience, and everyone wants to be so super-cautious. If you just look at the raw statistics, and I'll . . . accept the Burst Group's [first] assessment that it was a 12 percent chance that it was chance coincidence [here the very senior scientist is taking a very conservative view of the statistical outcome], that means it's an 88 percent chance of its having been caused by something. So roughly 8:1 in terms of the ratio of what the likelihoods were. And that's obviously not high enough confidence to make a claim of something new—I'll accept that—but in your heart of hearts you really ought to be believing, you ought to go with the odds, and we didn't do that. . . . The Burst Group actually carefully looked at it and really tried to understand what the systematics were. They did it, however, without making the best use of an experimental understanding of how the detector operates. I don't think they really ever got around to spending much time with the instrumental idiosyncrasies, where they might have found ways of either increasing or decreasing the probability of it being real.

The last remark implies that something might have been done to work out ways of lowering the background through a closer examination of the physical causes of the glitches, which may have enabled them to eliminate some of them. The interview was to take some interesting turns:

> Collins: So you didn't like [the Burst Group paper] because it was too dismissive of the possibility that this was a gravitational-wave?
> HRM: Yes it was.

The HRM then showed me a long paper produced by the Burst Group analyzing the Equinox Event and revealing that even though it looked like a glitch, its waveform would be what would be expected from certain real events after they had been molded by the "detector response function":

C: Yeah—I have heard people say that the instrument acts as a band pass filter, so what starts looking like an inspiral can end up looking like a glitch.

HRM: Yeah—because all of this stuff [the early part of the inspiral waveform] is at too low a frequency to show up, and so all that comes through is this last little bit, so obviously it doesn't match perfectly with the thing. But why should it? [given that we don't yet know for sure what such a waveform should look like as we haven't proved it experimentally and we have not calculated all the details].

And you know there are things like the reconstructed position that turns out to include the Perseus cluster—a super cluster. Which is one of the largest relatively nearby clusters of galaxies at a distance where this kind of an event, they actually reduce this signal by a factor of six to put it into the noise here.

So there are good reasons for you to say "it might not be real"—they put all those into the paper. There are equally good reasons why you might say "you know, there are some things that make it look pretty good." And those ought to have been in the paper also, and they weren't.

C: I think people have painted themselves into a corner, partly because of the heavy criticism of Joe Weber, and partly because of the really, really heavy criticism of the Italians.

HRM: Yes. There's that also.

C: They were never allowed to say anything tentative—you've either got to claim gravity waves or not claim gravity waves, otherwise you are illicitly trying to get a share in the Nobel Prize.

HRM: And [names another very influential figure] is one of the strongest from that position; he was very harsh on the Rome Group. I read what they did. I think statistically they fudged things a little bit. They didn't include the trials factor in it. But I didn't think what they said was so bad. They said we cannot rule out that this is a gravitational-wave. It has some marginal significance, and we cannot rule out that it is a gravitational-wave. Well, you probably could have actually from the point of view of the energy density and waves, I guess that's a . . .

C: But that's an astrophysical argument.

HRM: That's an astrophysical argument, and I think it's actually one that's very powerful, by the way, but beyond that I thought that what they had was not unreasonable. And of course the first evidence that we're going

> to have is going to be marginal evidence. It's going to be at the edge. . . . [These comments very lightly edited by HRM.]

Another very senior scientist who was known for his cautious approach told me:

> [T]o me the whole thing smells of "my god, are we going to demand before we start investigating seriously the candidates a perfection in the signals?" I'm not talking about publishing—please don't get that wrong. We have not taken things seriously, and things don't come forward through the collaboration. And so effectively what we've done is made our sensitivity a factor of two or three less good by the act of being so conservative. And that's scary. And that's my problem.
>
> C: Do you think it will change now?
>
> Well, people are going to be mad as hell about this and so things are going to change. I hope.

The clearest version of the other view I gained from discussing the actual wording of the abstract of the straw-man Burst Group paper with a scientist who had been heavily involved in its drafting:

> The point that I wanted to make was that a number of people, including myself, did not want the sentence to read like that Rome paper from whatever year it was—2001 or whatever—where what was infuriating about that was its coy playing with "we don't really have good evidence but we want you to think maybe." And that paper haunts us. And people draw different lessons from it, but what haunts me and some other people is not wanting to be accused of trying to straddle the line of winking at people when we aren't actually able to stake our honor on it. So we do not identify this—trying to stay on the sober side of coyness and winking. And I should say, that even having this much about the Equinox Event in the abstract was a heated debate, where a group of our colleagues, mainly the Virgo ones, felt that this was giving way too much attention to this, given that our conclusion was going to be that we were not making a detection claim. I thought they were wrong on that. I am proud that we moved it into the abstract, because goodness knows this was the thing we had to sweat so hard over in this search, but not just because it was hard but because it's important.

After more discussion of the kind of problems involved in calculating the exact significance of the event, this scientist stated:

> [A]t the end of the day I think it is exactly the right thing for us to hold out until we got the kind of evidence that wouldn't cause us to lose sleep overnight after this went out. So even though I'm now exposed to the world as extremely hard line on that issue compared with other people I respect, I still think it's the right thing to do.

The overall solution to these problems put forward by this scientist was to go for a level of statistical significance that would overwhelm all the unresolvable problems:

> I'm gradually coming to have more respect for the folklore that we could inherit from high-energy physics, which is "insist on 5 sigma before you claim a detection." And we've resisted that because, in part, our noise is so non-Gaussian that sigma is not necessarily a meaningful thing for us and that is probably why we haven't adopted it, but it has been put on the table as a kind of criterion. And then people gasp because the false alarm rate associated with 5 sigma in Gaussian noise is extremely small. But it does have the one virtue that it's why they don't agonize over the trials factor or anything else because they know they've got a huge trials factor thing, and I guess they've learned that it's really hard to account for that in any accurate quantitative way—there's too many ambiguities. And so you want to get to the point where no matter if somebody's wrong about the trials factor by a factor of two it doesn't change what is the, from a statistical point of view, nearly incontrovertible nature of the discovery. Whereas we are trying to be aggressive, at different levels—on the hard-ass side other people want us to tolerate more risk—OK. But I think we're in a very different regime than people who are used to, on a monthly or yearly basis, trying to decide is this a discovery or not?

This respondent is saying that, because it is a first discovery, one must be absolutely certain, and that is why a very strong statistical justification is necessary. Taking a broader perspective, however, one might say that because this is a new branch of science that has never made a confirmed detection, the first discovery will be something in which confidence rises slowly; it will at first be suggested by indication papers and then supported by steadily less tentative claims.[5] When I put this to him, however, another

5. A good model would be Joe Weber's series of increasingly confident publications in the 1960s! (See *Gravity's Shadow*, chapter 4.)

very senior scientist said: "I think that's abrogating our responsibility as scientists."

The respondent I am talking to in the long quotations above is one of the ones who pressed the argument that the Equinox Event looked too much like a glitch to be taken seriously. In a subsequent e-mail he told me:

> [T]he two seconds surrounding the E.E. in H1 data contained several actual glitches almost identical to the E.E. itself. They were slightly weaker, but only slightly. Separate from the weak statistics of the E.E., it would have been very difficult to defend the E.E. as a detection given that we'd have to explain its appearance in a "mini-storm" of actual glitches.

In our conversation I put to him the fact that we know that signals can look like glitches. He knows the argument but, as he says, is willing to "bare his soul": "This looks like the shit in our detector, and I won't face the world saying that the thing we found looks just like the shit."

I make the point that a signal can look like shit but can still be claimed as an event because of the statistical confidence in the coincidence. He says:

> That's right, from a statistical point of view that ought to be correct. So now I'm just telling you—I'm baring my soul as an experimental physicist. . . . Noise, when we talk about it in the abstract—we always have to live with it—it is what makes measurement hard. But there's the noise that will always be with us like the poor, and then there's the failures of craftsmanship, that people should be ashamed of, that's ill behaved. That doesn't have good statistics. That means that formal significance estimates should be more suspect than they otherwise should be. . . .

I interrupt and say that this means the noise is being counted twice—once in the statistical calculation and once when you look at the signal and say it looks like noise. My respondent agrees:

> I think you're on to it. And I think I agree with that. It does count it twice, and it counts it twice because it's bad noise because it's bad noise—it's crap noise. It's noise that shouldn't be there.

There is another problem that my respondent explains prevents him accepting the position I am representing. This is that we cannot fully trust the time-slide method for estimating background: "If I had confidence that we understood all the subtleties of our background estimates—if time shifts didn't sometimes dramatically give you the wrong answer for the background—then I would agree."

Why might time slides give the wrong or inconsistent answers? There are two problems. The first is the size of the time slides that are used. There is a lower limit on the period of a time slide. Thus, the Equinox Event lasted about 1/25th of a second, but some signals can be expected to last a second or two. If the time slides are shorter than this, it could be that coincidences between bits of the *same signal* could be counted as spurious and therefore part of the background. This would incorrectly reduce the apparent significance of any real event. On the other hand, if the time slides are too long, they might compare areas of the output of one machine with areas of the other that do not represent the background noise that was found at the time of any putative signal. As it is, the convention that seems to have been adopted, though I have not been able to find out exactly why, is that comparisons should be made between the output of one instrument and the output of the other subject to a succession of 3.25 second-long shifts. People worry, however, that changing this 3.25 seconds might produce a different result for the background. The choice of length of the time slides might give rise to an arbitrariness in the measurement of the background.

There is something of still greater concern to the community, however. The Equinox Event occurred in a short section of H1 data that was untypically highly populated by glitches. If one were to separate out that section of data and subject it alone to time slides in respect of the output of the other interferometers, one would expect the apparent background to be higher because, given that there were a lot of glitches in one of the data streams, there would be a higher probability of chance matches with glitches in the other data stream. The method as it was actually used—which compares a long stretch of data with another long stretch of data—might underrepresent the background for the short glitchy stretch. This is a powerful argument for not taking the background at face value when calculating the likely improbability of the event being due to chance—it makes it look a little less like double-counting. But, again, there is no certainty about how to use this "intuition" in a quantitative way.

These points taken into account, I again present the point that the whole design of the instrument has turned on the statistics of coincidence, but my respondent returns to an earlier passage of the discussion: "[W]e go back to the point that I keep saying: I think it is the morally correct stand for someone trying to move from knowledge that we don't have to knowledge that we do have to insist on a pretty high standard of evidence."

What is a heated argument finishes with a joke. He says: "And 3 sigma things happen all the time."

I say that ignoring 3 sigma results would rule out all quantitative social science. He says that the same would apply to medical research, and we both agree that maybe that wouldn't be a bad thing.

Now the scene shifts to the Los Angeles airport, and I begin to write this book.

7 Gravity's Ghost

What was the Equinox Event? As it happens, we now know exactly what it was—a blind injection with a waveform almost identical to the waveform extracted by the Burst Group. The astrophysical event imaginatively represented by those who injected the waveform was pretty well what the Burst Group said it would have been had it been real. The Burst Group did its calculations right!

The real Equinox Event—the blind injection—is much less interesting, however, than the metaphysical Equinox Event—the ghost that haunted the collaboration from September 2007 until March 2009. The existence of "Gravity's Ghost" makes it possible to unpeel the layers of argument and inference that will surround any marginal first detection by the collaboration. There is layer upon layer. First, however, an apologia.

Apologia: The Equinox Event and the Role of the Sociologist

As things stand, a sociologist is a guest in the house of gravitational-wave detection physics. Perhaps as physics moves into the twenty-first century, there will be more such guests whose role will be recognized. In my home university department, which houses half of the Centre for Economic and Social Aspects of Genomics, it is a standing joke that there are more social scientists and ethicists studying stem cells than there are biologists. Microbiology accepts that its social legitimacy depends on this phalanx of outsiders watching every move. Physics, that untypical corner of science, is still protected from this kind of scrutiny, which is why the sociologist's role remains so unusual and so delicate. In the case of gravitational-wave detection there are good reasons for privacy to do with making sure unscrupulous and untrained people do not get hold of data or proto-findings before they have been thoroughly analyzed, but there seems to be no reason why privacy should continue to be the default position in physics, an enterprise so heavily dependent on taxpayers' money. It would seem more natural that the right to privacy be justified on each occasion it was invoked. Still, a guest I am.

In this chapter and the next the sociologist's role gives rise to some reflection on how things might be done differently; this might not be thought proper, coming from a guest. For example, from the sociologist's perspective it seems that there is a tension between the model of high-energy physics, which many in the community endorse, and the essentially *pioneering* science of gravitational-wave detection.

In the spirit of the Anti-Forensic Principle, the subject of this sociological evaluation is not individuals and their intentions but the unfolding logic of "roles" or argumentative "positions" within the institutions of, first gravitational-wave physics and, second, science as a whole. Those roles are "illustrated" by the comments made by the individuals who are quoted here, any of whom is more than capable of switching from one role to another for the purposes of argument and analysis. Likewise, the sociologist's views that color the analysis should be seen as the product of the sociologist's role, not the particular sociologist who is writing these lines.

What is, or should be, special about the sociological perspective is that it retains a distance from the day-to-day activity of science—a distance that sometimes makes it possible to reflect more easily on strain or tension whereas full-blown participants are too busy *living* such tensions to reflect on them. To reflect and analyze properly requires, first, a journey as close to the heartland of the science as possible, where the participants have all the advantages, but then, and only later, a stepping back, which is not a natural or necessary activity for a participant. But for the sociologist not to step back and try to open the window on a larger perspective, even though it could be seen to violate the etiquette of a guest, would be to fail to fulfil the duties associated with the role.

There is a moral danger here nevertheless. Though gravitational-wave physics has occupied the larger part of my academic life, it is not the whole of my academic life. I do not spend the hours and hours every day of every week that the physicists spend, calculating, writing programs, analyzing data, and fixing bugs. I delete dozens of e-mails connected with gravitational-wave physics every day without reading them; the physicists have to read them and respond to them. I attend the occasional telecon when things look interesting; the physicists attend two or three telecons a week, often at unsocial hours. I do not spend nights away from my family in remote locations working shifts on the interferometers. Gravitational-wave detection is my respondents' world in a way that it is not my world. Though I put a lot of effort into observing and understanding that world, it does not compare with the effort, physical, mental, and emotional, that they put into making and living it. My reference groups are different—not high-energy physicists and astronomers but social scientists, philosophers, scientists who want to reflect on the meaning of their work, and perhaps a few general readers. Furthermore, when it comes to the algebra, the computer programs, and the calculations, I remain very much an outsider.[1]

Worse, for the time being I occupy, as Peter Saulson put it, the "bully pulpit."[2] Right now, duty or no, I am the only one who is writing a book about gravitational-wave physics, and this gives me more space and opportunity to talk about it to a wide audience than the scientists have themselves. In consolation, when the discovery of gravitational-waves is finally confirmed, the sociological commentary is almost certainly going to be trampled over in the triumph, a state of affairs that is nicely captured in the old joke: *Priest from pulpit*: "In the retreat the lame will be in the van." *Cynical member of congregation sotto voce*: "But not for long."

As a new science unfolds it is impossible to grasp the whole buzzing, blooming confusion of events. If there were no wrong choices, no company would ever go bankrupt, no general would ever lose a battle, and no cars, planes, or space-shuttles would ever crash. From the sociological perspective it can be seen that certain choices associated with the Equinox Event could have been made differently. But there is no culpability: it is just that the buzzing, booming confusion always intrudes in unexpected ways on the perfect world we are bound to believe we can create.

1. But on the relationship between mathematical understanding of physics and other kinds of understanding, see Collins 2007.

2. The term, coined by Theodore Roosevelt to describe the U.S. Senate, means a highly favorable platform for making a point. "Bully" was used in the original context as an adjective meaning "excellent" or "great."

The Layers of the Equinox Onion

We start to examine the layers of argument and inference surrounding Gravity's Ghost from the inside and work outwards. At the center of the science we see the still unresolved and seemingly irresolvable playing out of a set of tensions. The first tension is the freezing of protocols versus the application of common sense. The teams invent rules designed to prevent any overt or subconscious post hoc massaging of data that would lead to false statistical inferences. All development work must be done on the "playground" data or on the contrived coincidence data generated by a time slide. Only after the protocols are frozen is the "box" opened on the real data. But this procedure can fall foul of common sense, as the airplane event so dramatically demonstrates. If the frozen protocols have not anticipated everything, the unanticipated factor produces a tension between statistical propriety and the truth of the matter. It needed a vote in the case of the airplane event for the truth of the matter to triumph, but the vote was not unanimous—the tension remains unresolved; what, according the canonical model, should have been compelling and universal logic turned out to be a choice.

In any case, the idea of freezing protocols is placed under strain from the very beginning by the online searches. Common sense demands that online searches be carried out. Events loud enough to be seen prior to the application of the full panoply of refined statistical techniques have to be given special and immediate attention if only because they need to be brought quickly to the notice of astronomers working with neutrino bursts or the electromagnetic spectrum. As soon as an online search has spotted something, the clear distinction between playtime and real analysis is violated. There is no resolution except the very special vigilance that the logic of blind testing implies can never be sufficient.

The second tension, closely related to the first, is statistical purity versus craftsmanship. On the one hand, there is a stream of numbers that emerges from the extended causal chains of events triggered by the interferometers. The numbers include data quality flags which imply that we take "this" data more seriously than "that" data. In principle, a completely automated computer program operated by a trained pigeon should be able to analyze the data and, once the parameters are set, say "we cannot conclude there are gravitational-waves here" or "it can be said, with the following degree of confidence, that there are gravitational-waves there." In gravitational-wave physics as we know it, it seems that this cannot be done. Most of the gravitational-wave scientists hold the position that there is still a craft element to the work and that the data has to be looked at in concert with a close examination of the working of the machine.[3] In this case, the stretch of data which contained H1's contribution to the Equinox Event, on close examination, and only on close examination, was found to be populated by glitches that looked like the Equinox waveform. The role of the experimentalist, as opposed to the pure data analyst, gives rise to doubt about H1's contribution to the "coincidence." If this doubt gains the upper hand, there has been no coincidence. Seen from the different points of view associated with the many roles within the community, this was either the assiduous application of the experimenter's craft or the double-counting of noise driven by an overriding desire to find "reasons not to believe."

A complicating element is that experimental craftsmanship can be applied both ways. The most famous example of a positive application is Robert Millikan's analysis, in 1909, of his oil drop experiment, an experiment, incidentally, carried out at Caltech, the home base of the successive directors of LIGO. Millikan wanted to prove that the unit of electric charge was "integral"; that is, no electric charge could be subdivided beyond the unit of charge carried by the electron. To do this he needed to show that the charge on an oil drop was never less than a multiple of this unit—there were no "fractional charges." But Millikan's experimental notebooks show that he did find fractional charges—or, at least, apparent fractional charges. He applied his craft skills retrospectively to dismiss them as experimental artifacts—something that under the collaboration's protocols would be quite beyond the pale. But the judgment of history has confirmed Millikan's approach. He applied experimental craft to extract the right result where the actual data could easily have been seen to support his opponents, who said that charge was indefinitely divisible.[4] The history of science is full of similar examples; craft judgment is applied after the event to filter the data and extract a result which turns out to be right.[5] In gravitational-wave detection, however, experimental craft knowledge used after the event is nearly always used to make the event—a potential result—disappear. The approach is explicit: craft skills can be used in a proper way to reduce the salience of something potentially positive like the Equinox Event; craft skills cannot be used to reduce the salience of an event when an upper

3. And only very close examination would reveal whether completely blind analysis could be done in fields such as high-energy physics, even though from the outside it appears that it could.

4. The locus classicus for the discussion of Millikan's experiment is Holton 1978, 25–83. See Franklin 1997 for an opposite view.

5. See Collins and Pinch 1998.

limit is being set, however, because this would be to use them to make the outcome more astrophysically significant. The fuss over the airplane event arose because, for once, the opposite was allowed. The community is conservative—no mistake is being made by applying craftsmanship retrospectively so long as less science is being claimed rather than more.

Craftsmanship could be applied to enhance an event by using an understanding of the machine to explain and filter out more noise. If the causes of background events can be understood, they can be ruled out of the background. The less chance coincidences, the more do remaining events stand out—the higher their statistical significance. The Bison group's result were seen by almost everyone outside of Bison's immediate group to demand rejection because his method suffered from post hoc decision-making. But, to repeat, the glitchiness around H1 was allowed to be applied post hoc because it helped to eliminate an event.[6] One can, then, see the built-in technical bias in the procedures. The bias seems reasonable, but, assiduously applied, it would have ruled out many of the great pioneering results in science.

Not unrelated to this point is the judgment about the relaxation of vetoes. Too much relaxation could look suspicious, but some relaxation is necessary, as the events at Arcadia revealed. How should that choice be made?

These tensions are elements within the larger tension between the acceptance of Type I versus Type II errors—false positives versus false negatives. The strain is at the center of every statistical science, and, as we have seen in chapter 1, it goes right back to Joe Weber and the start of the gravitational-wave detection business. Contemporary gravitational-wave science has demonstrated a strong leaning toward the avoidance of Type I errors and away from the risk of claiming a false positive. Fear that this mindset had become pathological was the inspiration for the blind injections.

Moving out a layer, all of the above tensions come under the larger argument over objectivity versus subjectivity in statistics—the topic of chapter 5—which has been revisited throughout the book. Starting with the particular case of gravitational-wave detection, there are the uncertainties and choices over the interval, or length, of the time slides and also the uncertainty over how to handle the fact that time slides applied to

6. Ironically, the full extent and significance of this glitchiness was first spotted by Bison and his team.

short glitchy sections of data alone would show a bigger background—that is more likelihood of the Equinox Event being due to chance—than time slides applied to the whole data set. There is no clear, "mechanical" way to resolve the problem.

In chapter 5 it was shown that the meaning of any number reported at the end of a statistical process also depends on both knowing and understanding a large body of unknowable things about the history and contemporary activities of teams and individuals. It depends on what individual experimenters had been doing with the data prior to their reporting of the result—too much tuning activity now being the standard reason for explaining why Weber's results were wrong. It depends on what the members of the experimental team were thinking when they produced the number—that the Rome Group had *not* been thinking of the particular hour within the twenty-four that showed a peak prior to them finding the peak being a crucial argument against their claim to have found a marginally interesting peak. It depends on what other people in the team, and perhaps outside the team, and perhaps in private, have been doing with the data prior to the publication; and it depends not just on knowing but understanding the significance of what has been done—the seemingly irresolvable problem of how to calculate a trials factor. Lastly, it depends on what a community is willing to believe—the changing sociology of what counts as a reasonable belief in that community; this is the argument that extraordinary findings require extraordinary evidence, often referred to as "Hume's argument concerning miracles." As both Bayesians and Frequentists understand, what counts as "extraordinary" is a movable feast.

The 5 Sigma Solution and its Problems

One way to try to get round some of these problems is to *bury them* in statistical significance. Systematic errors and really large violations of statistical protocol aside, if the significance of the result is at the 5 sigma level—the level that has become the standard in the high-energy physics community, enforced by what I hear is sometimes referred to as the "5 sigma police"—then many of the conundrums and irresolvable issues discussed in the last paragraphs will be buried alive, as it were. Let there be an unresolved trials factor—it will never be enough to vitiate a confidence level equivalent to one chance in a million or so of being wrong. As a respondent put it, defending the 5 sigma standard:

> You want to get to the point where no matter if somebody's wrong about the trials factor by a factor of two, it doesn't change what is, from a statistical point of view, the nearly incontrovertible nature of the discovery.[7]

There are three problems with this approach. The first is that high-energy physics is different because beefing up the statistics is a matter of waiting long enough for more particles to be injected; it is all a matter of time.[8] Searches for individual sources are not like this. They are quintessentially unpredictable; one gets lucky or one does not, and one cannot control the source, which is the heavens.[9] Waiting for AdLIGO is one way to solve the problem, because there should then be a steady enough stream of sources to make the science look a bit like high-energy physics. However, if waiting for AdLIGO is too readily adopted as the only way to make sure no mistake is made, then Initial LIGO and Enhanced LIGO are being retrospectively redefined as machines which are much less able to make a detection than had been widely believed. iLIGO and eLIGO were built because they might discover something, and the likelihood of that discovery depends on their range. But, assuming that the first discovery is going to be of a weak event, too much caution means the calculated range effectively is cut by a factor of a few. For some, there was cause for concern even in the caution exhibited in the rejection of the Equinox Event because it occurred in a glitchy patch of data. As a very senior and very prestigious respondent put it: "And so effectively what we've done is made our sensitivity a factor of two or three less good by the act of being so conservative. And that's scary."

To go all the way up to 5 sigma, enormous luck aside, makes it much more likely that it is going to be necessary to wait for Advanced LIGO for a signal that can be talked about, and AdLIGO will not be producing

7. On the other hand, Allan Franklin (private communication) points out a case where a published 5.2 sigma result for the discovery of the "pentaquark" turned out not to be supported in subsequent experiments. See S. Stepanyan, K. Hicks, et al. 2003.

8. According to Krige (2001), Carlo Rubbia found a way of taking less time to reach a satisfactory level of statistical significance for the discovery of a new particle, and won a Nobel Prize, by drawing on the results of one of his rivals.

9. Actually, both the stochastic background and the pulsar searches could, in principle, just wait for more evidence if computer time was not so limited: the longer the observation time the more the signal, if there is one, builds up in terms of its statistics. In principle, if any vestigial signal were integrated for long enough it would become significant, though putting this fully into practice may await more sensitive instruments. One of my respondents argued that waiting for AdLIGO, given that it had been planned to be built from the outset, was actually like waiting for more particles to be generated in a high-energy physics experiment even though at first sight it seems like building a new instrument!

good data until about 2015 at the earliest. When Initial LIGO was being planned, an argument put by a critic was that there was no need to build two devices because, if there was little chance of an actual discovery, one site would be enough for all the development work to be completed until the genuine astronomical observatory was ready to be constructed. The 5 sigma level, in coming much closer to ruling out a discovery with the early devices, enhances the validity of this argument and intimates that some money could have been saved without significantly slowing progress—at least, if we lived in a world without politics or human emotions.[10] The missing entry from the "Downside" column of table 1 (see chapter 3)—on the advantages and disadvantages of blind injections—can now be filled in. The blind injections are forcing the collaboration to reveal the true working sensitivity of the detectors rather than their theoretical or measured sensitivity. In the absence of blind injections (or real events to come in S6), the difference would never have been exposed. As it is, there is a chance that we will see a continuing mismatch between the implicit promise of the early generations of LIGO detectors and the performance that they are actually permitted to achieve. For the sociologist, an upside corresponding to this is that the scientists are being forced into declaring their hands with respect to what is going to count as a detection, whereas without the blind injections the question could have been left unanswered until AdLIGO (putatively) makes the problem go away.

The third problem with the 5 sigma standard is that it may turn on a false model of gravitational-wave detection physics in the years leading up to Advanced LIGO. LIGO has been a success—in the sense of it being an apparatus of near miraculous sensitivity completed in a not totally unreasonable time in the face of enormous skepticism that it could be done at all. This success was brought about under the leadership of high-energy physicists. Virgo was largely driven by former high-energy physicists too. I argue in *Gravity's Shadow* that only the high-energy physicists could understand the subtleties, including the political imperatives, and the brutal and unsubtle mechanisms of big science, necessary to bring about this success. The understanding of these subtleties and the consequent application of a degree of cognitive and managerial savagery, were a necessity if LIGO was

10. *Gravity's Shadow*, 717. All the data analysis protocols could have been developed by working with time-shifted data sets from one interferometer. I argue in *Gravity's Shadow* that two sites had to be constructed to keep senior scientists interested enough to spend their lives on the project, and this still seems correct. Even this justification becomes shaky in retrospect if such a high a standard for a detection is set that no discovery becomes possible.

ever to survive and reach a reasonable level of sensitivity. But now that LIGO is "on air," contemporary high-energy physics might be the wrong model.

At the end of my discussion about statistical significance with Jay Marx (see chapter 5, p. 99), I asked him whether 5 sigma, though it been shown to the right standard for high-energy physics, was also the right standard for gravitational-wave physics:

> Collins: Do you think that this kind of standard [5 sigma] is OK for gravitational-wave detection as well?
>
> Marx: I have no idea—you don't know until you've had a sample of gravitational-waves and you can understand the statistical significance the analysis gives about background and how accurately information about the source can be extracted compared to what nature tells you. There's no experience yet. The comments about weak interactions in particle physics were based on years of experience by many people in that field with a significant number of experiments.

Here Marx says that knowing the right standard is a matter of experience and that there is too little of it in gravitational-wave detection to gauge the statistical standard that will nearly always precipitate correct findings.

The deeper point might be, however, that pre-AdLIGO gravitational-wave detection physics is not equivalent to technologically developed sciences like high-energy physics. LIGO and Virgo and the rest are right at the beginning of their scientific lives. To impose statistical confidence standards appropriate to a technologically developed science on a science that has not yet made its first, tentative, discovery could be to stifle it. Imposing these standards could have stifled high-energy physics in its early days too, as even it did not adhere to the "standards of high-energy physics" when it was first developing. As the extract from Alan Franklin's book shows (see chapter 5, p. 98), high-energy physics did not shift from using 3 sigma as its standard for a discovery until the 1970s.[11] There followed a period when 4 sigma was counted as satisfactory. Furthermore, high-energy physics admits and publishes papers, even in high prestige outlets such as *Physical Review Letters*, that are entitled "evidence for" rather than "observation of." "Evidence for" papers do not demand the 5 sigma standard—they accept

11. Franklin has found (private correspondence) at least one instance of a 2.3 standard deviation result mentioned in an earlier paper.

a lower standard.[12] The Equinox Event could have been considered as a potential candidate for "evidence for" even if it was never going to quite reach even that standard. That it was not seems, in a good part, to do with the "myth" of the "Italian" style of "indicazioni."[13]

Another science, which was stunningly successful as a pioneer of a whole new branch of physics, but which used statistical techniques that are very far from the standards of today's high-energy physics, was gravitational-wave detection! Joe Weber simply reported lists of events in the late 1960s and early 1970s when he was becoming one of the most famous scientists in the world. He did not report any levels of significance measured by standard deviations until well into the 1970s and then the levels were "all over the place." The obvious response to this is to say that Weber is a prime exemplar of how to do things wrong. But this is, perhaps, too glib. First, Weber did found the whole international billion-dollar science of gravitational-wave detection—he was an enormous success. Second, the major complaints about Weber's techniques were not to do with the absolute levels of significance reported but the way he generated them. Once one goes in for post hoc data analysis, any level of significance can be generated—statistical significance can never compensate for systematic error. Perhaps contemporary gravitational-wave detection science should be taking notice of the standards, if not methods, of its hugely successful pioneer; perhaps the wrong myth is being promulgated. Today's gravitational-wave detection is Weber in the early 1960s writ large.

There is a counterargument to the relaxation of statistical standards. As explained at the beginning of chapter 4, science has changed. In many of the earlier cases, such as that of Millikan, scientists had quite a bit to go on to guide their "instinct" when it came to choosing which bits of data to keep and which to throw out, or when it was wise to go with a claim even though the statistics were poor. Nowadays, because we are looking at more and more marginal events in observational science, there is nothing to go on but statistics—there is nothing, or almost nothing, to guide the scientific instincts when a judgment has to be made about whether this or that statistically unlikely concatenation of events signifies a real

12. For example, Allan Franklin points out to me that a paper by Abe, Albrow, et al., published in 2004 in *Physical Review Letters* is entitled "Evidence for Top Quark Production in $\bar{p}p$ Collisions at $\sqrt{s}$ = 1.8 TeV" and offers a 2.8 sigma significance. An extended version of the paper with the same title referring to the same data was published in the same year in *Physical Review*.

13. The 2.5 sigma associated with the Equinox Event is probably not enough even for "evidence for," but the point is that the Event was never even looked at as a possible "evidence for," and it never even reached "cannot be ruled out."

event. If there is nothing to guide the instincts, so that the numbers have to stand on their own, a higher standard may be justified—and, forgetting what Weber achieved in the way of founding a field, the demand for a high standard would have applied to Weber too.[14] On the other hand, since the standard of today's high-energy physics could well rule out evidence for gravitational-waves for many years, it may argue still more strongly for the use of statistics to produce "indications" and for the gradualist rather than binary approach.

As can be seen, there are two ways of looking at the question of whether gravitational-wave detection should treat itself like high-energy physics in respect of statistical confidence. There are, as one might say, the "statistical experience" argument and the "nascent science" argument. The statistical experience argument holds that since the Second World War a huge amount of experience on how to do physics has been gathered and that we now know that nothing below 5 sigma is reliable if statistics are all we have to go on in making a discovery—that is what physics has taught us. The nascent science argument holds that, while this might be a valuable lesson, it applies only to the well-developed sciences because their standards will stifle new sciences. It is hard to decide which of these two positions fits the case, but there is probably something to be learned from in both. The trick is, perhaps, not to buy entirely into one to the complete exclusion of the other.

Hammering the Equinox Event into Shape

At the risk of some repetition it is worth re-describing Gravity's Ghost as an ingot of knowledge forged by pressures coming from the past, the future, and the present.

The Past

The meaning of the Event is shaped by the whole history of failed claims to have seen gravitational-waves from Weber's first announcements to the Rome Group's 2002 paper. These events are treated as moments of shame

14. Weber took advantage of this argument in handling the spurious results he reported when he ran, as he thought, one of his own bars in coincidence with the detector of David Douglas at Rochester (*Gravity's Shadow*, chapter 11). He found a level of excess coincidences with a 2.6 sigma significance, but when he found that the bars were actually running out of coincidence by more than four hours, he claimed that 2.6 sigma was not significant by the standards of (mid-1970s) physics.

for physics which are not to be repeated. "Gravitational-wave detection is a science for flakes" is something that many members of the collaboration have heard before and do not want to hear again.

Ironically, of course, there would be no gravitational-wave detection were it not for this "flaky" history. There is now hardly a person who will not admit that without Joe Weber's crazy enterprise there would be no LIGO, no GEO, no Japanese detectors, no Australian effort, and probably no Virgo. Furthermore, there would probably be no LIGO without the widely promulgated theoretical calculations of the strengths of a variety of relatively speculative sources that, under favorable conditions, would fall within the sensitivity of LIGO—possibilities going well beyond an initial discovery of an inspiraling binary neutron star system or a supernova.[15] These calculations were easy to misread as probable outcomes; examine the small print and nothing was promised, but politicians do not read small print. Even the experimentalists' role left space for the expressed idea that something unexpected was bound to turn up, given that people were making claims to the effect that "LIGO was the first detector to reach a level of sensitivity that made it possible to see gravitational-waves," and "whenever a new instrument with an order of magnitude more sensitivity points at the heavens unexpected discoveries are made, and this one has two orders of magnitude more sensitivity."

Physics runs on optimism: the optimism of gravitational-wave detection science is indicated by the figures for the range of the interferometers recorded in the daily performance logs and represented in figure 2 (in chapter 1). Figure 2 shows the interferometers on a bad day, but on most days the range of L1 and L2 was reported as about 15 Mpc. That 15 Mpc did not take into account retrospective "craftwork" that could deal a body blow to a marginal signal. Once again, this is not in the "small print," and any of my respondents is in a position to claim that it cannot be *demonstrated* that there was a significant expectation that iLIGO and eLIGO could see an event. It cannot be *shown* that the caution associated with 5 sigma and the application of craft skills to a signal has effectively reduced the range of the instruments, since the range was never publicly defined in an exact way. Waiting for AdLIGO could be said to be consistent with any document in

15. Furthermore, at a number of meetings I have heard Kip Thorne describing LIGO's range in terms of is ability to see the inspiraling of massive black-hole binary systems rather than in terms of the more usual inspiraling binary neutron-stars (which imply a much shorter range), but there is no evidence that large black-hole binaries on the point of inspiral exist or that the universe is old enough to have given rise to them.

the archives. And yet the optimism associated with the new instruments was palpable. It can be illustrated, quantitatively, with the incident of the Ladbroke's bet.

In 2004, the British betting firm Ladbroke's, opened a book on whether gravitational-waves would be discovered before 2010. The criterion would be a report in *New Scientist*. The odds offered were 500:1 (rumor has it that Ladbroke's was advised by one of LIGO's old enemies), but within a couple of weeks the fevered betting among the scientists and those who knew what was going on inside the project dropped the odds to 3:1 and Ladbroke's closed the book.[16] One can say from this, and from the scuttlebutt, that most of the scientist-insiders thought the fact that a signal was going to be found was worth a bet even at low odds.

As far as outsiders are concerned, optimism may be the only fuel available for a big-spending science, given that the political system goes so much faster than the establishment of new knowledge. "Give us these hundreds of millions, with these opportunity costs for the rest of your constituency, so that we might deliver a bit of knowledge long after you are out of office" works for scientists—who are ready to devote their lives to future generations of knowledge makers—but it doesn't work so well for politicians. Ironically, then, the history of failure to achieve either detections or the fulfilment of promises might also be the condition of such success as gravitational-wave detection has achieved.[17]

So far so normal, but the power of the historical legacy of gravitational-wave detection has been amplified by the use of the history of failure in the internal politics of the field. In the normal way, incorrect or incredible claims are given a short "run for their money" and then ignored. In gravitational-wave physics, however, the failures were more salient. As soon as the big interferometers demanded one hundred times as much in the way of funding as the previous bar technology, the results coming from the much less sensitive bar detectors had to be shown to be worthless. Joe Weber forced the confrontation, were it not going to happen anyway, by writing to his congressional representative insisting that the interferometers were a waste of taxpayers money since his technique could detect gravitational-waves for a fraction of the cost. Joe Weber had to be shot

16. I squeezed in relatively late with £100.00 at 6:1!

17. Pinch (1986) points out that pioneering neutrino detectors were built on the back of promises of a high flux of detectable neutrinos. The promises were successively downgraded as the detectors were first funded and then built.

down in the most explicit way or LIGO could not be justified.[18] Thereafter, each positive result reported by the bars was likely to suffer the same fate, and this may explain some of the vigor of the rejection of the 2002 paper.

There is a danger that this history of forthright rebuttals of anything that looks like "indications" has painted gravitational-wave detection into a corner. To escape from the corner, it is necessary to understand the causes of the historical disdain for provisional results; that way the disdain can be to some extent discounted. The "highly respected member's" stated view could show the way out of the corner. Remember, he said of the Rome Group's 2002 paper:

> I read what they did. I think statistically they fudged things a little bit. . . . But I didn't think what they said was so bad. They said we cannot rule out that this is a gravitational-wave. It has some marginal significance and we cannot rule out that it is a gravitational-wave. . . . I thought that what they had was not unreasonable. And of course the first evidence that we're going to have is going to be marginal evidence. It's going to be at the edge. . . .

Even though no one now thinks the 2002 findings were right—and that includes their discoverers, since the findings were not confirmed by subsequent data—it is hard to be a pioneering science when to be provisional is to be despised.

The Future

The impact of the future is also such as to make the field—all except that part represented by its most longstanding members—risk averse.

At the September 2008 Amsterdam meeting a young but middle-ranking scientist, responding to the claim that the only way to be sure that one had seen a gravitational-wave was to have an electromagnetic counterpart, said:

> The optical counterpart could be argued that's the only way forward but it's not. There's a second way forward. The second way forward is to say "if it is a gravitational-wave that you're looking at, in fact, there is probably more than one per year if you look at the numbers. [This says] "maybe wait." And . . . depending on the actual run schedules, that could be right thing to do. If we

18. The history of this conflict and the evidence for its politicization is described in *Gravity's Shadow*, in the section around chapter 21.

> were going to shut down for five years before we had an instrument again, then I'd say you have no choice but do the best you can with what you have, but when you do know that you are about to go . . .

The implication was that if you know you have a new instrument coming on air soon that should see many more such events, you should wait for it.

Again, as reported, at the Arcadia meeting a respondent argued that gravitational-wave physicists can wait for a high-energy physics-like buildup of data when the more advanced detectors come on line:

> We're trying to do it now, but we know that if we can't do it now, it's *not* just a matter that well we wait and the universe may or may not get it. We're expecting to have, in a reasonable amount of time, such a dramatic improvement of the event rate, that if something like this is right, the event rate should go up by a factor of a thousand [he is thinking of Advanced LIGO] and they should be falling into our laps.

Likewise, as reported above, there was a question from the floor at the opening of the envelope session where the same sentiment was expressed:

> Are we too detection oriented? In other words, why do we need to make a detection in S5 or in S6? What is the rush? . . . What is the scientific basis for rush?

Another respondent pointed out to me that the younger people in the field, who would anyway not be earning Nobel Prizes for an early discovery, had nothing to gain and everything to fear from too early an announcement: it is bad for the chances of tenure if one is associated with an incorrect claim. The period of waiting is almost certainly no longer than ten years at the time of writing (2009). Unless something is very wrong, by that time Ad-LIGO will have shaken down and the data will be pouring out with around one event a day to look at. The future, as always, favors the young.

On the other hand, the old timers have everything to gain from an early announcement—a chance to enjoy or describe the detection of gravitational-waves before they become senile or die.[19] For some of the old timers there is also the chance for a Nobel Prize.

19. For the purposes of this analysis, I am an "old timer."

The Present

One force from the present is the matter of upper limits. Upper limits are the means of turning the lead of nondiscovery into the gold of a story with enough significance to be published. In the early days of upper limit publication, "the story" was that the instruments had simply been made to work. More recently, as described above, some of the upper limits have gained a degree of astrophysical or cosmological significance. LIGO has been able to set a not entirely uninteresting limit on the flux of the cosmic background gravitational radiation, a more interesting upper limit on the flux emitted by the Crab Nebula pulsar, showing that it cannot be hugely asymmetrical, and an upper limit that astrophysicists found interesting, showing that the gamma-ray burst emanating from the direction of Andromeda was not an inspiraling neutron star system within the galaxy. The majority of upper limit claims, however, have no astrophysical significance.

It is not immediately obvious why prestigious journals have been ready to publish so many upper limit papers, especially papers that bear on no astrophysical phenomena of real interest. Some of these upper limits were based on data from past runs where the devices had not yet reached full sensitivity even as new data was being generated and analyzed. The plethora of upper limit papers worries some of the scientists. At a meeting in March 2008, a plan was put forward by one group to publish fifteen to twenty upper limit papers based on S5 data. A scientist who had been with LIGO a long time remarked:

> When you are publishing that many papers, when you have not seen a source, it's a little bit like we're butterfly collectors who have not found a butterfly yet. . . . I think we are going to lose credibility if [we can't make a case for each new upper limit being] way, way better . . . People will get tired of reviewing our stuff.

Even the astrophysically significant events were not being received with unqualified acclaim by outsiders, as the extract in box 1 reveals. It is from the blog of "Orbiting Frog," someone who knows what went on at the American Astronomical Society (AAS) conference in June 2008.

A more subtle question about upper limits was not, so I understand from respondents, even being asked at the more general physics meetings where upper limit results were being unveiled. Analysis of the nature of calibration (see *Gravity's Shadow*, chapters 2 and 10) indicates the possibility that, since LIGO has never seen a gravitational-wave, it is impossible

My Beef with Gravity Waves

Posted on 3 June 2008

Yesterday there was brief moment when I thought that they had *announced* the *first detection* of a gravitational-wave by *LIGO*. Needless to say, this turned out to not be the case. If it were then you would have heard about it - most likely from a newsreader doing their very best.

The paper that caused this trouble describes how AAS has been used to place a low-limit on some properties of the Crab Nebula pulsar (you can *read it here* if you like). The way this paper was announced at the AAS meeting in St. Louis made it sound like they had a detection. But they didn't and don't (yet).

(http://orbitingfrog.com/blog/2008/06/03/my-beef-with-gravity-waves/)

Box 1. Orbiting Frog's comment on upper limit papers

to be sure that seeing the absence of waves actually sets an upper limit; it could be that there is something wrong with the process of detection. One senior respondent indicated to me (December 2007) that to point to this alternative explanation was "not to think like a scientist."

Another respondent suggested the following more nuanced analysis:

> Your attitude questioning our calibration method isn't entirely wrong—it depends how much one wants to "bracket out" for any particular purpose. We couldn't work if we didn't think we could correctly simulate (for calibration purposes) an incident gravitational-wave by moving the mirrors using magnets. But as soon as we succeed in seeing real waves, we'll claim it as an accomplishment to have validated the theory behind our having believed in our prior calibration. I think that what [. . .] was saying was that you are being impolite/anti-social in raising the question now. . . . We can always make a mistake in calibration, but the odds of us being completely wrong here are small—small enough so that, if we don't see something with Ad-LIGO and get around to questioning that aspect of our experiment, it would be interesting.

In this remark the sociology of scientific knowledge is encapsulated: it is not a matter of what is logically possible, it is what it is "polite" to ask at a particular juncture—that is, it is what is considered to be within the envelope of legitimate questions. When the first respondent said that to think that way was not to think like a scientist, he could be glossed as saying that my questioning indicated incomplete socialization into the way scientists belonging to this field are meant to think at this time. In the same sense, "the Italians" could have been said not to know how think like (interferometer) scientists when they published their 2002 paper.

To nail down the logic of the situation, consider what a flawless calibration of, say, LIGO's ability to detect inspiraling binary systems would be like. It would require one to contact, superluminally, a superhuman entity in, perhaps, the Virgo cluster, and ask it to have created and set in motion the final moments of a well-specified inspiral system at, say, a distance of exactly fifty light years exactly fifty years before 6 a.m. local time at some specified date. Then, at 6 a.m., a search could be made for its effect on LIGO.

It is "impolite," or demonstrates an inability to think like a scientist, to point out the differences between the way LIGO actually calibrates itself and this ideal way, and the potential created by the space between the ideal and the actual. For instance, inside this space between ideal and actual can be found disputes about the theory of the generation and transmission of gravitational-waves and their consequent detectability by interferometers. These disputes are not salient in the community's discussions—and to acquire the "interactional expertise" pertaining to the community is to know that these disputes are not worthy of discussion—but they have not entirely gone away; they have just spun out into the fringes of the field. For example, a paper claiming that LIGO and other interferometers are based on a flawed theory of the transmission of gravitational-waves was promulgated on arXiv, the physics electronic preprint server, as recently as 2008, and this paper was based on previous work of long standing, albeit discussion of that work within the community has died.[20] These claims do not have to be correct to have a life inside the logical space between ideal and actual calibration. The possible gap in our understanding of what happens between the generation and detection of gravitational-waves is not closed by calibration, it is assumed to be closed. And as my second respondent pointed out, this gap is acknowledged, in that the theory will of how they work will be said to have been validated when gravitational-waves are discovered. At that point the theory will no longer be something that is simply taken for granted; it will be treated as having been validated and worthy of a triumph.

20. I am grateful to Dan Kennefick for bringing this material to my attention: Fred I. Cooperstock, "Energy Localization in General Relativity: A New Hypothesis," *Foundations of Physics* 22 (1992): 1011–24; Luis Bel, "Static Elastic Deformations in General Relativity," electronic preprint gr-qc/9609045 (1996) from the archive http://xxx.lanl.gov; R. Aldrovandi, J. G. Pereira, Roldao da Rocha, and H. K. Vu, "Nonlinear Gravitational-Waves: Their Form and Effects," arXiv:0809.2911v1 (2008).

The second kind of gap between ideal calibration and calibration as it is carried out can be seen once we accept the assumption that the theories of generation and transmission are correct. In that case a gravitational-wave is encountered by an interferometer as a ripple in space-time that squeezes it one way while stretching it the other and then reverses the cycle. But a ripple in space-time impacts upon the entire earth and every part of the interferometer including not only the mirrors but the mirror mounts, and it affects every part simultaneously. For many years there were disputes about whether gravitational-waves could be detected in principle since, as one might say, the ruler one uses to measure the changing separation of the parts of the detector is affected just as much as the instrument itself.[21] Theory now assures us that gravitational-wave detectors can work, though the dispute about them does not seem to have been closed until the 1960s, and then only because Joe Weber actually built a detector and people started to argue about whether he had seen anything rather than whether he *could* see anything in principle.

That an interferometer is believed to be able to detect gravitational-waves is a matter of a theory that relates the movement of its solid parts—the hanging mirrors—to the effect on the light that bounces between them. It is not a trivial theory. It may be virtually inconceivable to most interferometer scientists that the theory could be wrong, but the *calibration* does not prove that the theory is right. When a gravitational-wave hits the interferometer, the signal is meant to be read off as the force required to hold the mirrors still in spite of the gravitational-wave's attempts to move them. But an injection of energy for the purposes of testing the collaboration's ability to detect a real wave is made by putting a signal into the coils that press on the magnets fixed to just one of the mirrors on just one of the arms, and, because of the theoretically deduced differential response of the mirrors to waves of different frequencies, quite a bit of calculation goes into determining what this signal should be. There is no reason to think this will not imitate the effect of a wave on the arms, and it may, once more, be almost inconceivable that it won't, but when gravity waves are detected, that all these inferences were correct will still count as a triumph because calibration does not test them, it assumes them.[22] In this passage I am being im-

21. These disputes are discussed in Kennefick 2007.

22. Within the collaboration the term "calibration" is used for something a little different—the rather more simple calibrating of the various individual components of the interferometer. It is from aggregating the results from all these individual tests that the range of the interferometers, as illustrated in figure 2, is built up. What I am referring to as calibration, the members of the calibra-

polite; it is impolite to talk of such things in the current state of the art—it is a sign of defective socialization, like bad behavior at a dinner party.

Actually, if one goes a little further outside the community, one can discover the questions being asked by others. They are being asked by people who seem to be physicists but are unschooled in the etiquette of gravitational-wave detection. What can be found in the physics blogs is illustrated by the remarks in box 2, which refer to the announcement of the upper limit on the Crab Nebula pulsar's spin-down energy.

That these considerations tend to float around on the fringe of the community and no longer enter its heart are the real sociological point. The critics—those bitter critics from the astronomical community, so active in the early 1990s, who were trying to stop LIGO being funded—have gone to ground.[23] Now that the political battle over LIGO's funding had been lost by the critics, they have nothing to gain from belittling LIGO's achievements. That no one deploys the potential of "the calibration criticism" shows how successful LIGO has been at building credibility.

This achievement of credibility means that upper limits can be announced and published with confidence in the absence of fear about the calibration question. But there is a downside. It means the younger segment of the collaboration can build a career on publishing risk-free upper limits and do not have to worry about risky positive discoveries. There is nothing in it for the old timers, but, ironically, the very success of LIGO—its credibility and legitimacy within the scientific community—is one of the indirect forces pushing toward a conservative interpretation of things like the Equinox Event—exactly the opposite of the forces experienced by the dying bar-detector community in the 1990s and early 2000s as the giant LIGO drew closer.[24] One can only wonder how LIGO and Virgo data analysis would feel today if the approaching advanced LIGO technology belonged to a rival group.

---

tion would talk of as "hardware injections" (thanks to Mike Landry for helping me through this thicket). My use of "calibration" accords with the more "philosophical" meaning.

23. See *Gravity's Shadow*, chapter 27.

24. Still more faux credibility can be generated when journalists report an upper limit results as a positive finding. The following headline from a journalist's blog glosses the result as a positive statement that the Crab Nebula pulsar is emitting gravitational-waves:

> Crab Nebula Pulsar Leaks Energy through Gravitational-Waves
> Up to 4 percent of the radiated energy is converted into gravitational-waves
> (http://news.softpedia.com/news/Crab-Nebula-Pulsar-Leaks-Energy-Through-Gravitational-Waves-87171.shtml).

jimmy boo June 06, 2008 @ 05:06AM

This could be a good opportunity to test the gravitational-wave hypothesis. If a method could be devised (perhaps by looking for irregularities/precessing of the radio pulse) to determine the shape of the pulsar, the expected gravity wave radiation from this could be calculated. If the mass distribution is sufficiently non-spherical such that the expected radiation is above the detection limit of LIGO it would provide a strong suggestion that 1) there is something wrong with LIGO, or (and more interestingly) 2) there is something wrong with our understanding/belief in the existence of gravitational-waves ...

(http://episteme.arstechnica.com/eve/personal?x_myspace_page=profile&u =8530045777 31)

Posted by *Iztaru* 06/02/08 15:23
Rank: 5/5 after 2 votes

They say the lack of gravitational-waves gives clues about the structure of the pulsar. But that is assuming the GWs exist and this pulsar has not produced them.

Their analysis of the structure of the pulsar is void until someone detects an actual gravitational-wave.

I remember a while ago they were very exited because they didn't detected any GWs from a gamma ray burst coming from a nearby galaxy because that ruled-out several scenarios for the source of the burst. But again, it is assuming GWs exist.

Is there anyone making analysis of these scenarios but assuming GWs do not exist? I think that is a must, until an actual wave can be detected.

(http://www.physorg.com/news131629044.html)

Box 2. Two examples from physics blogs

The other force from the present has already been discussed. It is the scientists' peer groups in related areas of physics and astronomy working with the canonical model of science, and/or working in established, technologically well-developed, fields. The binary model of discovery, and the 5 sigma imperative, are also imported via the recent experience, in high-energy physics—the paradigmatic big science—of many of the collaboration's members. If the nascent science argument has some truth in it, so that the statistical experience argument does not hold the field unopposed, it might be better to think of LIGO as a small science in big science clothing. Gravitational-wave detection was a small science in the days of Joe Weber and even the cryogenic bars. To build LIGO and the other interferometers took the political, economic, and organizational techniques of big science. But now that the interferometers are collecting their first data, it may have become, effectively, a small science once more. Big sciences are characterized by advances of around an order of magnitude in sensitivity that are justified by predictions of what will be found based on the past successes of the previous generations. There are always surprises, but there is some order to the progress. But there is not so much as one past success in terrestrial gravitational-wave detection on which to build experience, and the increase in sensitivity of the detectors has been two or three orders of magnitude. This kind of risk—this degree of floating free of the past with no lifeline of past success—is normally the prerogative of small sciences. To build a science one needs results, and, if every one of these has to be perfect, maybe the new shoot of a science will wither and die before it reaches the full glare of sunlight above the canopy of its competitors. *Horribile dictu*, it may be best for the big-but-really-small science of gravitational-wave detection to act like "the Italians"—at least in some respects. These respects include endorsing "evidential collectivism" rather than "evidential individualism"—exposing provisional findings to the wider scientific community rather than keeping every disagreement or uncertainty behind closed doors. Remember that at the Equinox Event discussion in Arcadia we heard the following:

> . . . for the most part we are almost certainly going to be in a case where we are not going to have the kind of confidence that some of us would perhaps like to have, and I think we need to get used to the idea that we may have to, as a group, say we have seen something and put ourselves on the line over it.
>
> I think we have to decide are we willing to live with the possibility of not seeing something that is really there in order to be conservative or not.

Conclusion to the Main Part of the Book

That the Equinox Event was an artifact made some difference to the way it was analyzed by the scientists. Some analysts guessed what it was and put less effort into trying to show whether or not it was noise than they might otherwise have done. Indeed, the whole blind injections exercise illustrates the difficulties of "social engineering": an effort that was supposed to change scientists' mindset away from rejecting potential discoveries to embracing them had the opposite effect on at least some and sowed confusion among some others. Overall, however, the exercise seems to have been a scientific success because so much was learned from it, as the postscript below will show.

Sociologically, the blind injections have been a huge success. Gravity's Ghost was almost as good as a genuine liminal event in the way it revealed both the true meaning of what it means to discover something in the "exact sciences" and also the tensions and judgments that surround the process even in an esoteric field of physics. In this it continues the work of Wave Two sociology of science and continues the application of it to gravitational-wave detection, which is exemplified by much of *Gravity's Shadow*. In this particular case something new has been done. The observation of the proto-discovery process has made it possible to explore the day-to-day meaning of statistical analysis of data in a depth that has not been done before so far as I know.[25] There is nothing missing from the account of such a statistics-based discovery except the still more burning problems that might be associated with a full-blown announcement and, of course, with the process of reception in the wider scientific community. Statistical analysis has been shown to be a social process in a very concrete way. Most of that was summarized in the first part of this chapter, which was organized on the principle of "stripping away layers."

Discovery and statistical analysis being social processes, it has been possible, in the second part of the chapter—the one employing the metaphor of the ingot of knowledge—to consider the social and cognitive forces that press upon the meaning of the Event. Throughout, the underlying idea is that what is happening in a social group is revealed by what its members take seriously enough to talk about and what they reject without consciously thinking about it. Thus it has been possible to see something like the "calibration question" ceasing to have a place in debate within the core of the science though it lives on in logic and can still be found to be discussed among the poorly socialized, such as those at the fringes of the field (and those taking the role of the poorly socialized—namely myself). I have tried to show how what is considered seriously and what comprises meaning is affected by pushes and pulls from all sides. Sometimes these forces will be almost palpable. The force of history is an example in this field, and it has been seen how its power is maintained by the recounting of myths—stories about historical incidents which are taken to embody vital lessons.

25. Which is not to say that there is no sociology of statistics: see for example Mackenzie 1981, which analyzes the social roots of the correlation coefficient and its basis in questions of race.

Less powerful influences have also been discussed in the final section. These include the existence of "escape routes" circumventing the felt need to make immediate discoveries, such as upper limit publications and the prospect of more powerful detectors to come. It is impossible to say how strong these less powerful influences are, or even if they make much difference at all, but the point is that the close examination of the data analysis process reveals how they could be having an effect. The analysis shows how any such effect would be mediated by the discourse of the field. Snatches of that discourse even suggest that these considerations are playing a role.

The book goes beyond Wave Two, however, in that it tentatively explores the possibility that the reflective sociological stance might contribute to the discourse as well as record it. It has been suggested that from the distanced position it appears that there is a tension between gravitational-wave detection as a pioneering science and gravitational-wave detection as a big well-established science. With great trepidation it has been suggested that clarifying this matter might help resolve some of the tensions in the statistical analysis caused by, for example, the desire to meet the 5 sigma criterion while not having the means to do it with the current generations of detectors.[26]

In the next section of the book the distance from the field is going to be suddenly increased—to use a metaphor from cosmology, a kind of "sociological inflation" is going to take place. That is why there is no chapter 8 but rather something that I have called an "envoi." An envoi, according to the *Chamber's Dictionary*, "is the concluding part of a poem or a book; the author's final words, especially the short stanza concluding a poem written in certain archaic metrical forms."

26. Allan Franklin points out that some of the processes and problems described here do have precedents in high-energy physics. Franklin is preparing a paper, which he kindly refers to informally as a "footnote" to this book, which will show where similar problems are to be found in the recent history of high-energy physics. The paper will also show how the problems were debated and resolved in those cases.

ENVOI
Science in the Twenty-First Century

This book has dealt with the problems of detection of liminal events. But maybe this is not the future of gravitational-wave detection. Maybe there is some big surprise or a colossal stroke of astrophysical luck to come. A big close event may be visible in many different ways via electromagnetic or other effects. With the detection of such an event, the science would be propelled from provider of upper limits to almost incontrovertible certainty-maker, without the in-between stage of "indications."

From a sociological viewpoint, it would be a pity if the interferometers had the stroke of scientific luck that would suddenly put an end to the debates and uncertainties that have lasted for nearly fifty years. There is more to be learned sociologically from the dilemmas surrounding marginal events and from a gradual phase-transition from nonbelief to belief than from dramatic discoveries. That is one reason for writing this book now, before the arguments can be overtaken by a strong event of this sort. The arguments will survive such an eventuality, but gravitational-wave physics would no longer exemplify them so nicely, and the agony of the Equinox Event would be quickly forgotten. In sum, the sociology of knowledge is best served by an uncertain science. There is, however, a bigger question. The topic now shifts from gravitational-wave detection in particular to science as a whole and to its role in social life. Here gravitational-wave detection becomes merely an illustrative example rather than the central topic.

Science as a Reflection of Society

Science no longer has the unquestioned authority it had when it was winning wars and promising power too cheap to meter. Nowadays it is beset by raids on its epistemological grounding from academe, a backlash from religion, an attack on its professionalism from free-market ideology, and scorn from those who think a simple life without technology is the only salvation for the human race. The attack on experts and expertise that has come with the academic movement known as postmodernism actually does the same work as the attack on the professions begun by Margaret Thatcher and Ronald Reagan and takes the same view on experts as religious fundamentalists. The academic movement holds that expertise has no epistemological warrant, the political movement holds that the judgment of professionals has to be replaced by quasi-markets in which every aspect of performance is measured and compared so that it can be properly priced, and the fundamentalists hold that expertise is worthless in the face of revelation. The arms of this grotesque pincer are squeezing science from three sides.

But is science worth preserving? What is science? Nowadays it sometimes appears that science has no unique cultural identity left. Like a teenager, today's science is continually borrowing from others' cultural repertoires to gain attention. We have science-as-Hollywood with its superstars, their vanity bolstered by the media industry and a new style of popular science publishing. We have science as a religion, with Stephen Hawking and Charles Darwin as its figureheads. Hawking sells millions of copies of a completely incomprehensible text which ordinary people revere as they might once have revered the Latin Bible. Meanwhile the media treat Hawking and his mystic utterances as revelation, while other physicists see, or are said to see, the "Face of God" in the skies. Richard Dawkins and his colleagues attack organized religions with the gospel of Darwin, borrowing religion's rhetoric, beatification, and iconography. There is also science as the slick player in the market, from Silicon Valley to the muscular-capitalism start-ups of the new biology and its instant millionaires. If this is science in the twenty-first century, then there is nothing to learn and nothing to preserve, at least not in the way of values. As far as cultural values are concerned, this kind of science is all reflection and no substance. Such science cannot claim to be a source of transcendent value, since there is nothing transcendent about it. This way, even without the pincer, science will garrotte itself as a cultural movement.[1]

The pressures to act this way are, of course, enormous. Economic pressures lead governments to demand visible short-term economic pay-offs from the public funders of science. Big science has to compete for funds in the political arena, where publicity is a dominant force. The massively costly astronomy and particle physics need the front pages linking them to the story of our beginnings and our ultimate fate—previously the agenda of religion. Individual scientists find a ready justification for pushing their noses into the trough because to glamorize is to survive.[2]

Maybe this is the fate of science—to be a secular religion servicing the economy and the entertainment industry. But science, at least as exemplified by the story of gravitational-wave detection, has the potential to lead not just follow. It has the potential to provide an object lesson in how to make good judgments in a society beset by technological dilemmas. For more than three hundred years the old-fashioned values of science have seeped into Western societies like the air we breath. Imagine a society without any place at all for the cultural authority of science. It would have surrendered all responsibility to politics, market forces, or competing modes of revelation, and it would be a dystopia—at least as anyone who prefers reason to force would see it. This is not to say that science is the only cultural institution the removal of which would lead to a dystopia, but it is central to the kind of society most of us prefer to live in.

The Central Values of Science as a Cultural Institution

It is hard to list the special values of science, because it is an activity only vaguely defined by the "family resemblances" of its different parts.[3] One

1. For what appears to be an alternative perspective on contemporary capitalism-linked science, see Shapin 2008. Shapin makes no distinction between derivative and essential values and appears to treat with equanimity a science-capitalism nexus which, while blind to matters of race, gives differentially favorable treatment to young, physically fit, and competitive persons of local origin.

2. The political maneuverings of scientists have been wonderfully documented over the years by Dan Greenberg (e.g., 2001).

3. The term "family resemblance" was introduced by the philosopher Ludwig Wittgenstein, who argued that the idea of "game" has no set of clear defining principles but that games were linked by family resemblance—which can be thought of as a set of overlapping sets, each containing a subset of game-like qualities but whose extremes need have little or nothing in common; games, after all, run all the way from professional football to not stepping on the cracks in the

can make progress, however, by imaginatively taking away different elements and seeing if what is left can still be called science. Thus, one can take away the ability to see the face of god and still have something recognizable as science; one can take away the best-selling books that no one understands and still have science; one can take away the religious war against religion and still have science; one can take away the venture capitalists and still have science; and one can take away the front page stories and the superstars and still have science. These features of science as a cultural institution are merely "derivative."

One the other hand, one cannot take away integrity in the search for evidence and honesty in declaring one's results and still have science; one cannot take away a willingness to listen to anyone's scientific theories and findings irrespective of race, creed, or social eccentricity and still have science; one cannot take away the readiness to expose one's findings to criticism and debate and still have science; one cannot take away the idea that the best theories will be able to specify the means by which they could be shown to be wrong and still have science; one cannot take away the idea that a lone voice might be right while all the rest are wrong and still have science; one cannot take away the idea that good experimentation or theorization usually demand high levels of craft skills and still have science; and one cannot take away the idea that, in virtue of their experience, some are more capable than others at both producing scientific knowledge and at criticizing it and still have science. These features of science are "essential," not derivative.[4]

Three qualifications are in order. First, not every one of these values belongs to science alone. For example, most religious institutions, every professional institution except those to do with magic, illusion, and crime, and indeed every society as whole, must value honesty if they are to last. But honesty seems more logically integral to science than to the others. Everywhere else honesty must be the default or average position; in science it seems to be always vital or science simply is not being done. Therefore science provides a kind of "sea-anchor" for the value of honesty and integrity.

---

sidewalk. (Whether there is "little" or "nothing" in common is not so clear; here I am trying to identify some things that all, or nearly all, science has in common.)

4. The essential/derivative distinction is radically at variance with the popular "Machiavellian" approaches to science found in the field of Science and Technology Studies, which, laying all the stress on detailed observation of science as it is encountered, see the achievement of scientific success as essentially a political process.

Second, not every value listed in the "essential" category is equally central. There is still an argument going on about whether the theory of the historical evolution of species satisfied the falsifiability tenet. But even if it does not, we know that the overall theory has been strengthened by small-scale demonstrations of the mechanisms in the laboratory and that these do satisfy the tenet. Thus, overall science needs the falsifiability value even if there are violations here and there.

Third, generalizing the second qualification, honesty and integrity aside, one can still have passages of scientific research that produce valuable findings where one or more of the values is violated.[5] The list of values, contrary to the way they have sometimes been thought of, does not comprise a set of necessary *conditions* for the production of good scientific work.[6] Rather, the overall "institution" of science is formed by these values, and it follows that the day-to-day "form-of-life" of science is built up through actions driven by the "formative intentions" of scientists who are guided by these values. The idea of family resemblance shows how such a set of values can hang together to form a cultural institution without always needing to be manifest in every one of its corresponding activities all the time.[7]

Given these qualifications, it could be said that the essential values of science are far more important than science's substantive products and findings. The essential values of science are worth preserving so that they can continue to seep out into society as a whole and help to shape the way we live our lives. That is why the pincer needs to be resisted even if, in the short term, the maintenance of the conditions for continued production of scientific substance and findings seem to rest with the derivative values. Either one sees that this is the case and chooses the central values or one does not; if one cannot see these values as good in themselves and prefers not to choose them, argument ceases and force prevails.[8] As far as I have been able to observe, the heartland of gravitational-wave detection science exemplifies these values quite closely.

5. As the Second Wave of science studies has clearly demonstrated.

6. Arguably, Merton's (1942) "norms of science," which overlap with the values listed here, were initially justified as a set of conditions for the efficient functioning of science.

7. One of the persistent errors of some critical approaches to science is to take it that discovering lapses from the central values of science—something that is easily done—shows that there are no central values.

8. The choice to make the essential scientific values (not findings) central to the life of a society can be called "elective modernism" (see Collins 2009).

Another thing that cannot be taken away and still leave science—another essential value—is the idea of the replicability of findings; replicability is a corollary of the idea that there is anything stable to investigate. Replication often cannot be carried through in practice, as when unique events are observed or apparatuses are so costly that only one can be built, but the idea of replicability still informs what is going on. Thus, should there have been only a unique observation, the idea is that anyone who was in a position to see it would have seen the same thing. If there is only one apparatus, the idea is that another similar apparatus would make the same discoveries or that anyone else who knew how to run the machine would find the same things.[9] Adherence to a value does not always mean that the value has to be instantiated there and then, but it does mean that the value remains an aspiration even when circumstances cannot allow it to be realized.

Including replicability, another consequence of the essential values of science is that secure claims cannot be based on the authority of individuals or other unique sources. Holy men or women cannot "reveal" the truth in virtue of their unique relationship with a deity. If the relationship is unique, then it cannot be replicated (not even in principle) nor criticized, so it cannot belong to science. The same goes for books of obscure origins or with obscure authorship. What is written in such a book cannot command scientific respect in virtue of the fact that there is something special about it as an object: its contents have to be open to criticism and investigation. Inter alia, these principles rule out creationism as a part of science, because the story of the creation depends too heavily on a particular book of obscure origin and its interpreters. These principles also rule out the more technical version of creationism, "intelligent design," because it too is heavily dependent on a book of obscure origin for its ideas and because it seems impossible in principle to think of any falsifying observation or experiment—that is, one that would show that any still unexplained state of affairs could *not* be counted as the work of an intelligent creator.

Another thing that follows from these principles is that works of science should be as clear as it is possible to make them, failing destructive simplification. The simpler and better the explanation, the easier to criticize; to make a work obscure beyond necessity is to make it, effectively, more private, and privacy is incompatible with freedom to criticize. Note that at the other end of the spectrum of valuable cultural institutions, creating

9. Replicability is, as one might say, a "philosophical" idea. To see how it works in practice, see Collins [1985] 1992.

works that are open to competing interpretations and debate by those that consume them is an essential feature of the arts.[10]

The principles also indicate why the imitation of major features of non-scientific cultures can be antithetical to science—the means destroying the ends. If millions of people are being encouraged to treasure, for their scientific content, books that are so far beyond their ability to criticize as to be completely incomprehensible, then they are being encouraged to think of scientific worth as like religious worth—based on the authority of the author or the text. The same goes for all "glamorization" of science and scientists: as soon as the person, rather than the ideas and findings, becomes important, then the very idea of science is being subverted. It is the same when the virtues of science are advertised for their commercial potential. The balance of scientific worth is not the same as the balance of commercial worth. That science generates more economic value than it costs is probably true, but that science is worth keeping primarily because of its value to the economy is not true.

Science as an Object Lesson in Judgment

Contrary to what some commentators claim, quantitative exactness is another thing that *can* be taken away from science without destroying it. It is a shame that the idea that exact quantitative analysis is crucial to science is so widespread because it does great harm. For example, a social-scientific finding has little chance of influencing government policies unless it is expressed in quantitative terms, whereas many quantitative social science results, when they are not simply wrong, are of no social significance. In social science qualitative findings are often far more robust and repeatable than quantitative findings and often more socially significant, yet they make small headway with policy-makers. As has been seen in chapter 5, even in the physical sciences the expression of a result in statistical terms is often the tip of an iceberg of hidden judgments, assumptions, and choices, and yet, to those who consume rather than produce them, the numbers still have a force quite out of proportion to their true meaning.

Certainty, along with the binary model of discovery and the Nobel Prize, is another thing that can be taken away while leaving science intact.

10. The argument is worked out in detail in chapter 5 of Collins and Evans 2007, under the heading of "Locus of Legitimate Interpretation" (LLI). If the LLI is forced pathologically close to the producer in the arts and humanities, it is "scientism." If forced pathologically close to the consumer in the sciences and technologies, it is "artism."

Indeed, one might say that certainty is the province of revelation rather than science, and I have heard even potential Nobel laureates claiming that lust for prizes distorts and damages science. Even a philosopher of science such as Popper can point out that all scientific claims are essentially provisional.

A shallower but more important point about the false allure of certainty was made in the introduction: most science is applied to such messy domains that good judgment rather than certain calculation is quite obviously the best that can be done. If certainty and quantitative exactness were the keys to science, then science would be restricted to that small corner described at the outset as comprising the products of Newton, Einstein, quantum theory, and so on—the so-called exact sciences. But there is a much larger domain of "inexact sciences," which impact much more immediately on our lives. This is the domain where society needs its object lesson on how to make decisions as well as how to live its life.

Of course, gravitational-wave detection fits naturally into the realm of the "exact sciences"; its precursors are Newton and Einstein, and, as far as its place in the spectrum of sciences is concerned, it takes its "reference group" to be high-energy physics and the like. I suspect that when the final announcement of the terrestrial detection of gravitational waves is confirmed, it will be impossible for the scientists to resist the retrospective redescription of their field as one of heroic point-discovery, and I suspect they will be unable to resist the glorification of the enterprise in culturally derivative ways. At that point, as I have suggested, "the lame" (that is, the gradualist and uncertain description of discovery) will be trampled under the stampede; the "Drums of Heaven" will be said to have sounded and "Einstein's Unfinished Symphony" will be said to have been completed.[11] Perhaps more important to the mass of working physicists who have dedicated their lives to the project will be the right to face their critics from the other exact sciences with pride; making the most of the newly accomplished similarity between gravitational-wave detection and the sciences of their peers will be irresistible.

But what we have seen both here and in much preceding work involving detailed analysis of the sciences is that even the exact sciences are inexact when examined under the microscope. Here we see that this inexact

11. With apologies to David Blair, whose (so far as I know, unpublished) manuscript on gravitational wave detection was given the first title, and to Marcia Bartusiak, whose very good, popular introduction to the field was given the second.

face is visible even to the metaphorical "naked eye" when the science is engaged in true pioneering. The analysis of the Equinox Event, though the event was an artifact, was a moment of true pioneering as far as the procedures of science are concerned. As we have seen, judgment after judgment had to be made. Indeed, as *Gravity's Shadow* reveals, the entire multi-decade history of gravitational-wave detection has consisted of judgment after judgment.

I want to suggest that the mythical hero of the science of gravitational-wave detection should be not Einstein but Galileo. Einstein may have conceived of the idea of gravitational waves, but Galileo represents science as a leader of social understanding—in Galileo's case, the end of the geocentric universe. I have already suggested that the essential scientific values encountered in the heartland of gravitational-wave science are a self-evidently good model for social conduct. What I am now suggesting is that the *inexact* science of gravitational-wave detection (not the triumphalist and *exact* version almost certain to come) is an excellent model for judgment wherever technology and society intersect in an inexact way.

The science from which we can learn to live our technological lives is hard, frustrating science full of flawed judgments which are nevertheless the best judgments there can be. Barring massive luck, the first indication of gravitational waves will be "the best guess about what this funny looking blip in the statistics might mean." Such a display of how to understand the world and what it is to understand it as well as possible is a vital thing that science has to offer society. It is something that belongs to science alone, something which allows it to lead, not something derived or reflected.

One of the senior scientists in LIGO is quoted above as saying that to offer only gradualist and tentative results in the search for gravitational waves would "abrogate our responsibility as scientists." He was reflecting a very widespread view among the collaboration's members as at the beginning of 2009. From the sociological and political perspective offered here, the opposite is the case. Scientists' responsibility lies in making the best possible technical judgments, not in *revealing* the truth. To represent every judgment as a calculated certainty is to abrogate *social* responsibility. To be a producer of certainties is, at best, to consign oneself to the nonexemplary sciences—the corner of the scientific world that has dominated, and distorted, Western thought with examples of what it claims to be a perfect and, worse, *attainable* mode of knowledge-making. To strive too much for certainty is to abrogate the responsibility of taking that leading role in Western societies that only science as a cultural activity can assume.

If scientific judgment is not applied to the uncertain world that faces us, the role of finding solutions will be relegated to populism, fundamentalism, force, or what amounts to the same thing, the market. If we want to maintain the society we live in, the technical element of judgments had better continue to be based on technical experience. By exemplifying scientific values, and judgment under uncertainty, gravitational-wave physics could be among the sciences that play a still more important role in twenty-first century political and social life.

The view put forward in this last section of the book is uncompromisingly sociological—perhaps a violation of etiquette characteristic of the improperly socialized. One of its bizarre implications is that the more incorrect results physics makes public the better, so long as they are meant to display the process of a gradual move toward better understanding and better judgments. It follows that all the rejected incorrect or partial claims described in *Gravity's Shadow* could be more valuable in social terms than the point-discovery and even the gravitational astronomy to come. The trouble is that the physicists are ashamed of all those incorrect results. But why should they be? They all arise from physicists doing their job with a burning passion. It is likely that the shame is generated, not by a disappointed public, but by close rivals in the fight for resources—one set of physicists damning another in their attempt to get a share of their funds. That kind of competition is not an essential value of science—scientists can get things wrong and science very definitely still goes on. In the science in the twenty-first century, more valuable than the truth is the demonstration, and the guidance this can give to society, of how experts with integrity judge when they do not know.

POSTSCRIPT
Thinking after Arcadia

Dates can be important in contemporaneous work. The first draft of this book was written in about five weeks. The first key was struck in Los Angeles airport on 19 March 2009, and, after a week or so of primping after completion, a confidential copy of the manuscript was sent on 12 May to my gravitational-wave physicist friend, Peter Saulson, for his comments.[1] The manuscript has undergone continual polishing since then, partly in response to the remarks of those mentioned in the acknowledgments, while the envoi gained a life of its own. But up to and including chapter 7, the *substance* of what appears here is very close to the draft that was completed as April turned into May in 2009. Those chapters represent the best I could do in the way of recording physicists' thinking and talking from the first intimations of the Equinox Event to the end of the Arcadia meeting in the middle of March 2009.

Things started moving very fast after that, and this postscript is an attempt to capture the way the blind injection experience affected things in the few months that followed. There were two further large meetings, one in Orsay and one in Budapest, held in June and September respectively. I did not attend either of these meetings but accessed them by other means: being by now a kind

1. The five weeks of composing was, of course, based on a store of knowledge and experience built up over nearly forty years, on some preexisting draft papers on statistics, and in particular on hundreds of pages of notes referring to the blind injections.

of "honorary" member of the LIGO Scientific Collaboration, and having been given the passwords that allow me access to almost everything that is put up on the web by members of the community, I was able to view the slides presented at the meetings and read e-mails exchanged in response to them; I was also able to access some of the discussion in Budapest via a telephone link, and colleagues reported additional goings on.

It is hardly worth mentioning that the Inspiral Group agreed that they must find a way to move faster and do all the checks on the data as soon as possible. Once more, there is nothing more of sociological significance to say about their response. Of greater potential sociological significance were the reactions to the Equinox Event. The task that confronted those who wished to be in a position to make a claim about the detection of a burst of gravitational waves from statistical likelihood alone—without the reassurance of correlated optical effects or a well-defined inspiral waveform—was being reassessed. In particular, some were now suggesting that, if events demanded it, "evidence for" claims with a much lower level of statistical significance could come before "observation of" claims. The notion of what could be presented in the journals was becoming less binary and more "Italian." Nevertheless, the divide between the Virgo and LIGO teams was, if anything, becoming sharper, with the Virgo people in general pressing a much harder line. But even one member of that group, who had previously been of the conservative persuasion, displayed slides at both meetings which included the following sentiments:

What if we have to deal with a single candidate without any counterpart?

Seems hard to reach the "discovery" criterion.

The "evidence" criterion is not out of reach.

It is likely that our first detection paper title will be "Evidence for . . ."

Then "Strong evidence . . ." etc. . . . as we are able to increase the significance of candidates.

Continuum of possibilities between first evidence and a clear and convincing detection.

First paper likely to be "*Evidence for blabla . . .*"

A long (and painful) (and exciting) road in front of us.

Though these remarks show that something less than discovery was now being countenanced by one or two Virgo people, a hard line in respect of discovery itself was still being pushed by everyone from Virgo whose remarks or presentations were accessible to me. Most were in fact pressing an even harder line, which was being described as the "Virgo position"—a position, incidentally, that is consistent with the account provided by 'S' (see chapter 4).[2] This Virgo position was that nothing at all could be announced about a single event unless it (a) correlated with optical effects or (b) demonstrated a well-defined waveform. A third possibility was that (c) statistics alone would be sufficient so long as there were a series of repeated events—effectively a call to wait for AdLIGO. The criterion of coincidence between separated detectors—the detection principle that has informed gravitational-wave detection from the days of Weber—was simply no longer to be counted as good enough on its own.

Some LIGO people agreed with this thinking, but others suggested that the caution in respect of statistical analysis that was being argued for—leaning over backward to take into account all possible contributions to the trials factor, assuming that the non-Gaussian nature of the data made any extrapolation from the actual false alarm rate generated from time slides impossible, and refusing to accept a single event until similar events had been seen—took the gravitational-wave community well beyond what was normal practice in astrophysics and perhaps even of high-energy physics itself. A specially interesting intervention came from Jay Marx, the executive director of LIGO. He represented the detection of gravitational waves as normal science, given the theory, the astrophysics, and their indirect observation by Hulse and Taylor via the slight decay in the orbit of a binary system observed over many years (see *Gravity's Shadow*). Marx suggested that 3 or 4 sigma should be enough to support an *observation* claim because an observation of this kind was to be expected.

My reading of the overall state of discussion at Budapest was that there was still considerable confusion and disagreement, sometimes exhibited by what looked like contradictory statements within individual talks. In particular, a couple of talks from Virgo spokespersons insisted, on the one hand, that no discovery could be made with a single candidate and, on the other, that whatever was found might have to be reported (just as "the Italians" argued). The apparent contradiction is made clear in the two sentiments expressed one after another on the same slide presented by a senior figure in Virgo:

2. Some LIGO scientists were surprised that they could be such things as "Virgo positions" or "LIGO positions" in a unified team, questioning once more the exact nature of the LSC-Virgo collaboration.

> We cannot afford "Evidence for . . ."
> WARNING: If the situation is borderline with respect to this we might be forced to publish our data WITHOUT claiming a detection.

The resolution of this apparent contradiction is that the putative publication flagged by the "WARNING" would discuss all the possible sources of noise in the detector that might have led to a coincidence that, though statistically significant, could not be presented even as "evidence for" in the absence of other kinds of confirmation.[3]

Pressed further by people from LIGO about the apparent contradiction in the Virgo position, another senior Virgo spokesperson, himself, on the face of it, a hard-liner, said:

> I think that this is a very difficult moment we are passing through because it is difficult to find the first event and so we may have to take a chance in accepting such an event. It is very difficult to say what is the right thing to do. Honestly, I don't have the answer. I am very worried that we might publish something that is not a real event. The solution could be repeatability, but that's also difficult because we are looking for rare events. I don't know what to do in this case—honestly. [Edited for English style by HMC]

The debate, and the dilemmas it embodies, can be summarized along three dimensions:

Dimension 1: What kind of science is gravitational-wave detection?

This question has already been discussed at length in the main body of the book. GW continually compares itself with high-energy physics, but that standard may be too high for a pioneering science. In any case, the standards of high-energy physics were more relaxed when high-energy physics was developing. Furthermore, GW may be setting itself standards even higher than HEP when it tries to take into account every imaginable aspect of the trials factor problem and when it refuses to countenance the calculation of a background which goes beyond what can be directly measured with time slides. It may also be going beyond other sciences in the way it applies craft understanding of the instruments in an asymmetrical way.

3. Private communication with the author of the slide (October 2009). The author added in this private communication: "I know several cases in other fields where publication of unexplained behavior occurred, ending into nothing. Measurements continued and the effect didn't show up again."

Finally, it may be astrophysics that is the better model for the discipline; astrophysics, so at least one commentator said, operates with less conservative standards than high-energy physics.

Dimension 2: "Evidence for" or discovery claims only?

Again, this dimension has been much discussed in the body of this book. At the Orsay meeting there seemed to be a new willingness to countenance "evidence for" (something that, for what it is worth, is argued for throughout this book). But at the Budapest meeting some very strong sentiments were expressed that ran against this idea. But can "evidence for" claims be avoided if some marginal event is found in S6? We do not know what will happen in S6, but we can try a thought experiment. Imagine that four "events" are seen in S6, all of which have around a 3.5 sigma significance level. Imagine that when the envelope, or envelopes, is/are opened, three of these events are found to correspond to blind injections but the fourth correlates with nothing. Would it be credible to say nothing about the fourth event? Would a refusal to say "this looks like evidence for" demonstrate integrity or lack of integrity?

Dimension 3: Setting the rules a priori or responding to events?

For years, members of the community have looked for ways to set detection criteria in advance. The counterposition has also been argued. In Budapest both positions were once more advanced. The debate was nicely expressed in the concluding remarks of a senior member of the LIGO team:

> I've tried and I think every time I've looked at it I've failed. And that is to do what you guys have been trying to do for the last hour-and-a-half. What are the criteria you would apply, given our current state, [given] the fact that we have many different inputs—multiple detectors, external triggers, waveforms that have been calculated? How do you take that whole mix and make something that's sensible out of it? Can you make a set of rules? And that was what I heard—people making rules about how they were going to approach this. I don't think we should do it that way. I think we should make a set of concepts . . . write down a set of concepts as guides. . . . I know of no other way to go at this. . . . The Detection Committee, if they ever get exercised, will look at each case on its own merits. I don't know any other way.

Again, for what it is worth, I have argued in this book that the last sentiment is the correct one. While it is worthwhile trying to imagine what criteria one might apply in the future, these exercises are best looked at as a kind of "training" in thinking about the many possible scenarios that might arise rather than as a way of developing a rigid set of rules. The reason for saying this is set out above (in chapter 2), and it is the deep philosophical point made by Ludwig Wittgenstein that rules do not contain the rules for the own application. Mostly, the rules for application of most of the rules we use in day-to-day life are so deeply sunk in the sediment of ordinary practices and cultures that we do not even notice that we are performing a philosophically miraculous act of interpretation every time we use a rule. In cases like the one discussed here, however—the true pioneering sciences or in any radically new practice or culture—the problem of the proper application of rules becomes more salient. In such cases the only way to know how to apply a rule—for it is a certainty that every eventuality cannot be foreseen—is to work it out as events force the arguments to unfold.[4] The outcome of these lived arguments sinks down through the ocean of history to form the sediment of society—in this case, scientific society. Calculation cannot produce that sediment; only the cycle of culturally innovating life can create it. It is possible to make some attempt to rehearse the future, as war-games rehearse battles, but, as generals continually rediscover, war games are not war.

Conclusion

The discussions that took place in the post-Arcadia meetings demonstrate that the blind-injection exercise was a success. It was not a success in the way it was intended to be: it did not result in the injections being identified as pseudogravitational waves, and it did not give rise to a complete exercising of the detection procedures. But one may be sure that the approach to the next round of data analysis will be influenced by the impetus of blind injections much more than the last round was affected. The success is being realized one experimental run later than expected.

4. The "airplane event" illustrates this deep point of principle in almost comic form.

APPENDIX 1
The Burst Group Checklist as of October 2007

Burst Detection Checklist for GPS 874465554 event
The LSC-VIRGO Burst Working Group
1 October 2007

1. Zero-level sanity Convert GPS times to calendar times and check for suspiciousness.
   Sep 22 2007 03:05:40 UTC not suspicious

2. Zero-level sanity
   Read e-logs for the times/days in question.
   Was there anything anomalous going on?
   seismic activity and loud/long transient at H1; nothing at L1

3. Zero-level sanity
   Check state vector of all instruments around the time of the candidate.
   Data correctly flagged?
   FOM shows all IFO's in science mode

4. Zero-level sanity
   Identify inspiral range of the instruments in order to set the scale of sensitivity. Is this typical/low/high?
   Good H1/L1 range, typical H2 range. Range drop 6 min after

5. Zero-level sanity
   Identify nearest in time hardware injections (also type/

amplitude of them). Were there ongoing stochastic/pulsar ones? When did they start and what was their amplitude?
No nearby hardware burst injection, pulsar on

6. Zero-level sanity
How close to segment start/end for all instruments does this event occur?
Science segment boundary is over 10,000s away for all instruments

7. Data integrity
Check for undocumented/unauthorized/spontaneous hardware injections.
Nothing in injection log. Nothing unexpected in any excitation channel

8. Data integrity
Examine all possible test points recorded in frames or saved on disk. The latter part might be time critical if data are overwritten.

9. Data integrity
Establish if there was any data tampering. O[s' responsibility]

10. Data integrity
Check integrity of frames; check raw/RDS/DMT frames.
data-valid flags are OK on the L1 RDSs for both observatories
no CRC mismatch over range [874463712, 874467392)

11. The event
Run Q-scan on RDSs/full frames and on all available instruments in the network of detectors.
Some mildly interesting things (see link).
H1 is very glitchy across IFO channels, must understand
H2 is at inconsistent Q of 23, while 4km IFO's have Q of 5

12. The event
Run Event Display on RDSs/full frames and on all available instruments in the network of detectors.
no obvious glitches. some spikes in AS-AC,REFL-I/Q about 1 secs later.

13. The event
What is the overall time-frequency volume of the event in each detector? Does it look consistent?
minimal uncertainty, looks very consistent

14. The event

    What is the expected background for such a candidate?

    What is the significance of the observation given the background?

    Compare background estimated from time slides and from first-principle Poisson estimates. Is it consistent?

    background depends on algorithm/search. at the predetermined thresholds

    for the online search we have a significance of about .03 for KW+CP.

    However the event is well above the set thresholds. This is for the probability of observing N events on Sept 21, 2007

15. The event

    How robust is the background estimate? Do randomly-chosen shifts as opposed to fixed shifts—is the result consistent? How about other steps?

    currently the background is only estimated for Sept 21

16. The event

    How robust is the significance of the event to the threshold chosen? Do a "stair-case" analysis (varying threshold) to appreciate this.

    Naturally the significance of the event depends on the threshold

    For KW+CP it varies from .03 (fg 3, bg .68) to .01 (1, .01) depending on Gamma threshold chosen (5 or 9.5)

17. The event

    If more than 2 IFOs are involved in the event, would any 2-IFO pair be able to identify it as well and with what background/significance? **easily visible in H1L1.**

    probably visible in H1H2 Q analysis (guessing from qscan)

18. The event

    If only 2-IFOs are involved, why it was not detected by the rest of the detector network?

    below KW threshold in H2, but clear signal in all 3 IFO's

    G1 and V1 too poor in sensitivity at 100 Hz

19. The event

    Examine the status of the detectors not involved in the event and establish why this is so.

    H1H2L1G1V1 all in Science Mode (wow)

20. The event

Examine frequency content out to the KHz range—is there anything there? Will a broader bandwidth search pick it up?

There is nothing in h(t) up to 8192 Hz. However the high frequency excess in AS_Q_0FSR for both H1 and L1 visible in the qscans should be understood (aliasing at nyquist?)

21. The event

Run parameter estimation code.

22. The event

Identify the sky ring/sky patch using noncoherent methods, i.e., signal timing and amplitude O['s responsibility]

23. The event

Compare source reconstruction between coherent and incoherent methods

24. The event

What fraction of the network's acceptance comes from this direction?

25. The event

How robust is the candidate against the stride size/start of the analysis?

Seen in KW and QPipeline, so it should be robust

26. The event

How the peak times established by the method, parameter estimation and the various coherent methods compare? Is this as expected?

using QPipeline as a reference, KW peak is 9ms later. Both algorithms give dt=4ms between H1 and L1. CorrPower peak is at −38ms with a 50ms window

27. The event

Could it be an artifact of the prefiltering?

Q Pipeline and CorrPower use zero-phase filtering. The positive offset of the KW trigger is due to an uncorrected filter delay

28. The event

What is the detection efficiency for the search method calculated near to the candidate event? average/low/high w/r/t the S5?

29. The event
    Could there be any effect from lines not filtered enough and/or any other artifact? Could it be violin mode excitations? Any other mechanical resonances sneaking in?
    does not look like a line in qscan. short duration (few ms)

30. The event
    How stationary were the instruments around the time of the event?
    Quantify this both in terms of singles counting and PSD.

31. Vetoes
    What could have caused this event other than astrophysics? blind injection.
    faulty injection. common low-freq glitch

32. Vetoes
    Are there obvious environmental disturbances in the Q-scan/Event Displays?
    nothing obvious in the initial RDS qscans. however, we might want to look at qscans which display all channels regardless of significant content immediately around event time. There is a nearby low freq glitch in H0:PEM-BSC3_ACCX at −200ms, and noise in H0:PEM-BSC6_MAGZ from −6 to 1s about event time

33. Vetoes
    Are there obvious interferometric disturbances in the Q-scan/Event Displays?
    H1 shows a variety of IFO channels glitching nearby stopping 1s after event time. The glitches are not directly in coincidence with the event, but should be understood

34. Vetoes
    Examine what known earthquakes occurred around the time of the event.
    fairly quiet around time of the event

35. Vetoes
    Contact power companies and obtain known power line transients around the time of the event.

36. Vetoes
    If available, check in/out records for trucks and heavy equipment to the Lab that might not have been recorded in the ilog. Check for overpassing airplanes (airport flight logs).

37. Vetoes

    Check for any switching of major electrical equipment around the time of the event that might not have been recorded in the ilog.

38. Vetoes

    What are the KleineWelle (KW) triggers in auxiliary channels around the time of the candidate?

    Nothing much immediately around event time (ms within .714/.718)

    Glitches within the same second are weak. BSC3_ACCX is missing in H0 as KW triggers are only generated from 10–512 Hz.

39. Vetoes

    If there are any overlaps with KW trigger from auxiliary channels, what is the expected background of such coincidence and what is the significance of that channel as a veto channel?

40. Vetoes

    Which of the overlapping channels are safe, which are not?

    Analyze most recent hardware injections.

41. Vetoes

    For PEM/AUX channels with measured transferred functions, is the signal present in the them consistent with the one in the GW one?

42. Vetoes

    If nothing in the non-GW channels in the RDS, proceed with scanning full frames and repeat above checks.

43. Vetoes

    Any known data quality flags overlapping with event? How is this dependent on DQ flag thresholds? What is the coincidence significance?

    None of the DQ flags that have been evaluated so far overlap, but many flags have not yet been evaluated for this part of the run

44. Vetoes

    Examine minute trends/Z-glitch/glitch-mon data.

45. Coherent Analysis

    Run the H1-H2 Q analysis; anything in the H1-H2 null stream?

H1-H2 stream has norm energy of 7. H2 has energy of 15.
So H1 and H2 seem more or less consistent.

46. Coherent Analysis
    Run the r-statistic cross-correlation over all detector pairs (involved or not in the trigger); how significant each is?

47. Coherent Analysis
    Run coherent analysis/null stream burst analysis on the available detector network.
    RIDGE pipeline finds maximum statistic at location of event

48. Coherent Analysis
    Run the inspiral multi-detector coherent analysis on the available detector network and compare to the burst one.

49. Coherent analysis
    What is the best fit waveform extracted from the data?

50. Other methods
    Do other burst ETGs find the event(s)? If yes, compare extracted event parameters, including background/significance.
    The event is seen in all 3 IFO's using QPipeline, however no coincidence is done, nor thresholds set for a detection
    Event also seen in Block-Normal in H1 and L1 only
    Event seen in CWB at FAR 1/598 days over S5A.

51. Other methods
    What is the outcome of the Inspiral and/or Ringdown search around the time of the burst event? If something is present, what is the background/significance?
    No online BNS event (1–3 Msun)
    Frequency range suggests BH mass objects

52. Calibrations
    What is the calibration constants and errors around this event?

53. Calibrations
    How robust is the event analysis against calibration version?
    KW uses uncalibrated data. May affect CorrPower

54. Calibrations
    Could there be any calibration artifact?
    such a strong signal is not present in the excitation channel

55. Calibrations
    Is the event identified when analysis is run on ADC data?
    Compare findings of an analysis starting with ADC(t) vs h(t).
    event is found on DARM_ERR with KW and h(t) with QP/CP

56. Miscellanea
    Check timing system of the instruments (well in sync?).

57. Miscellanea
    Check for any recent reboots, software updates/reloads.
    Any suspicious acquisition software changes?

58. Miscellanea
    Check recent logins to the various acquisitions computers.

59. Other GW detectors
    Any signature in the non LSC-VIRGO detectors? TAMA/bars online?

60. Other GW detectors
    What is the expected signal size given what we know for the event?
    undetectable in V1, G1 and anything else due to low freq (100Hz)

61. Non-GW detectors
    Any known or "sub-prime" event in E/M or particle detectors around the globe? O['s responsibility]

62. Astrophysics
    Any known sources overlapping the ring/patch on the sky corresponding to the direction of the candidate event?

63. Astrophysics
    Examine events (other than the candidate) reconstructed at the same direction. Perform a directional search; if a point source is behind this, more, lower SNR events might be in our data.

64. Astrophysics
    How the extracted waveforms compare to astrophysical waveforms?
    What is the energy scale going into GW, assuming galactic distances?

65. Vetoes

Create a hardware injection starting with signal waveforms corresponding to the best fit waveforms extracted from the instruments.

66. Other methods

Take the extracted waveform per IFO and run matched filtered search in order to establish how often the specific morphologies appear in the data.

67. The event

Run the Q-event display

68. The event

Run the Coherent-Event-Display (CED)

69. Vetoes

Play audio files corresponding to GW, H1+−H2, auxiliary channels

70. Vetoes

Check that signal is the same in all photodiodes.

71. Vetoes

Check wind speeds

Wind speeds normal and less than 15 mph

72. Vetoes

Check for fluctuations in power levels of TCS laser

Normal: no mode hops or big jumps in TCS power levels

73. Vetoes

Do a seismic Q-scan

Some seismic noise in L0:EY_SEISY and H0:EY_SEISZ need further study.

APPENDIX 2
The Burst Group Abstract Prepared for the Arcadia Meeting

We present the results from a search for unmodeled gravitational-wave bursts in the data collected by the network of LIGO, GEO 600 and Virgo detectors between November 2006 and November 2007. Data collected when two or more out of the four LIGO/Virgo detectors were operating simultaneously is analyzed, except for a few combinations which would contribute little observation time. The total observation time analyzed is approximately 248 days. The search is performed by three different analyses and over the entire sensitive band of the instruments of 64–6000 Hz. All analysis cuts, including veto conditions, are established in a blind way using time-shifted (background) data. The overall sensitivity of the search to incoming gravitational-wave bursts expressed in terms of their root-sum-square (rss) strain amplitude hrss lies in the range of 6 × 10−22 – 6 × 10−21 Hz−1/2 [tentative] and reflects the most sensitive search for gravitational-wave bursts performed so far. One event in one of the analyses survives all selection cuts, with a strength that is marginally significant compared to the distribution of background events, and is subjected to additional investigations. Given the significance and its resemblance in frequency and waveform to background events, we do not identify this event as a gravitational-wave signal. We interpret this search result in terms of a Frequentist upper limit on the rate of gravitational-wave events detectable by the instruments. When combined with the previous search using earlier (2005–2006) data from the fifth science run (S5) of the LIGO detectors, this is at the level of 3.3 events per year [tentative] at 90% confidence level. Assuming several types of plausible burst waveforms we also present event rate versus strength exclusion curves.

II

BIG DOG

INTRODUCTION
Big Dog Barks

September 2010: Great Excitement

There is great excitement and activity in the gravitational-wave business. For the first time, it really seems as though a gravitational wave has been directly detected. The event that is being feverishly analyzed and written up as a paper for submission to the prestigious *Physical Review Letters* happened on 16 September 2010. The event was a big one and therefore different than anything anyone has seen. In the nearly forty years up until September 2010, during which I have been following the science of direct detection gravitational waves, there has been nothing other than marginal proto-signals, all of which were eventually discounted.

As luck would have it, there is a meeting of the gravitational-wave detection collaboration beginning on 20 September, just four days after the event was registered on the detectors. The meeting is in Kraków and I am there; this is where my account begins. My story ends more or less with a second meeting almost exactly six months hence on 12 March 2011, in the Californian hamlet of Arcadia. Arcadia is about twenty miles from Los Angeles and, not coincidentally, directly inland from Pasadena, home of Caltech, the unofficial headquarters of gravitational-wave physics. So this book should report on six months of activity, sandwiched by two meetings. As I will discover, the filling in the sandwich for the scientists is thousands of e-mails and dozens of telephone conferences, or, telecons. For me, the filling is those same thousands of e-mails plus a few more personal ones, about a dozen of the telecons and a few phone calls.[1]

The Bombshell

On the first full day of the Kraków meeting, the events of 16 September are introduced as follows: "This is a detection and we are going to start writing a detection paper"; it was not a marginal event that had appeared on 16 September.[2]

The pattern of the event—the "waveform"—was consistent with the final moments in the life of a binary star system. It looks as though two massive stars rotating round each other faster and faster in a frantic death spasm had finally merged. The signal came from gravitational waves generated in the last few seconds when the stars were spiraling round each other hundreds of times a second; it ended a second or so later as the stars coalesced. Because of the strength and dynamics of the signal, one or two black holes must have been involved.[3] A rough, initial estimate was made of the direction of the source: the constellation Canis Major. Within a couple of days the gravitational event assumed its irresistible nickname: "Big Dog." Within a week, Big Dog became as familiar as a family pet.

Like the Equinox Event, Big Dog might be a "blind injection." The envelope opening that would reveal the truth, would, it turns out, happen in Arcadia on 14 March 14 2011. The Equinox Event took eighteen months and many meetings to resolve itself; Big Dog will take six months and two

1. One methodological lesson is that reading e-mails as a way of maintaining contact with a field is pretty good but not quite as good as being immersed in conversations. As I point out in the course of the book, I missed a few things, and it is also the case that the importance of certain matters turned out to be more obvious when they came to my attention through spoken discourse than when I was simply reading a huge stream of material on the screen.

2. The "relevant detection group" was tasked with analyzing signals with a waveform putatively coming from the inspiraling of a binary star system comprising neutron stars or black holes. This group is also known as the "Inspiral Group" or "CBC (Compact Binary Coalescence) Group." The large event, however, was first noticed by the Burst Group, which had been tasked with looking for signals with no predicted waveform. The Burst Group does not have the best tools for investigating an inspiral in detail. I learned later, however, that members of the Burst Group could immediately see from their "Q-scan" graphical analysis that the CBC Group were going to find a loud inspiral, and this informed much of their attitude to the subsequent analysis.

3. The shape of the signal is worked out by continually matching a bank of preexisting templates to the data stream as it pours in. Each template represents a different model of a binary system—each with different masses for each component and, in the more sophisticated templates, different spins on the individual masses. One of the scientists pointed out to me: "We do 10,000 experiments [the number of templates we search] every millisecond or so."

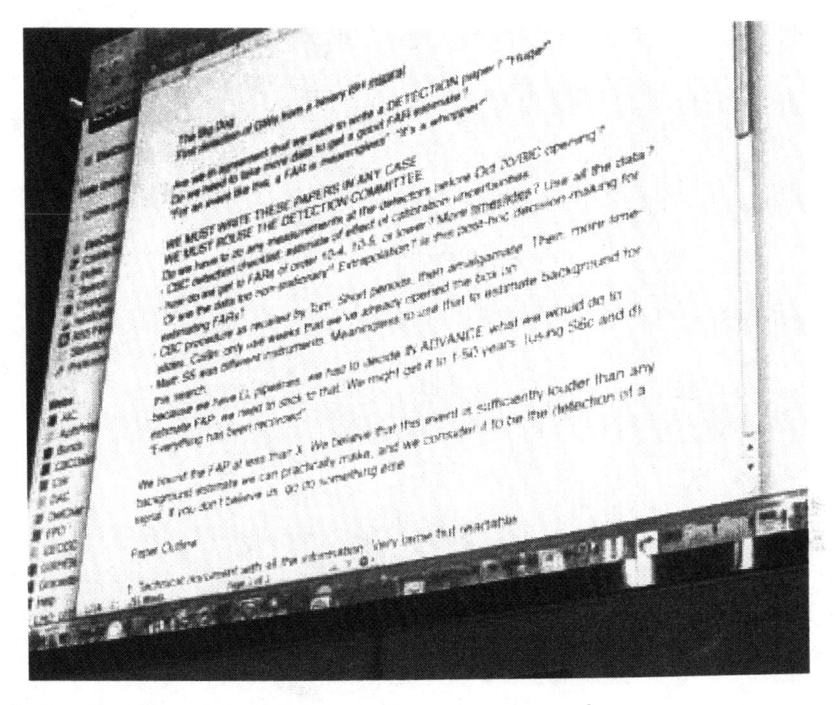

Figure 15. Kraków, 21 September 2010—the event is announced

meetings. Otherwise, history is repeating itself. Until 14 March everyone is going to act as though the event is real. As one of the senior scientists put it in an e-mail circulated on 4 October:

> The B[lind]I[njection] challenge . . . is supposed to be a "system" test of the L[IGO] V[irgo] collaboration to make a detection; to get us all the way to point of making a public claim of a detection (papers, press release). In as much as we can, we should just forget about blind injections—they don't exist—and treat the Big Dog as a real event . . . One thing I really do like about [the process] is the BI envelope is only opened after we're ready to submit a paper. It forces people to go all the way—deciding what's essential for the first paper and what can be covered in a follow up paper.

The signal being introduced is big and beautiful. The Equinox Event had been weak and weedy, and the scientists realize almost immediately that it could not give rise to a discovery announcement even if it was a real event. In the case of the Equinox Event, suspension of disbelief and its consequences for action came hard—the paper that was meant to be written never got further than the abstract. This was different. Whether as a result of firm conviction or successful role taking, drafts of the entire Big Dog paper were being refined almost from the word "go." There were going to be sixteen promulgated drafts before the envelope was opened. Every single sentence, diagram, and even word was to be the subject of often heated and prolonged debate consistent with true belief or firmly suspended disbelief. *Physical Review Letters* papers are limited to four pages, and the entire community was involved in making sure that not a word was wasted and that there was not a single phrase that could give rise to misunderstanding. As one of the scientists reportedly said, every word of the paper that would emerge would cost a thousand dollars.

Physics and Philosophy

The climax of this book for those most interested in the physics story is chapter 15, but for those interested in the nature of scientific discovery, the crux is chapter 14. That chapter was written in Arcadia, only a day or two before the envelope was opened. Looking back though the chapters I had already written, I found that around twenty-five major decisions that the physicists had made in deciding how to write the Big Dog paper were not really "physics" decisions at all—that is, they were not physics in the sense that they had answers that could be calculated. These twenty-five puzzles had been or were being resolved by debate, by default, by invocation of tradition, by anticipation of the future, by invoking points of philosophical or stylistic principle, or some combination of these things. Chapter 14 assembles these decisions in a numbered list.

The Material

It is astonishing how much there is to think and talk about in the buildup to a discovery claim. Having the personal contacts, knowing where to look, being put on the mailing lists, and being trusted with the passwords, I could read every draft of the proto-paper; I could talk to my gravitational-wave physicist friend, Peter Saulson, and one or two others on the telephone or via e-mail; I could listen in to the occasional telecon, I could read the minutes of meetings; and, above all, every day I could read the e-mails that circulate around the community. On 11 January, I deleted 65 e-mails relating to gravitational-wave detection, the vast majority of which had to do with Big Dog. It was a heavy day; the day before I had deleted 33, but that was not unusual. On each of those two days I also collected five of the e-mails and stored them in my "Big Dog" folder. By 12 January 2011, the Big Dog folder contained 684 e-mails, which means I had collected about 6 per day since the beginning of the Kraków meeting; in the same period I must have deleted around 4,000. By the end of the debate, in the middle of March, the Big Dog folder contained around 1,200 e-mails, and I had deleted about 10,000. It is through this storm of interchange that the scientists pulled together the meaning of their discovery, or proto-discovery.

Much of the work of "authoring" this book has been more like editing—selecting and quoting the words of others from the storm of e-mails and recorded discussions. As with the other books I have written on the history and sociology of gravitational-wave physics, I can only *hope* that my own analytic contributions will be lasting, but I can be pretty certain that what the scientists said in the face of these early events is likely to become more rather than less interesting as gravitational-wave astronomy establishes itself. Writing contemporaneously allows me to quote speech (and e-mails) that are normally lost to historians; it encourages the use of more direct quotations rather than less, rendering a service to the future. The quotations also act as an anchor for my particular obsessions and interpretations not allowing them to float completely free of what the scientists were saying at the time. Of course, I am still responsible for the editing—what I delete, what I collect, what I quote and don't quote—but there is some constraint which I have chosen to maximize later in the book by quoting scientists critical of my circulated draft.

8 Black Holes Observed?

The form and tone of the discovery paper was already being anticipated in Kraków. On 21 September the PowerPoint shown to the collaboration (figure 15) said what is shown more clearly in figure 16.

Before the end of September a draft title had already been written by one of the senior members of the collaboration and leader of the Compact Binary Coalescence (CBC) analysis group [Acacia].[1] The draft title shows that the debate to come about the strength of the claim has been anticipated; indeed, it had already started. I wrote to Acacia about the title and his e-mailed response (11 October) begins to capture the positions that would be staked out:

1. The first occurrence of pseudonyms will be in square brackets with a brief explanation of the scientist's role. I will work through the alphabet for initial letters of names of trees or shrubs trying to find those that do not coincide with ordinary surnames so that the pseudonymous nature of name should be obvious even after the first occurrence. A glossary of pseudonyms with descriptors is provided on page 293. Thanks to Susan and Emily for suggesting this scheme and finding the names of trees. I also sometimes use less elaborate signifiers for different scientists when I do not want, or there is no need, to tie them into the tree-pseudonym scheme.

The Big Dog
First detection of GWs from a binary BH inspiral
Are we in agreement that we want to write a DETECTION paper? "Huge!"
Do we need to take more data to get a good FAR estimate?

MUST WRITE THESE PAPERS IN ANY CASE
MUST ROUSE THE DETECTION COMMITTEE
Do we have to do any measurements at the detector before Oct. 20/BIC opening?

. . .

We bound the FAP at less than X. We believe that this event is sufficiently louder than any background estimate we can practically make and we consider it to be the detection of a signal. If you don't believe us, go do something else.

Figure 16. The tone of the paper to come as anticipated in the Kraków slide of 21 September

Hi Harry,

The original title, which I fed to the students and take full responsibility for, was:

"The First Detection of (or First Evidence for) Gravitational Radiation From Black Hole Binary Coalescence"

I was anticipating a debate over whether we would be claiming "discovery" or "evidence for" (and honestly, I don't care, because in the long run, it will be the first discovery, no matter what we say in the title).

I certainly agree with [Balsa, a senior theorist] that "Direct" should be used.

[Quince], a very senior scientist] suggested we change "Radiation" to "Waves," since the public is afraid of radiation; that's fine with me.

The students [who were tasked with writing the very first draft] then insisted I change it from ". . . from Black Hole Binary Coalescence" to ". . . from a Black Hole Binary Coalescence." But I don't like it. Yes, it is one instance of a Black Hole Binary Coalescence, but it is also the first direct observation of that entire class of phenomena, and I think that's more important.

I also expected someone to object that, as [Balsa] said, "The deletion of the qualifying phrase 'from a black hole binary coalescence' is meant to show that this is first (direct) detection of any kind not just from a black hole binary."

My response to him was: Yes, it's the first detection of GWs [gravitational waves] from ANY source. So we COULD stop there. But the fact that it's from BBH [black hole–black hole] (assuming that the parameter estimation converges convincingly on that) is also very significant (*it's the first direct*

> *evidence of black holes, without the complications of matter*) so I actually think it deserves equal weight in the title. Adding the qualifier detracts from the "First Detection," but adds more specificity and significant information. (I like paper titles that are as specific as possible) [my emphasis].

To clarify, on seeing Acacia's initial title, Balsa wrote to him on 1 October:

> Thanks for kick starting the draft.
>
> My first, and only, comment now is that we should change the title to simply "The First *Direct* Detection of (or First *Direct* Evidence for) Gravitational Radiation."
>
> There are two changes: the inclusion of the word *Direct* to show that we do recognize the *indirect* detection by the binary radio pulsar observations. The deletion of the qualifying phrase "from a black hole binary coalescence" is meant to show that this is first (direct) detection of any kind not just from a black hole binary.

And on the same day, Quince wrote:

> I believe the title should be "Evidence for the First Direct Detection of Gravitational Waves."
>
> "First" should apply to "Detection," not "evidence," since there has certainly been "evidence" for detections before, just not good evidence.
>
> Also, this paper is going to be the subject of significant popular press. The public likes "waves" (witness the hoards that head to the beach each weekend, But they are suspicious and fearful of "radiation." This is a case where pandering to popular opinion is justified.

Acacia, in his e-mail to me, was agreeing with the inclusion of "direct" because that acknowledged that gravitational waves had already been detected "indirectly" but was unhappy about leaving out "from a black hole binary coalescence" because that was also of great importance.

The flavor of the growing debate can be gauged by further e-mails. Another scientist [A] wrote on 1 October:

> I don't like using both of the two words, "detection" and "evidence." It's probably just a matter of taste, but it seems redundant to me. How about:
>
> The First Direct {Detection of/Evidence For} Gravitational {Radiation, Waves} from a Binary Black Hole Coalescence
>
> My personal preference would be (Detection of, Waves).

A different scientist [B] wrote on the same day:

> Hi [Quince], I would prefer a little bit different title: "Evidence for the First Direct Observation of Gravitational Waves"
>
> However such paper may have a significantly different text than a paper with this title: "The first direct detection of gravitational waves" Perhaps we need to start with this one and later downscope it to the "evidence" or even "blind injection" paper.

This attracted the following immediate riposte from [A], which makes a more philosophical point:

> We didn't observe gravitational waves. We observed a blip in our detectors, which we have inferred was due to gravitational waves. The observed signal is direct evidence for the presence of a gravitational wave.
>
> The phrase "Evidence for the First Direct Observation of X" does not make sense to me.

Scientist [C] chimed in:

> "A chirp from an astrophysical binary black hole merger: the first direct detection of gravitational waves"
>
> This is slightly long, but covers *astrophysical,* *bbh,* first *direct* detection of any kind of GWs.

While [B] replied to [A]'s "philosophical" point:

> The phrase "Evidence for the First Direct Observation of X" does not make sense to me.
>
> makes perfect sense to me—all observations are blips in the detectors which can be our eyes, ears, ATLAS detector at LHC, etc.

[B] is claiming that this is all you ever get in modern science—a statistical blip which has to be interpreted. Thus, even in an observational science, matters of "meaning" are always lurking just under the surface.

Features of This Account

Each word in the four-page *Physical Review Letters* draft was worth a thousand dollars and my book cannot do justice to them all. I have chosen to organize the account around two related "rallying points": the title and the statistical significance of Big Dog. It is only fair to say that not all of the physicists were equally obsessed with these two features of the debate. Some physicists told me that they didn't really care about the title—it was the content that was important; they said it would be the wider scientific community who would decide whether this was a discovery irrespective of the eventual title. Furthermore, they considered that the exact level of statistical significance was not important, whereas from the point of view I adopt here, reinforced by an analysis of the journals, 5-sigma is a crucial hurdle. My obsessions are not, then, an exact match for the physicists' obsessions, but I will use the evolution of the title and the negotiation over what was to be agreed as the statistical significance of the event as touchstones for what would otherwise be logistically beyond capture.

Black Holes

Concerning whether "black hole" should be in the title, there are three related arguments. First, there is the question of exactly what was being seen—one black hole or two? Second, there is the question of whether the black hole was really being seen "directly." Third, there is the matter of anticipating how big a claim can be presented to the physics community without attracting criticism or even scorn.

Was There One or Were There Two Black Holes?

It turns out that it was not entirely clear that two black holes were involved in generating the signal. There was a chance that one of the masses might be a neutron star. Working out what the signal represented depended on comparing the gravitational wave with a set of theoretically derived models known as "templates," or "waveforms." These models depend on different assumptions, such as whether the stars involved are spinning and in what plane they might be spinning. It appeared that most modeling exercises favored two black holes of around 4 or 5 solar masses (a solar mass is the mass or our sun), but there was one wild scenario based on a certain kind of model with the stars spinning that gave a system with one star of 1 solar mass and another of 23 solar masses—and there were other possibilities too. As [Dogwood], a young but extremely active analyst put it in an e-mail of 1 October:

> Given the wild variance in m1/m2 numbers [the masses of the individual component stars] between [statistical analysis] pipelines [which rest on different models], how strongly can we support the phrase "Binary Black Hole" in the title?
>
> What the . . . search gives us "detection" of is not necessarily BBH [binary black hole]. It is GW from a compact binary coalescence [jargon for two stars coalescing]. Parameter estimation studies can give us a better handle on what type of binary, but don't (at present) do the type of "detection" or "evidence" we require to write the title. . . .
>
> In any case, it's probably misleading to imply we could distinguish a BBH system from a heavy NSBH [neutron star–black hole] by eye.
>
> My suggestion for the title is
>
> "First Direct Detection of Gravitational Radiation from Compact Binary Coalescence"
>
> and in the abstract:
>
> "a transient (. . .) with a time-frequency evolution strongly suggestive of compact binary coalescence"
>
> and at the end
>
> "the source was most likely to be the coalescence of a (X,Y)Msun [solar mass] binary black hole system (etc. . .)."

The 1/23 mass model was not to last, but a problem remained. It could be that under certain assumptions one of the masses was near the upper limit for a neutron star.

Stars are formed when matter—hydrogen gas in the first stages of formation—clumps together under the influence of gravity. As more and more gas, and possibly other matter, is sucked into the nascent star, the gravitational field gets stronger and stronger and the matter is compressed more and more by its own gravity. At a certain point the hydrogen gas is sufficiently compressed for nuclear fusion to begin, and a star in same family as our sun is formed. Still more gas or other matter can be sucked in, however, or one star might merge with another, increasing the mass still further, and, under the right circumstances, there will come a point when the gravitational field is so strong that even the atoms of which matter is built cannot withstand the gravitational pressure. A crisis is reached at around 1.4 solar masses. At this point the electron orbits that give ordinary matter its relatively low density are crushed out of existence and only the nuclei of the atoms are left. The result is a neutron star, in which, it is said, a thimbleful of matter would weigh as much as Mount Everest and take as much effort to climb. A neutron star that is about 1.5 times the mass our sun would be about 20 kilometers in diameter, compared to our sun's 1.4 million kilometers. That, however, is not the end of the story. If matter is squeezed close enough together its internal gravitational field becomes so strong that even the neutrons of which it is made can no longer stand the strain. At that point black holes are born, those mysterious objects which cannot be said to have a diameter at all—they are so-called singularities. The only kind of diameter-wise thing that can be said about black holes is that they have a "Schwarzschild radius"—defined as the distance from their center, inside which anything, including light, cannot escape their embrace but must be sucked in. They are called black holes because light or any other kind of electromagnetic radiation cannot escape from inside the Schwarzschild radius—which is why they cannot, in the normal way, be seen directly.

Unfortunately the maximum possible size for a neutron star is not known exactly. It will depend, for example, on how fast it is spinning; if it is spinning very fast the centrifugal force generated by the spinning can counteract the inward pull of gravity so heavier neutron stars can survive without collapsing into black holes. The latest thinking about these issues among astrophysicists was summarized by [Elderberry], a scientist who mediated between the gravitational-wave world and astrophysics. Elderberry wrote in an e-mail dated 12 January 2011:

> within the astro[physics] community a compact object is not observationally claimed as a black hole (in x-ray binaries) [binary system detected by their emission of x-rays] unless the best mass estimate exceeds 3Msun [3 solar masses]; this is considered the safe, EOS-independent upper limit [EOS means "equation of state" and refers to the stiffness of the material from which the star is composed]. This is adopted regardless of the fact that no existing EOS predicts a max NS beyond ~2.2Msun for non-rotating objects and ~2.6Msun (if I remember correctly) when rotation is high (close to break-up). [All calculation of stiffness and strength suggest that a non-rotating neutron star cannot be heavier than 2.2 solar masses before it collapses into a black hole and even if it spinning so fast that it is on the point of flying apart, the theory says it cannot be heavier than 2.6 solar masses.] . . .
>
> I think we should stick to the general practices of the astro community for this kind of announcement, but I would still be comfortable with this

> statement above; at the low-end our secondary masses extend down to 2.5Msun.
>
> Note the most massive neutron star known so far is at 1.97 ± 0.04Msun [reference given].

To summarize, Elderberry is saying that even though the top limit of the mass of the most massive neutron star that has been observed is no greater than 2.1 solar masses, and even though the heaviest neutron star that is compatible with any available theory is 2.6 solar masses, the astrophysics community still don't call anything a black hole unless it can be shown to be at least 3 solar masses. As certain of the models that remained in play and fitted the Big Dog allowed one of the stars to be as "light" as 2.5 solar masses, according to this tradition, it could not be claimed that more than one black hole was involved.

But why did it matter if there was one black hole or two if it was the first direct discovery that was at issue? As [Foxglove], a very senior scientist, wrote in an e-mail dated 4 October:

> Please forgive me if I'm confused on this point, but isn't it the case that we can determine the chirp mass reasonably well, and thus that we can tell that we aren't dealing with a NS-NS system, but instead with either NS-BH or BH-BH? If this is so, then we would be in a position to state that we have observed a BH. Until we had a high SNR signal with unambiguous ringdown, I'd agree that we are still looking for even better tests of the existence of black holes. Still, aren't we in the situation where the evidence would show that way more than the Chandrasekhar mass was contained in a pretty small volume? There are ways to write a paper to properly deal with uncertainties, and thus it is good to have these discussions (and all of the subsequent ones that will ensue before we've agreed on a paper draft!) But I'm worried that we seem to be stepping back from what I thought one could claim with reasonable confidence, based on the evidence before us. What am I missing?

Somehow this sentiment was never discussed at length until near the end of the story. On 1 February, one could hear the following sentiments expressed during a telecon:

> S1: Are we confident that this is two black holes? Across all the models that have been run, across all the codes that have run them, the lowest mass we can have at the 95% confidence level is 2.6 solar masses.

. . .

Several people have made the suggestion, which I think is quite nice, that if we are confident we make that the title . . . [putting] black holes in the title, I think would be kinda cool. . . . Are we confident we can do that?

S2: if you want to follow the astronomy convention you don't call a black hole anything with a mass below 3 solar mass. So in the fully spinning runs . . . the ranges have been getting narrower and narrower. If they cross 3 solar masses then I would say "no doubt" we should say "binary black hole in the title." . . .

And so on. So far as I know, the most ambitious titles were not discussed and, at this point, it seemed that for some reason black holes were not going to reappear in the title at all unless there were two of them.

Directness

As for "directness," every other "detection" of the existence of a black hole has been made by inference from the behavior of the visible material close to it. This close matter could be another star or stars circulating in an orbit around a compact point so fast that it could not be explained in any way other than that the point was a massive black hole with a correspondingly huge gravitational pull. Alternatively, the matter could be more diffuse, but emitting such energetic electromagnetic radiation (in the x-ray region of the spectrum), that it must be being manipulated with the violence that could be offered only by something with the concentrated mass of a black hole. No form of radiation can escape from the near vicinity of a black hole other than gravitational waves which, being oscillations in space-time itself, are effectively emitted by a black hole when it moves. Thus, it seems to be the case that looking at a black hole via the gravitational radiation it "emits" is just like looking at a visible star via the electromagnetic radiation it emits. That is why, potentially, Big Dog could be the first *direct* observation of a black hole—or so I thought.

But what is direct and what is indirect? This is another philosophical problem. Turning away from black holes for a moment, nearly all accounts of Nobel Prize–winning first "detection" of gravitational waves in 1993 by Hulse and Taylor refer to it as an "indirect" detection. The observation was of the slow decay of the orbit of a pair of stars, one of which was a pulsar, enabling the orbit to be monitored over a number of years. It was found that the rate of decay was consistent with Einstein's theory if the loss of energy was due to the emission of gravitational waves. So the observation in that case is a matter of timing whereas in the case of the interferometric detectors the observation can be said, by me and at least a few of the physicists, to be the effect of the direct impact of the waves on the instrument itself.

I had heard it reported, however, that at an important meeting, Joseph Taylor claimed that his work should be counted as a *direct* detection. I wrote to Taylor and another of the scientists involved, Thibault Damour. In 1983, Damour was responsible for the first complete calculation of gravitational-wave interaction in binary systems and had worked with Taylor. Damour responded first. On 1 March 2011, with some slight amendments on 2 March, he wrote:

> Several binary pulsars' data have provided direct experimental evidence that the gravitational interaction between the two bodies is not instantaneous, but has propagated between them with the velocity of light (and with the tensorial structure predicted by Einstein's theory of gravity). This constitutes *direct* evidence for the reality of gravitational radiation.

Damour went on to explain that he believed (along with Taylor) that the reason these results can count as a *direct* observation has to do with the details of the theoretical derivation of the interaction of the stars which is invoked in interpreting the pulsar data; this derivation either makes explicit use of the wave-like propagation of gravity between the two stars (as in Damour's calculations) or resorts to a more indirect and heuristic argument related to the gravitational waves emitted far from the system.

Damour went on to say:

> Evidently, there is here some sociology at work: people like Joe Taylor or me have written in several papers such statements . . . while people who fought for the funding of LIGO etc. tended to downplay the value of the pulsar-experiment/GR-theory agreement as a *direct* evidence, probably because they wished to insist on the novelty that will represent the first direct detection of GW's arriving on the Earth (in addition to the important scientific prospects opened by GW astronomy).

Taylor, prompted by Damour's response, wrote on 3 March that he agreed. He added:

> In the binary pulsar experiment, and also in a LIGO-like experiment, one infers the presence of gravitational radiation based on effects it induces in

a "detector." If a ruler could be used to measure the displacement of LIGO's test masses, I would grant that detection to seem rather more "direct" than one based on timing measurements of an orbiting pulsar halfway across our Galaxy. However, LIGO can't use a ruler; instead they use servomechanisms, very sensitive electronics, . . . and finally long sequences of calculations to infer that a gravitational wave has passed by. Such a detection, like the binary pulsar timing experiment, is arguably many stages removed from being what most people would call "direct."

There is one significant difference. The "detector" in the binary pulsar experiment is the pair of orbiting neutron stars—the same thing as the "transmitter." "Detection" involves measuring the back-reaction of the emitted gravitational waves on the transmitter itself. The time and place of "detection" are the same as the time and place of "emission."

These things will not be true of gravitational waves detected by LIGO. In that case, the waves will necessarily have traveled a very large distance from transmitter to receiver.

These thoughts suggest that a better distinguishing characteristic of the two experiments would be something like "in place" and "remote" rather than "indirect" and "direct."

Whether the "sociology" referred to by Damour is correct I do not know but it is clear that there is disagreement over what counts as direct. Thus, on 1 March I also encountered the following e-mail on the lists as part of the debate over directness that was still rumbling:

I disagree that the Hulse/Taylor pulsar was the first "direct" detection. It was most definitely an indirect detection, as the waves themselves were not observed. In fact, I think this is exactly why the paper title should use the term "direct detection." That's kind of the whole point of this enterprise, and we should make that clear in our first detection paper.

But Taylor is also saying something similar to what I wrote in *Gravity's Ghost* (see chapter 5). When we come to the interferometers, it is not absolutely clear that even they are making "direct" detections, because all the analysts do is make statistical inferences based on strings of numbers that measure electric impulses, which are needed to hold the mirrors still so that the interference pattern made by the reflected light in the interferometer does not change. That is a lot of nongravity-based steps between the impact of the gravitational wave and what counts as an observation. It is not like looking at the heavens and having light from a star hit your eye.

Nevertheless, seeing a gravitational wave with an interferometer is called a direct detection, and none of the scientists involved would call it anything else. Furthermore, if one calls everything that isn't seen with the naked eye and a ruler "indirect" detection, then modern physics never makes a direct detection; the argument has to be about whether Taylor's detection was as direct as Big Dog and, while it seems to me that it wasn't, there is obviously plenty of scope for argument. Once more, this is not something that can be settled with a calculation.

Going back to black holes, the idea that Big Dog could be the first direct detection did not survive for long. The strongest counterargument made to me was that there was a difference between looking at the waveform of the circling stars as they spiraled into each other and looking at the waves emitted by the quaking—the "ringdown"—of the single black hole resulting from the final coalescence. Thus Foxglove, who was sure that there must be at least one black hole involved, wrote to me on 26 October:

> The deeper point may be this: when we say that GWs will give us a qualitatively new/better way to demonstrate that BHs [black holes] exist and obey GR [general relativity], what we were referring to was seeing ringdowns of the characteristic "quasi-normal modes" of a BH. We haven't seen that. So what we've done is inferred the existence of a BH (or two) by measuring masses in a binary. This isn't SO different from more classical methods of study of binary star systems. It is different, but not our most distinctive result.

What can be said is that by 22 October the title of the versions of the draft paper that were being circulated had lost any reference to black holes at all, never mind "direct detection" of black holes. The 22 October title read:

"Direct Detection of Gravitational Waves from
Compact Binary Coalescence"

Black holes were not to reappear in the title until early March, though the view that black holes were a likely source could still be found in the abstract.

9 Evidential Culture and Time

Evidential Culture-Space

"Evidential culture" is, I believe, a useful sociological concept. What counts as an acceptable claim in a science or any other realm of activity is a matter of tradition and the context within which the actors live and work. For example, the evidential culture surrounding newspaper astrology is a loose one: newspaper astrologers can put forward strong claims based on very little evidence and need not fear for their reputations. Modern physics is at the other end of the spectrum: claims must rest on detailed evidence, and a mistake can lead to a permanent loss of standing in the community.

There is some variation among even physicists' evidential cultures, which can be captured in a diagram. Different groups of physicists can be thought of as occupying different positions in a three dimensional "evidential culture-space," as shown in figure 17.[1]

Evidential threshold is the most straightforward dimension of the three: it represents the level of evidence that is currently demanded to confirm various categories of finding, culminating in *discovery*. For example, in high-energy physics since the 1960s, the expected level of statistical significance for anything that is to be announced as a discovery has increased from 3-sigma—three standard deviations from the mean—to 5-sigma. Very roughly, to

1. This analysis can also be found in *Gravity's Shadow*, chapter 22.

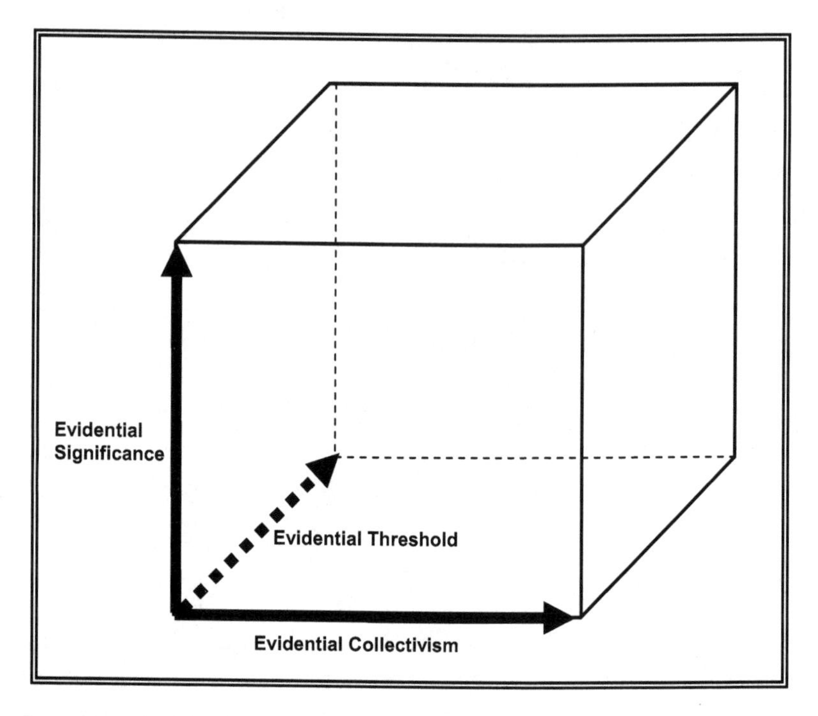

Figure 17. Evidential culture space

announce something as a discovery in high-energy physics in the 1960s one had only to have proved that there were no more than three chances in a thousand that the result could be due to the play of chance whereas in more recent times one has to argue that there is no more than one chance in a million.[2] One can, however, announce "evidence for" at a lower evidential threshold.

The dimension labeled *evidential collectivism* is a little less straightforward. It concerns the extent to which it is acceptable to publish a provisional finding. Convinced *evidential individualists* believe nothing should be published until the result is certain; the development and debate that takes place prior to the public "announcement" should be conducted behind closed doors. Convinced *evidential collectivists* believe that suggestive

2. To be more exact, this is what we know from analysis of papers in the journal *Physical Review Letters*; its publication policy will be analyzed more closely in chapter 5. We may be sure that it reflects such a change across the whole of this kind of physics even if the rule is not always applied in quite such a rigid way in other journals.

findings should be published as soon as possible so that the wider community can begin to prove, disprove, or develop a claim as early as possible.

In the case of Big Dog, the evidential collectivism dimension played out in a still more subtle way. There was never any question that the Big Dog result (if it wasn't a blind injection) would be published, but the debate, as has been seen, was about whether it should be called a discovery or merely "evidence for" a discovery. If it was to be called only "evidence for," then a degree of evidential collectivism was going to be endorsed, because a relatively insecure finding was going to be published inviting the wider community of physicists to treat it as unfinished business and therefore relieving the gravitational-wave community of the responsibility of having made a firm discovery—something that convinced evidential individualists believe to be a moral obligation. In the years before the establishment of interferometry as the sole method for gravitational-wave detection, members of the community had been scathing in their condemnation of "suggestive" results published by groups using earlier technologies. Thus, acceptance of an "evidence for" title could be construed as a *volte-face* in respect of this dimension of evidential culture-space—a point to which we will return.

The idea of *evidential significance* is also quite subtle. It is the extent to which any finding—any set of numbers with a satisfactory *statistical* significance—is said to indicate something more or less *scientifically* significant.[3] The way evidential significance plays out in this debate has already been seen in the discussion of black holes. The paper was going to be the first direct discovery of (or evidence for) gravitational waves, but was it also the first direct observation of a black hole? To include the first direct detection of a black hole represents a positive advance along the evidential significance scale, whereas keeping black holes out of the title is to go back toward the origin.

Time

As with the Equinox Event, the Big Dog *Physical Review Letters* paper can also be thought of like an ingot in a forging machine being shaped by the past, the present, and the future (figure 18). The idea of evidential culture-space and the notion that there are pressures on the debate from three directions representing time will act as organizing principles for much of what follows.

3. The idea of evidential significance was first set out by Pinch 1985.

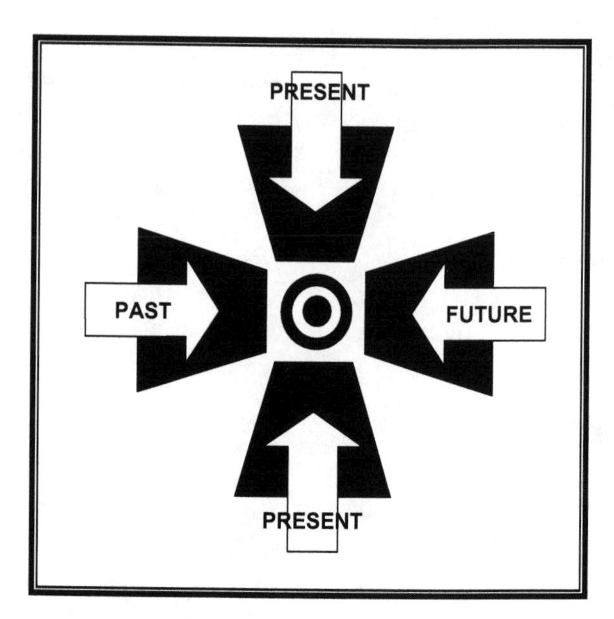

Figure 18. Forging the shape of the paper

10 Time Slides and Trials Factor

What was happening in the present? In most modern observations in physics and astronomy, all that is "observed" is a set of numbers. The numbers carry no trace of the thing they are supposed to represent. There are no fingerprints on the numbers that represent a gravitational wave, they are just numbers and everything is imputation.[1]

Suppose, then, you are looking for an example of a new kind of thing—let's call it a "tharg." No one can recognize it from its appearance. All that can be said is, "This apparatus has been set up in such a way that if X occurs in the apparatus it will be consistent with the apparatus having been affected by a tharg." But given that thargness itself cannot be felt, smelt, or otherwise experienced, the only way to know something really is a tharg is by showing that it could not be anything else. If it can be shown that it is sufficiently unlikely to be anything else then the 'observers' feel entitled to claim that a tharg has been 'seen.'" Two questions follow: How unlikely is "unlikely"? How does one show how unlikely it is to be anything else on any particular occasion?

One shows how unlikely it is by trying to think of all the other things it could be and showing how unlikely it is to be any of them. But this leads to another question: How likely does it have to be

1. We will see when we get to the discussion of the waveform of Big Dog that it is not quite true that a gravitational wave leaves no traces on the numbers—it does have fingerprints—but "no fingerprints" is the right thought to hold for the time being.

that some particular thing created that tharg-like effect, or gravitational wave–like effect, for the scientists to have to take it seriously in the first place? This is a matter of evidential threshold and, of course, is very much tied in with history and tradition—the past.

Thus, here are some examples of potential creators of gravitational wave–like effects on interferometers that physicists do not take seriously because their traditions tell them they don't have to bother and the evidential thresholds for their existence are very high.[2] (1) The scientists are not worried about the possibility that aliens from another planet are having fun with the apparatus at the expense of terrestrial beings. (2) The scientists are not concerned in case the force of desire to see a gravitational wave is so strong in some of the physicists that it is producing a psychokinetic effect on the apparatus. (3) The scientists are not worried about the possibility that the theory of relativity is so deeply flawed that it could be that, for a moment, someone's wristwatch emitted some local gravitational waves that affected the interferometer in such a way as to mimic the inspiraling of two massive stars. All of these could be creating a gravitational wave–like effect on the interferometers but in none of the discussions is anyone checking out whether they could actually have happened or trying to eliminate the possibility that they did.[3]

An example of what the physicists are somewhat belatedly finding that they have to take seriously is the possibility that someone has entered the laboratories and secretly injected a false signal into the apparatus—not an official blind injection but a kind of hacker's prank. They have to take this seriously because a couple physicists confessed that, before the official blind injection program had been invented, they had at one time thought of doing this as an unofficial test of the readiness of the team to detect gravitational waves. Furthermore, they had gone so far as to work out how to do it so they knew that doing it was possible. In an e-mail of 26 October, Acacia wrote that people e-mailed him with this concern:

2. One can say the prior probability of their existence is too low if one prefers the Bayesian formulation or one can use the metaphor from *Gravity's Shadow*, chapter 6—the Dam of Disbelief from which one can only remove the upper stones if the water of physics is to stay in the reservoir.

3. It was not always thus—traditions change. In the heady days of the 1960s and 1970s, it was seriously considered that a psychic, Ingo Swann, had paranormally affected a sensitive magnetometer in the Stanford laboratory of Bill Fairbank, a pioneer of gravitational waves among other things (see Collins and Pinch 1982, 139–40). For an account of the "heady" period in physics, see Kaiser 2011.

> Some people feel strongly that "Direct Detection . . ." is too confident for the title (the main worry is sabotage!), and it should be softened to something like "Evidence for the Direct Detection of Gravitational Waves . . ."

The possibility of sabotage was taken seriously enough for it to be agreed that the logs that keep track of who is in the building should be examined around the time of the event and that a small committee should be set up with the following brief (see the following e-mail from Quince dated 22 November):

> "Are we confident that the event is not the result of malicious tampering?"
>
> This is a difficult one. . . . Has anyone looked at the possible excitation channels? Are there other things that we can/should do to rule out tampering?
>
> I would suggest that . . . chair a small subgroup . . . to look into this point. I, for one, would like to know if this group can figure out a way that they believe that they could inject a signal into the detectors without it being caught.

This small committee reported to the Detection Committee on 18 January. It concluded that, because of the immense complexity and frequent modification of the analysis programs and other software associated with the experiments, it would be extremely difficult if not impossible for someone from outside the collaboration to create anything that looked like a signal but was not immediately detectable as a fake. As far as internal tampering was concerned, they were unable to examine every channel where such a false signal—a "blind-blind injection" as it were—could have been inserted because that would mean unblinding the Big Dog. Nevertheless, they recommended that should the envelope prove to be empty and Big Dog real, such channels would have to be examined retrospectively with great care. In the meantime, members of the collaboration were to be trusted:

> The action of a rogue trusted colleague can't be ruled out, but even among collaboration ranks, the required skills are rare. The necessary commitment of time and resources would also be large and difficult to conceal. Depending on the specific mode, we could name about 10 colleagues with sufficient technical expertise to "easily" mount an attack. To our knowledge, of course, none of these individuals would find anything to gain from sabotage. Indeed, our experts' vigilant attention to machine performance and data integrity can be seen as a powerful deterrent.

Finally, what the physicists have always had to take seriously is the possibility that the apparent signal is just a coincidence of two very loud noises of the right kind of waveform that had nothing to do with gravitational waves but that just happened to occur on both LIGO detectors at the same time. Here, it is not only that a gravitational wave—like a tharg—doesn't have fingerprints, it is that a gravitational wave–like thing cannot even be said to be a gravitational wave–like thing unless something coincident is seen on two widely separated apparatuses at the same time. With rare exceptions, such as when there is a very well-observed and documented event in the sky which happens in coincidence with an exactly matching wave, there cannot be a gravitational wave–like thing on one apparatus alone because there would be no way to know it was not "noise." So, and this is a point that will prove important later, the scientists are not really looking for gravitational waves, they are looking for coincident "excursions" from low-level background noise between two detectors. I suggest that they are not gravitational wave hunters but "coincidence hunters." When they find the right kind of coincidence they will, however, call it a gravitational wave.

The False Alarm Rate

Ever since the first gravitational-wave detectors went on air in the 1960s, the basic technique has been to look for coincident bursts of energy in two widely separated detectors. The effects being looked for have always been so small in respect to the sensitivity of the apparatus that they have been comparable with the background noise. The background noise is caused by disturbances in the air, the ground, and the electromagnetic environment and is also created by the restless movement of the atoms and molecules in the materials of which the apparatus is made. In spite of the ingenuity put into suppressing these noises, without which the whole enterprise would never get started, noise can never be eliminated. Further ingenuity is required to separate the signal from the remaining background. If the separation between two detectors is wide enough there are only a few outside disturbances other than gravitational waves that could give rise to a coincident jump in output. These few include earthquakes centered near the midpoint of the detectors, electric storms, or everyone switching on their kettles at the same moment when the commercials appear during *Monday Night Football* and creating a coincident change in the characteristics of the electricity supply at both detectors. Such disturbances, however, should register on the "environmental monitors" and result in a warning "data quality flag" in the signal channel; putative signals that are coincident with such flags can be discounted. Actually, the environmental monitors have to keep a sharp watch for disturbances even at individual sites because if even one of the coincident signals is known to have been caused by a truck hitting a curb, or a tree falling in the nearby forest, then there has been no "real" coincidence caused by external forces.

The environmental monitors cannot detect chance coincident peaks in the movement of the atoms and molecules of the materials and mirrors so there is always a risk that these, or some other coincident "glitches" of unknown origin, are the real source of something that pretends to be a gravitational wave. Luckily, the majority of the random noise coincidences were small. But it remains necessary to work out how unlikely it was that any putative signal was caused by chance effects that have not been eliminated in nonstatistical ways. This is called working out the "false alarm rate" (FAR) and the related "false alarm probability" (FAP).

The traditional way to work out these false alarm likelihoods in gravitational-wave detection physics is to carry out "time slides."[4] Imagine taking the output of the two detectors (think of them as squiggly lines on ribbons of paper with peaks and troughs a small fraction of a second apart) and laying them out, one above the other, as in figure 19.

The ribbons in this figure exhibit one coincidence, as indicated by the vertical line (normally one would expect not a single peak but a few coincident cycles of data in each ribbon when a coincidence occurs but for simplicity, just one peak is shown). The streams of data—"ribbons"—are, of course, much longer that this; they extend for weeks, months, or even years, depending on how long the detectors remain "on air," but the principle remains the same. Now, suppose a promising coincidence is found that is not vetoed by a data quality flag. What happens next is that one of the data streams, one ribbon, is "slid" along—offset—by, say, a few seconds. We now have a section of overlapping ribbons of length "1." (An offset is also known as a "delay" so the starting point is known as "zero delay.") Once more the number of coincidences is examined. In this case any coincidences that are found—call them "offset coincidences"—can only be spurious. They must be the result of coincident noises; they cannot be result of a signal from outside, because any such outside signal will show up only at zero delay.[5] This is how a "background" is artificially generated.

4. The basis of the old "delay histograms" discussed in *Gravity's Shadow*.

5. For the purpose of explanation I am ignoring niceties, such as the light travel time between detectors, which means that coincidences don't have to quite coincident.

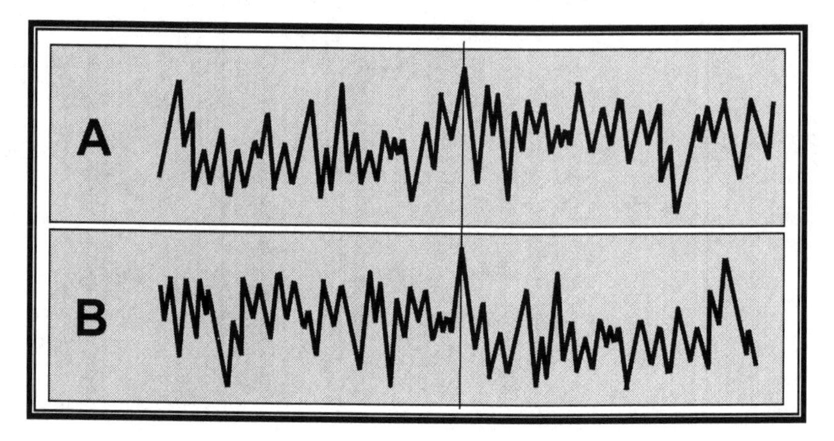

Figure 19. The outputs from detectors "A" and "B" laid out above one another

Therefore, if one counts up the number of offset coincidences and finds "n" of them, one can say that in a time period equal to 1. N coincidences might be found purely by chance: dividing the length of the ribbons by n enables one to say that the apparatus us likely to exhibit one spurious coincidence every $1/n$ seconds. We will know from this time slide exercise that a purely chance coincidence is likely to occur only this often. This exercise shows how likely it is that a promising zero-delay coincidence was actually caused by chance alone.

But that is only the beginning. To get a sufficiently strong "unlikelihood" for causation by chance alone, more background is needed. It is built up by carrying out many more time slides. If the first offset was 5 seconds the next could be 10 seconds, the third 15 seconds, and so on, with each new delay generating a new contribution to the artificial background. If each one imagines each ribbon is joined at the ends to make a loop, one can rotate the top loop all the way round in five second intervals until one reaches the beginning again. Each time a new offset is created it is as effectively a fresh analysis on new streams of data so if the ribbons are 2 months long and 100 time slides are carried out 100 × 2 months of data will have been examined. This is 16.7 years' worth of noise data. If "p" offset coincidences are found in all this data the FAR is p per 16.7 years.

More and more offsets can be made until certain limits are reached. One limit is that if the offsets are too small—say they were one second instead of five—the cycles associated with, say, a two-second excursion in the two ribbons would overlap so certain apparently discrete coincidences would be the same coincidence repeated twice or more. This would give rise to a mistakenly elevated false alarm rate. The offset settled on for the Big Dog analysis was, as it happens, five seconds.

Another limit is that the lengths of the ribbons cannot be too great, because the interferometers are not stable over a very long time. Compare a piece of data at time "t" with something that was happening a year later and, effectively, the output of two different machines is being looked at. To provide a fair estimate of the unlikelihood of a signal it is necessary to create the time slide background from ribbons with roughly similar noise characteristics to that present at the time of the putative signal. As will be seen, for this reason, the time slide exercise was limited to the period known as "S6D"—which is the continuous, roughly sixteen-week run during which Big Dog appeared. More remote stretches of S6D date were excluded from the time slide exercise.

It is also worth bearing in mind that to create a lot of time slides and count the coincidences takes a lot of computing power on big computers. In the case of S6D a maximum of four million time slides would be carried out but it took a month or so to do it.[6]

The threshold of what is taken to count as a coincidence will also affect the time slide analysis. In nature, small noises are more frequent than large ones so there will be far fewer large noise coincidences than small ones. If the potential signal that occurs at zero delay is small it is known in advance that there will be many such coincidences in the offset data so there is no point in doing the time slide exercise—the false alarm rate will be so high that the signal will not be credible. If, however, the potential signal is large, then the time slide exercise becomes worthwhile because it should generate few offset coincidences.

By the middle of the debate scientists had come up with innovative ways of doing effective time slides. The output of the plain old time slide method was slowly becoming useful as the computers hummed toward the four million mark, but there were two other methods on the table. One, to which we will eventually return, was an extrapolation from what was known assuming that the noise followed a fixed statistical distribution; one could say that if you extrapolate from the distribution of noise that

6. S6D was 16 weeks long. Sixteen weeks consists of 1.93 million 5-second intervals, or, roughly 2 million. But one must imagine the two ribbons end to end to start off with the right-hand end of ribbon "A" opposite the left-hand end of ribbon "B." The end of ribbon "A" can then be shifted along 2 million times in 5-second intervals until the ribbons are exactly alongside each other. But then one has to make a further 2 million shifts until the left-hand end of ribbon "A" reached the right-hand end of ribbon "B." That is how it comes out to 4 million time slides altogether.

is already plotted with the data in existence then more data would have produced more of the same distribution and a false alarm rate could be computed from this.

The other method was made possible by the sheer size of the coincident signal. It made it reasonably economical to look through the whole data set and count up how many noncoincident noise excursions were loud enough to create a pseudosignal of similar magnitude. Two of the scientists embarked on this method at roughly the same time. One champion of the method (the second to report the work), called it the "find all coincidences" method:

> The basic idea is to construct all possible coincidences from any time slide—rather than actually doing slides.

This method gave a promising result. The find all coincidences method also drew attention to the fact that if you removed Big Dog from the data there was simply nothing in the background that could produce an offset coincidence that would match it in strength; this was to give rise to a very lively debate. This aside, discussion of these alternatives was to die down as the number of real time slides approached four million. The first results were, of course, less impressive. One hundred time slides produced a FAR of "less than one every seventeen years." As everyone seemed to agree, this was "not interesting"—not good enough to form the basis of a discovery claim.

11 Little Dogs

At the heart of the time slide method is an intriguing philosophical puzzle with real and consequential effects on the meaning of the Big Dog coincidence. A putative signal like Big Dog comprises a big excursion in both ribbons of data at zero delay. When coincidences are looked for in the time-shifted data should these excursions be included or excluded? Dogwood put it as follows in an e-mail of 27 October:

> If we include the event's triggers, we are effectively assuming them to be background and there is a finite probability that we are overestimating the FAR by some amount. This is a conservative procedure and is correct in the limit of small event rates.
>
> Conversely if we exclude the event's triggers from the slide distributions, we are effectively assuming it to be signal and there is a finite probability that we are underestimating the FAR.

To explain, if the putative signal is really a signal then its component excursions should be removed before the time slide exercise is carried out, because when the component in ribbon A finds itself opposite loud noises in ribbon B, and when the component in ribbon B finds itself opposite loud noises in ribbon A, it will appear that there are offset noise coincidences being produced. But they aren't really offset noise coincidences, because half of each coincidence is not noise but a real signal. The net result of this will be that the false alarm rate comes out higher than it should—all those places where offset coincidences were part produced by the real signal in A and B should have been removed. If, on the other hand, the putative signal is only noise, then its components should be left in. When the component in ribbon A finds itself opposite loud noises in ribbon B and vice versa, they really are offset noise coincidences and *should* contribute to the false alarm rate. This is what Dogwood is saying with the claim that assuming the putative signals are noise "is correct in the limit of small event rates"; Dogwood means that when there are only very few signals the conservative assumption is that they are not really signals at all and they should not be treated as signals at the outset. This, however, came to seem less clear and more paradoxical as the debate unfolded. One determines whether a signal is real by setting it in the context of the false alarm rate but to know the false alarm rate one has to know whether the signal is real! It's the "time slide regress."

For about a week spanning the end of October and the beginning of November there were a flurry of e-mails debating the paradox. I collected around seventy. A new piece of terminology rapidly developed. An offset noise excursion that paired up with one of the components of Big Dog to produce a false alarm comparable with Big Dog in terms of combined signal-to-noise ratio became known as a "little dog." The question at the center of the debate, then, was whether or not to ignore the little dogs. The following severely edited extracts from those e-mails provide some of the flavor of the argument:

> [From scientist A, 20 October:] Some of the armchair analysts at the DAC [data analysis council] call were opining that it is legitimate to remove the little dogs in this plot (but I don't think there's unanimity about this). On the chance that the legitimacy of this is convincing, [we need] extrapolations in which, e.g., the 5 minutes around the big dog are excluded from the background estimate?

> [From Ginkgo, 2 November:] I am against choosing a procedure [involving cutting time either side of the Big Dog—which works because, as it happens, there are no medium dogs in that area] that works just to improve the big dog's significance, that does not work in general. This sounds too biased to me. Can you convince me otherwise? . . . I actually think the result WITHOUT the big dog in there will be closer to the "correct" FAR, and feel it is OK to state which way think is better. But I don't feel comfortable about not giving the FAR with the big dog in the background in the paper, because

I want to avoid any appearance of "confirmation bias." . . . What worries me about "confirmation bias" is that while I know we are all smart enough to try to avoid this, it seems really smart people are not able to avoid it.

[From scientist C, 2 November:] I realize that that creates a slippery slope. . . . However, I don't think it's something we have to worry about now. Currently, we're in a regime where we expect at most one gravitational wave in our data.

[From Ginkgo, 3 November:] I think the key point you make is, we can remove the "big dog" AFTER we are confident it is a detection, but not to make ourselves confident it is a detection (which risks appearing as "confirmation bias"). [I was confused about this: what counts as AFTER we are confident—is it 4-sigma or 5-sigma? It seems to me that if you have even 4-sigma then the input of the little dogs to the background should be very much reduced. I wrote to Ginkgo on this point and the discussion is reproduced below this exchange.]

[From scientist D, 3 November:] I'm one of the armchair analysts from the DAC call who wondered why one would include the . . . big dog triggers in the background estimate. Why don't we want a background estimate that is absolutely statistically independent of the foreground event? To the extent that including a single-IFO big dog trigger distorts the background estimate, its inclusion is contrary to the underlying assumption that it's part of the background.

[From scientist E, 3 November:] When the time slides are done to estimate the FAR, this is to answer to the question: what is the significance of any event? The same procedure should be applied to every event, including the loudest one. Otherwise we are increasing artificially the significance of this event because we have already assumed that this is a real event.

[From scientist D, 3 November:] What I'm suggesting is that the "correct" blind analysis should have excluded every zero-lag coincidence's single-ifo trigger from its own background estimate. The analysis would then still be blind, but would no longer have a background estimate that is biased high—as the current little dog estimate is biased. If I understand correctly, the current default background estimation is more or less guaranteed to be biased upward for very loud true signals. The louder the signal, the greater the bias. . . . In this instance, I think [we have] uncovered a conceptual flaw in the background estimation in the presence of a strong signal. (I could imagine the flaw has been known for a long time, but neglected under the assumption that no true GW would be strong enough to cause trouble.)

[From scientist D, 3 November:] It's only a distortion if the big-dog event is already assumed to be real, otherwise the original background distribution including the triggers is correct.

I disagree. I'd like to know if the big dog is consistent with the rest of the data taken. Something extraordinary happened in a short interval of time on Sept 16.

[From scientist F, 3 November:] We have a bug in the analysis when computing the FAR . . . as we don't exclude the trigger from its own background computation.

[From Ginkgo, 3 November:] I do not think either leaving or removing triggers generated from a foreground trigger in its background is right or wrong. I think the former can over-estimate the background and the latter could under-estimate the background. A priori I see no unbiased perfect way to find the correct background. I see no "bug" in what was done, just a choice that was made.

[From scientist C, 4 November:] To label [certain excursions] "glitches," you have to appeal to some outside information that you don't have access to, unless you have God's phone number.

You're accusing me of introducing a bias, but look at your own assumptions carefully. For any foreground trigger, you have no way of knowing whether that trigger is from a gravitational wave source or a noise source. If you are going to do a true un-biased search then, following the principle of indifference, you should assign a uniform prior when calculating the probability that that trigger came from a specific distribution. If you allow that trigger into your background, however, you are not assigning a uniform prior. You're skewing the prior toward believing it is noise, which will in turn skew your posterior. That is not an unbiased search.

This issue was not to go away. As January turned to February and the paper was being finalized, one could still hear the following (from a telecon of 1 February 2011):

> We don't really have a clear answer to whether the [little] dogs should be included or not. I suspect that if you talk to 100 physicists, half will say one and half will say the other.

As more and more time slides were completed and analyzed the solution that seemed to be emerging was to somehow include both versions of the calculation in the paper.

To save confusion one should note that in the slides as actually executed, the data ribbons were not arranged as loops but were left as straight stretches of data. This meant that the end of one ribbon slid off the end of the other as it ratcheted along. So while 4 million slides were completed this was equivalent to only 1.75 million full-length slides with the ribbons arranged in loops. What this means is that to calculate the full background time that was generated via time slides one has to multiply the length of the run by 1.75 million. This produces an effective background time of 200,000 years rather than the 1.25 million years that would result from 4 million time slides applied to 16 weeks of data arranged in continuous loops.[1]

With Big Dog left in, there were 5 false-alarm excursions found in the 200,000 years of time slide data giving one false alarm in 40,000 years. But there were no false alarms at all if Big Dog was removed—all the false alarms with Big Dog left in were little dogs! So if the little dogs were removed, and this is important, only *a limit* to the false alarm rate could be calculated. What could be said with little dogs excluded was that the false alarm rate was *less than* 1 in 200,000 years. This is a lower limit and the, unknowable, true false alarm rate could be much more favorable to the detection claim.[2]

The "Kill Your Grandfather" Paradox

As noted above, I was confused about Ginkgo's remark: "we can remove the 'big dog' AFTER we are confident it is a detection, but not to make ourselves confident it is a detection (which risks appearing as 'confirmation bias')."

I wrote to Ginkgo on 15 February:

1. 6 (weeks) × 1.75M / 52 = 201,923 years

2. Thanks to Collin Capano for sending me these results in a clear form (private communication, 2 February 2011).

> What I wondered was what your criterion would be for "being confident that it is a detection." Thus, suppose you do time slides with little dogs in and you get 4.3-sigma. Is that "confident that it is a detection" or do the time slides with little dogs in have to reach 5-sigma before "we are confident."

Ginkgo replied the same day:

> I was arguing that we needed to report the false alarm rate (FAR) with the big dog included in the time slides, because that is the way the background estimation algorithm was designed before we opened the box. After we opened the box, we saw that perhaps we should have removed the time around the big dog, but since we were no longer blind to the result doing so introduces bias. But we can always go ahead and remove the big dog related events from the background, as long as we are clear that this is done after the fact. So that is what we have done. . . . The exact number of sigma shown (~4.5) is somewhat arbitrary . . . but so is the particle physics 5-sigma criteria. So whether this is defined as a confident detection or not is a judgment call. Regardless, we can see in the current draft of the paper, if you post-facto remove the time of the big dog from the background estimation, the FAR would be much less than 1/(40000 yr), and perhaps (I am completely guessing here) it is 1 in [1] million years. It is then up to the reader to judge for themselves what this means. The point I was making above is that we needed to give the 1/(40000 yr) number. If we did not give that number, but just gave the 1 in a million year number, doing anything post-facto like this (i.e., changing our background estimation algorithm after the box is opened) risks appearing as "confirmation bias," or would make us guilty of "confirmation bias."

So Ginkgo is saying that the potential sigma value with little dogs removed is huge (remembering that any number we have is only a lower limit) but that there was a worry about confirmation bias and that is why both values have to be given.

Ginkgo's suggestion that sigma could be recalculated without little dogs after it had been confirmed that Big Dog was real leads to interesting consequences. We know that the standard for a discovery in physics has moved from 3-sigma, through 4-sigma, to 5-sigma. Imagine Big Dog had been found when the standard was 4-sigma, so it definitely would have been a discovery, and therefore, its true significance was recalculated with little dogs removed. And imagine that the recalculated significance was equivalent to

5-sigma. Assuming no other evidence about the existence of gravitational waves has been forthcoming in the meantime, the question one must then ask is whether Big Dog is still a discovery in the era of the 5-sigma hurdle; it has reached 5-sigma so that suggests "yes," but it reached it only because at the time it surpassed the hurdle the barrier was set at 4-sigma. Should we be going back in time and killing the "grandfather" that gave birth to the 5-sigma result? Discovery in physics shares its genetic inheritance with social convention. Indeed, Ginkgo himself says, "The exact number of sigma shown (~4.5) is somewhat arbitrary . . . but so is the particle physics 5-sigma, criteria." We could ask a similar question for any result that was established at a time when the statistical significance hurdle was lower and which has not subsequently jumped the more recent higher hurdle.

When I sent my e-mail to Ginkgo I accidentally copied it to the whole analysis group (an easy mistake given the way the e-mail is set up), and Quince responded with a much stricter interpretation. Quince said that there was nothing to be done except present the answer with the little dogs *included*: once this had passed the threshold (or not) it was otiose to do any other calculation.[3] Quince did not want to leave it to readers to make up their minds but was continuing in his current role as a conservative interpreter of the event.

During the six-month period before the envelope was opened the sentiment for leaving the little dogs in or taking them out seemed to wander about with, as noted, the consensus around the end of January that both results should be in the paper. Later I detected a move to report only with the dogs in place. Thus Quince wrote to me on 1 February (my comments are in square brackets):

> Time slides will always fail when the number of real events begins to approach the number of noise glitches in your data. [I think this is my starting position in figure 16]. It is our inability to block the GW signal to get a pure background that forces us to use the artifice of time slides. [As argued in *Gravity's Shadow*, the time slide is, effectively, a way of removing the GW signal given that there is no other way of turning it off. In the case of Big Dog, however, we have discovered that it does not work perfectly.] . . .
> What I object to is the bias of removing the thing you are testing from its

3. This was a clever argument that took a long time to dawn on me. Quince was saying that if the signal passes the significance threshold then it has passed it—it cannot pass it a second time by being recalculated as still more significant, it either passes or does not pass. Therefore, only the first calculation is needed—a second calculation serves no purpose.

> background estimate. Fortunately, we are in the position where I don't think we have to remove the little dogs to make a good case for the Big Dog being a GW event. My preference for not even showing a background with the little dogs removed comes from the strength of our case. Removing them makes it look like we think we need to in order to make a credible case. If skeptics see any evidence that we are making (after the fact) choices to lower the background, it will just fuel their doubts. We don't need that.

In this book the little dogs will reappear in two places. At the end of *Big Dog* the result of a calculation with little dogs removed will be reported. In the third part of the book, "The Trees and the Forest: The Sociological and Methodological Reflection," there is a technical discussion turning on my own attempt to solve the philosophical/technical problem of the little dogs. I examine the scientists' reactions to this adventure and the methodological perils and benefits of trying to become so closely involved in the world of the actors.

The Logic of the False Alarm Probability

All this discussion of sigmas turns not on the false alarm rate (FAR) but another measure of how likely it was that the event was caused by noise. This other measure is the false alarm probability (FAP), and it is subtly but importantly different. It is the false alarm probability that is converted into a sigma equivalent. The full relationship between FAR and FAP and its consequences took me a long time to understand, though it is pretty easy once one "gets it." The real difficulty is that, in this case, the logic of the FAP depends on the fate of the little dogs.

False alarm *probability* is generated by multiplying false alarm *rate* by the length of time the instrument has been on air. Thus, if the FAR is 1 per 40,000 years and the apparatus stayed on air for 40,000 years, it is likely that it would see one false alarm. In those circumstances that false alarm probability would be 1. If, on the other hand, the apparatus was on air for 6 weeks, or roughly 0.12 years, the FAP would be 0.12/40,000 = 0.000003 (it is this figure that is converted into a sigma equivalent). As can be seen, given a fixed FAR, the longer the apparatus stays on air the less significant does the FAP become and the more the sigma level falls.

Applying this to some hypothetical lengthening of the run is complicated. The following is my own reasoning but it was verified by one of the scientists and I was later to win a beer following a light-hearted argument with another of the physicists who said I had it wrong.[4] My reasoning depends on certain assumptions. (1) Big Dog was a very lucky event that would not be repeated if the S6D run were to be extended for any reasonable length of time. (2) The noise in the detectors found in S6D would remain at roughly constant level even if the run was extended. Given these assumptions, (a) if the length of the run were, say, doubled, there would be no more events as striking as Big Dog but (b) there would be double the number of little dogs. If Big Dog were taken out of the time slide exercise, however, then (c) the number of false coincidences would still remain at zero.

What would happen under these assumptions if the length of the run were indeed to be doubled? If that were the case it would be possible to complete twice the number of time slides before the limits were reached and the ribbons used in each slide would be twice as long, meaning that four times the background could be generated—800,000 years instead of 200,000 years. But there would be 10 little dogs instead of 5. The false alarm rate would then be 10⁄800,000, or 1 every 80,000 years instead of 1 in every 40,000 years. This is twice as good. But since the run would be twice as long, there would be twice as much chance of seeing such a false event so the false alarm probability would remain unchanged and so would the sigma equivalent! In sum, under these—not-unreasonable—assumptions, increasing the length of the run does nothing for the sigma value or the significance of the signal.

If, on the other hand, the little dogs are removed, the limit that can be reported under my assumptions does change because the number of false alarms—false coincidences that are comparable with Big Dog—remains at zero. The limit, then, becomes less than 1 in 800,000, which is *four times* better than it was rather than twice as good. This still has to be multiplied by two to get the false alarm probability because the length of the run has been doubled but it is still twice as good as it was. Under these assumptions, if Big Dog is removed, the FAP *limit* improves proportionately to the

4. The point I want to make when I explain that I won a beer over the little calculation is not that I can do physics better than physicists, because any of them could have worked this out much quicker than me. The point is that I had the opportunity to work it out first because the physicists weren't thinking about the question, or had given up thinking about the question when they learned that the little dogs were definitely going to stay; they weren't, or were no longer, asking themselves how they could get up to 5-sigma whereas significance level was one of my two "rallying points." Thanks to Tom Dent for some initial help and the confirmation.

increase in the length of the run (if the maximum number of time slides is carried out) and the sigma value will improve accordingly.[5]

To me it seems right that in the case of a very lucky event it should be possible to improve one's confidence that what has been seen is real by staying on air longer so as to demonstrate that nothing in the background is comparable with the event itself. This is a "philosophical" argument for saying that the right way to do the time slide exercise is by removing the little dogs. In high-energy physics the apparatus is run for as long as it takes to get the sigma value up to the desired 5—so why shouldn't this be possible in gravitational-wave detection?[6]

Whatever, the instrument was not run for longer, and no more than 4 million time slides were carried out, so the FAR remained at 1 in 40,000 years with little dogs.[7] In that case the longer the instrument *was counted* as running the less impressive would be the FAP (and the sigma level). I stress "was counted" because, as we will now see, it turned out that how long the instrument was counted as running was not necessarily the same as how long a period was used for the time slide exercise.

How Long Was the Instrument Running?

The question of how long the instrument was running goes back to the notion of an "experiment space" or "sample space," as discussed in chapter 5. A recent paper promulgated by Louis Lyons, an authority on statistics for the physical sciences, shows in still-clearer form that the problems discussed there were not philosophically "contrived." I came across this paper because, on 18 January, it was circulated by Gingko, who was looking for a solution to the problem.[8] Gingko remarked:

5. Though it increases less than directly with the length of the run.

6. Admittedly the logic of the increase in the sigma value is somewhat different in high-energy physics; in that case they are seeing more events rather than making it less likely that the one event is a false. See Krige 2001 for the unusual way that Carlo Rubbia reached the necessary 5-sigma level to win his Nobel for the discovery of the W-boson.

7. Could more time slides have been carried out—perhaps if the interval was shortened from 5 seconds to, say, 3.25 seconds, a convention that had been used in earlier runs?

8. Louis Lyons, "Comments on "Look Elsewhere Effect,'" http://www.physics.ox.ac.uk/Users/lyons/LEE_feb7_2010.pdf (last accessed 5 March 2011). In fact there are flurry of web discussions from the beginning of 2010 onward of what is called in earlier chapters of this book "the trials factor." These can now be found by web-searching for "look elsewhere effect." I knew nothing of these when I wrote chapter 5 so the discussion in that chapter is independent of Lyons et al (and with somewhat more of a philosophical slant). Louis Lyons, by the way, sits on the CDF (collider detector at Fermilab) Statistics Committee and has published *Statistics for Nuclear and Particle*

> it seems a bit surprising [but] it seems there probably is not a consensus [in the high-energy physics community] on how to present results like this.

Lyons is asking the same question about how widely one should look in order to calculate the trials factor (or "look elsewhere effect," as he calls it). Lyons argues that when one is searching for similar experiments that need to be taken into account when calculating a trials factor—when one "looks elsewhere" to see what else has been done—there is a range of possibilities. He talks in terms of results that come out in the form of a histogram and he says that there is an "ever widening list of possibilities" about where one should look:

> 1) In the particular location of the histogram that shows the effect.
> 2) Anywhere reasonable in the histogram that shows the effect.
> 3) Anywhere reasonable in the analysis I performed here.
> 4) Anywhere reasonable in any analysis performed by our Collaboration.
> 5) etc.

He then comments on this list as follows:

> Choosing 1) is over-optimistic and unrealistic. Option 2) is a more reasonable alternative. My favourite is 3), but this can be hard to define in a non-blind analysis. It allows for other binning of a histogram, alternative choice of cuts, other histograms that were looked at, etc. Suggestion 4) and anything further down the list is casting the net too wide for my taste. (2)

Ginkgo was right to be surprised that there is no consensus among physicists about such an extraordinarily important matter—how to decide on the trials factor that goes into the calculation of the statistical significance of a result. But the reason there is no consensus is that, once more, a philosophical conundrum lies at its heart, which makes the lack of consensus less surprising. Note Lyons's last phrase, and this is coming from someone who is such a high-ranking authority on statistics that he sits on important advisory committees: he is saying that it is a matter of *taste*! He is, in short, helping to develop the culture of physics in respect of a matter, of real

Physics (Cambridge: Cambridge University Press, 1986), so he can be accounted an expert of considerable standing. His ideas were further discussed at a conference on statistical methods in high energy physics as is reported at length in the *CERN Courier* of 26 October 2010, http://cerncourier.com/cws/article/cern/44115 (last accessed 5 March 2011).

significance (pun intended) to the reporting of the validity of physics results; and he is doing this because there is no clear answer that can be calculated; and he believes a reasonable input into the creation of the culture is good "taste." Choosing a good trials factor is like choosing a good wine!

What is essentially the same "philosophical" problem as applies to the trial factor arises with respect to the question of how long the experiment is said to have been running. If one carefully examines figure 15, the PowerPoint slide presented in Kraków on 21 September that introduced Big Dog to the community, one will see a couple of sentences that were not reproduced in figure 16. They are:

> S5 was different instruments. Meaningless to use that to estimate background for the search.

It is being said that the length of the run should not include the consolidated yearlong period of observation made with a slightly less sensitive instrument between September 2005 and November 2007 known as S5, or Science Run 5 (the basis of the events reported in *Gravity's Shadow*). S5 was to be counted as *not* belonging to the experiment space, because the instruments belonging to S5 were not counted as being the "same." But when is an instrument the same as another instrument? It is the case that instruments are being improved all the time. For, example, the instrument was improved between all the separate runs of S6. These comprised: S6A, which began on 7 July 2009 and ran for eight weeks; S6B, which ran from 24 September 2009 for 120 days; S6C, which ran from 28 January to 25 June 2010; and S6D, which began on 26 June 2010 and ran for 16 weeks.

Figures 20 and 21 show S6C and S6D. The vertical scale shows sensitivity expressed as the range in megaparsecs at which an averagely orientated inspiraling binary neutron star would be detected (a megaparsec is about 3.26 million light-years). One can see the difference in the sensitivity in these two runs (and in figure 21, one can also see the relative insensitivity of Virgo).[9]

On 18 January one could hear the following from a member of the Detection Committee:

9. Throughout I contrast "Initial LIGO" with the "Advanced LIGO," which is expected to be gathering good data in another five years or so. The LIGO interferometers that detected Big Dog are an in-between generation, which should be called "Enhanced LIGO." They were meant to have double the range (and eight times the detecting capacity) of iLIGO but, as can be seen from the figures, achieved only 20 megaparsecs instead of the intended 30 megaparsecs. Because of this I have not used the eLIGO label but lumped the first two generations together.

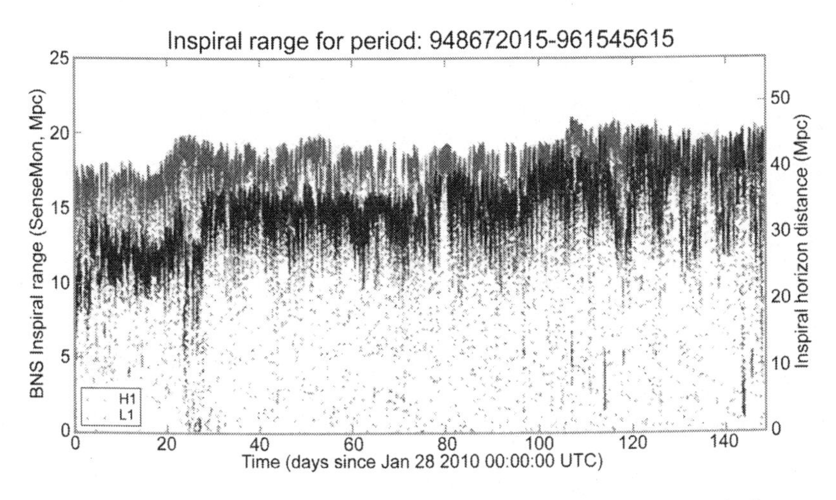

Figure 20. S6C run (28 January–25 June 2010) ranges in megaparsecs. The Hanford detector is in light gray, the Livingston in dark gray

> I have heard people say we should use all the science runs and some people say we should use something very local. I guess all of S6 is the right compromise here.

On the nineteenth:

> How to convert this [FAR] to [FAP] is ambiguous because—it is unclear what total length of time is appropriate—e.g., should we look at S5 too?—should we only consider times when H1L1 were operating?

So what we are getting is a compromise—a matter of choice (or even taste) once more. The Detection Committee argued long over this choice, concluding at a meeting of 1 February that they could not reach consensus. In the end the whole of S6, not just S6D, went into the calculations.[10]

But the trials factor and the length of the run, in one way of looking at it anyway, are not independent. One scientist explained:

> To obtain a false alarm probability over the whole of S6, we account for a trials factor of 12, due to 3 mass bins times [and] 4 different types of coincidence we look for over all of S6.

10. I was later told that a consensus was reached about using the whole of S6, but this happened in meetings of which I have no record.

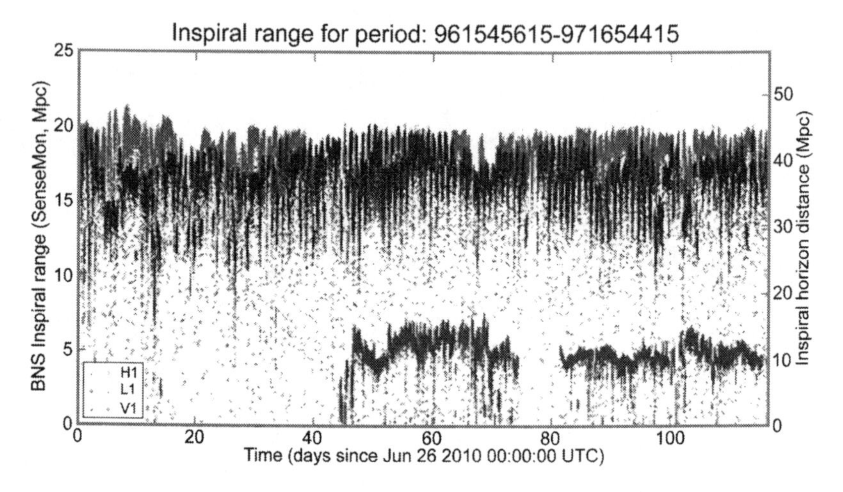

Figure 21. S6D run (26 June–20 October 2010) ranges in megaparsecs. The Hanford detector is in light gray, the Livingston in dark gray, and Virgo is at bottom.

Another wrote on 18 January:

> I understand the arguments for using 6 for the trials factor. It is correct for the time the event occurred, and this is what the code uses to rank triggers. But if we us the entire time of the run to get the FAP, I would prefer using a trails factor of 12, which is the upper limit on the trials factor for the entire run.
>
> Using 12 fits in better with the above suggestion (iii) of giving the p-value for the trigger (in our case big dog) occurring "anywhere in the analysis being reported in the publication."
>
> My previous e-mails also have advocated for using 12 to be conservative, and to make the explanation simpler (in my opinion). But I have also seen there is a argument for using 6.In other words, I am not giving up on advocating on 12, but I am not "dog"-matic about it either. We do have to be very clear about whatever we do in the paper. Using the time for the entire run in one calculation and the trials factor for just part of the run in another calculation I think could be confusing, though I am not saying it is impossible to be clear about this. But I still like using 12.

First note, once more, the word "choice" and the tendency toward conservatism. The trials factor of 12 stood for a long time with a few people unhappy with it. On 19 January one could read:

I don't like a trials factor 12 at all, it seems far too conservative, but I did not feel comfortable making it 6 [in the draft I was preparing].

In fact, the idea of a trials factor of 12 when applied to the entire S6 run was eventually to be dropped in favor of 6. This was because the "four types of coincidence" did not apply to S6D—the period when the observation occurred; at that point the number of types of coincidence was only two. Somehow one had to choose whether 2 or 4 (and therefore 6 or 12) was the right number because it was not self-evident. In the end, the 2 that applied to the actual period of the Big Dog—S6D—was chosen and, along with it, a trials factor of 6.[11]

The choice about counting the sample space as being S6 or S6D makes a difference to false alarm probability which, with dogs included and expressed as a sigma value, is 4.3 versus 4.5; with dogs excluded it is 4.7 versus 4.8 (and, as noted, it could be potentially much higher with dogs excluded because that is a limit). So there are two choices that, made conservatively, provide a sigma value of 4.3 (a long way from "discovery") and made the other way provide a sigma value of 4.8 (right on the edge of "discovery").[12] (According to my calculation, and I have reason to think it is about right, to get up from 4.8- to 5-sigma one would have to lower the FAP by a factor of a bit more than 2, which would have had to be accomplished by running for a bit more than twice as long or squeezing more than double the number of time slides from the same run or, more feasible, some combination of both approaches.)

It seemed to me that counting the whole of S6 as the time period was a kind of "double jeopardy." If little dogs were going to be included, then, assuming the noise was constant across whatever run was chosen as the basis for the time slides, the false alarm rate would increase linearly with the length of the run (assuming that the maximum number of time slides were carried out). The false alarm probability, however, would decrease linearly, leaving it the same however long the run went on. But here it was being proposed that the instrument was too *different* between S6 and S6D to make meaningful time slides across the whole duration but was

11. The number of types of coincidence depended on which detectors were on air and running in a satisfactory manner. Four types of coincidence are HL, HV, LV, and HLV, where the letters are the initials of the detectors: Hanford, Livingstone, and Virgo. The two types of coincidence during S6D were HL and HLV, since Virgo was not running well enough to provide a satisfactory HV or LV. The three mass bins refer to the three separate collections of templates used to model low-, medium-, and high-mass events, respectively.

12. Thanks again to Collin Capano for supplying me with these numbers.

sufficiently *similar* to allow the whole period to be counted for the sake of calculating the FAP. Instead of FAP being constant irrespective of the length of run, this double interpretation of what it means to be "the same," at a stroke reduced the FAP by 2. As far as I can see no one made a formal *decision* to do it this way, it just came out that way because the question was no longer at the forefront of the attention of the wider community.

Finally, there is an argument about what the 5-sigma criterion means. According to [Juniper], contributing to a Detection Committee telecon of 1 February, the convention in high-energy physics is to quote FAPs without trials factor. The point is that everyone knows there is a trials factor and hence the demand is for such a sigma value as high as 5. Physicists just know that the true significance is lower because of the trials factor but the "threshold value for publication as "discovery" is 5-sigma *before trials factor*. I would discover that some of the physicists were confused about this.

12 Discovery or Evidence Revisited

"Detection," "Evidence for," "First Result," "Observation of"

On 26 October, a long e-mail came round from [Hornbeam], a young but energetic scientist. This began the more extended discussion of whether Big Dog was a "discovery" or only "evidence for."

> Many have argued about whether the title and abstract should say "detection of" or "evidence for," or something in between, but I do not think we can even make an informed decision about that until we have a point estimate of the false alarm rate. If it is really as high as 1/(17 years), well, that is pretty weak—something that I think we would have an obligation to publish but with a description of "evidence for," at best. On the other hand, if it turns out to be 1/(200 years), then I think that our readers might not complain if we say "detection" as our interpretation of the evidence. But without having pinned that down, I do not see the point in pretending that we have settled on a title and abstract wording.
>
> Actually, my personal opinion is that we might as well be conservative in the title and abstract and describe our result as "evidence" even if it IS in the 1/(200 years) range. We will then go on to quantify the strength of that evidence, and readers will be able to appreciate for themselves what the data leads them to conclude about the reality of the event. We can even offer our own interpretation of the evidence in the body of the paper. I do not think such a conservative approach to titling and abstract-writing

> would detract in any way from the result. (People are going to read the paper eagerly in either case!)
>
> In a conversation about this earlier today, someone suggested that "evidence for" papers lose out to "detection" papers when Nobel Prizes are given. I asked if there was any example of an "evidence for" result having been published, but the Nobel Committee picking some later "detection" result instead to recognize with a Nobel Prize. I couldn't think offhand of any case in which a conservatively stated initial interpretation has led to that. Someone said they thought that the discovery of the cosmic microwave background was such a case, with earlier hints from someone other than Penzias and Wilson. But I followed up this evening and can relay that Penzias and Wilson's paper was entitled "A Measurement of Excess Antenna Temperature at 4080 Megacycles per Second," and their abstract was modest in its claims: "A possible explanation for the observed excess noise temperature is the one given by Dicke, Peebles, Roll, and Wilkinson (1965) in a companion letter in this issue." It is a conservative approach that appeals to me, at least.

Here we see the group beginning to feel out the positioning of their finding in evidential culture-space. The *evidential threshold* for a discovery is, as everyone agrees, well above a FAR of 1/17 years. Here it is suggested that 1/200 years justifies discovery but this will be questioned later. The *evidential significance* of Big Dog turns, in part, on whether this was the first direct detection of a black hole but that is not discussed by Hornbeam. Evidential collectivism is discussed, however. Hornbeam prefers to leave much of the interpretation to the wider community even if, as Hornbeam believes, a 1/200 FAR would justify a discovery claim. A new and important theme is introduced—the reaching back to physics tradition for guidance and the invocation of tradition to support an argument. In the most palpable way, physics culture is pulling itself up by the bootstraps.

Incidentally, Hornbeam's e-mail mentions the Nobel Prize, and it is a theme that surfaces in other parts of the discussion. Sometimes presenting extracts from a much more extensive body of text can give a false impression and, in this case, it would be quite wrong to imagine that the physicists were really much concerned about the Nobel. I do not know if anyone at the top of the organization was seriously thinking in terms of "the Prize" but it was not going to be awarded to anyone who was discussing Hornbeam's e-mail. Rather, "the Nobel" along with phrases like "book your ticket to Stockholm," are used in a jokey way. It is a useful way to symbolize what is meant by a discovery. When people refer to the Nobel in these extracts they are nearly always just using a shorthand to refer to the solid but highly significant scientific accomplishment that they hope to achieve.

The famous paper by Penzias and Wilson that is cited here is the Nobel Prize–winning first discovery of the cosmic background radiation. Penzias and Wilson could not eliminate what they thought was a source of noise in their horn aerial which they were using to scan the heavens for microwave sources. Their first thought was that the source must be dirt such as pigeon droppings, but they were eventually persuaded that it might be something more significant. It is a famous story and its modest initial presentation is a powerful resource for Hornbeam to draw upon.

Acacia replied to Hornbeam, referring to a meeting of the CBC Group—the group primarily responsible for the analysis of Big Dog. Acacia also referred to the Detection Committee, an overview group including the very senior people from across the collaboration who are tasked with critically reviewing the discovery paper and asking for changes in approach or such new data as might be thought necessary:

> Thanks for weighing in. I agree with you that it is unlikely that we won't get our Nobel if we say "Evidence for" instead of "Detection of."
>
> It's true that we haven't yet firmed up the post facto FAR estimate; 1/17y is only a bound on the FAR, it is certainly significantly lower than that. But here in CBC land we're all pretty convinced that the FAR is [will turn out to be] sufficiently low to be quite confident of detection.
>
> We had yet another discussion of this in the well-attended CBC call today, and I think it was fair to say that *everyone* on the call was in favor of "Detection of" and *no one* wanted to speak in favor of "Evidence for." . . . Some people were quite strong in their opinion that "Detection of" was preferable to "Evidence for."
>
> It may well be that the detection committee will ask us to soften the title and abstract. I'd prefer to leave the bolder statements in, until they do.

A new philosophical theme now entered the discussion. This was in the form of a sentence added to the conclusion of the draft by Quince:

> However, we cannot completely rule out other possible causes, and look forward to more sensitive detectors being brought online to observe rates and distributions of numerous gravitational-wave events.

The "philosophical" point is that it is impossible ever be sure exactly what was being seen; a coincidence could always be due to a chance concatenation of noises of unknown origin that just happened to mimic a gravitational wave. Let us call this the "epistemological sentence" since it captures something irrefutable about all knowledge. Quince, as we will see, had strong backing for this sentence from some members of the Detection Committee.

It should be noticed that Quince was reaching toward the future. The importance of Big Dog seemed to be reduced in the minds of many of the scientists because they knew that in a few years much more sensitive detectors ought to be detecting many events so there was no need to risk reputations on this one. Toward the end of the book I will suggest that were it not the case that the new detectors are already funded and, on 20 October 2011, the process of building them had begun, the pressure to pin Big Dog down as a real "discovery" might have been greater.

Blas Cabrera

Now another element of physics tradition was drawn in. This was the introduction of the precedent of a famous paper by Blas Cabrera. Quince circulated the following on 15 November:

> I would like to ask that each of you read or reread the attached paper, as I think it is another real world example of the complexity of a discovery claim. I am sure that many of you will remember it. In 1981–82, Blas Cabrera was operating a search for magnetic monopoles, and he reported on 151 days of observation, with a single event that stood out from his background. It is in many ways strikingly similar to the situation that faces us. A single event in a long run. In some ways, Cabrera's event stood out sharply from his background, five times larger than any other event he saw (compared with the 50% larger event we are seeing). It matched a predicted particle type (Dirac monople) with rather good accuracy (5%). This paper doesn't say it, but as I recall, I think he was not expected to see a monopole, given the expected fluxes of the day and his observation volume (area times time), but there was a finite chance that he could have. One difference was that without a coincidence method, Blas had very limited ways to measure a false alarm probability.
>
> Blas published this paper, which was extremely restrained in its claims. Subsequent searches with much greater sensitivity failed to find any additional candidates, and the theoretical basis for expecting to see monopoles eroded away. It is my impression that the caution which he showed in writing this paper, if anything, served to enhance his reputation in the phys-

> ics community, rather than harm it for publishing a result which was not subsequently confirmed.
>
> To this day, I think no one knows what caused the event that Blas saw. I don't know that anyone can say it wasn't a one in a million monopole, but the odds are against that. The lack of subsequent confirmation at any level has largely eroded any belief in the event. You might find it interesting to speculate what the community would think if subsequent experiments had confirmed the existence of a monop[o]le, but perhaps at a rate that would only give Blas a 1% chance of having seen one in his integration time.
>
> In any event, prepare for some interesting times over the next several weeks.

To fill out the story, in 1982 Cabrera published a paper in the same *Physical Review Letters* that was the target outlet for the gravitational-wave discovery paper. Cabrera's paper was entitled "First Results from a Superconductive Detector for Moving Magnetic Monopoles." Read deeper into the paper, however, and a momentous discovery was being claimed: the first detection of the theoretically conceivable but empirically elusive "magnetic monopole," a subatomic particle with only one magnetic pole instead of the usual "north" combined with "south." This was such a major event that Cabrera's paper was an object lesson in how to eliminate every other possibility, and a highlight of the paper is a list of deliberate attempts to reproduce the effect such as by hitting the apparatus with the handle of a screwdriver (the possibility of a continued series of such tests in the case of Big Dog was shut off by the decommissioning of LIGO on 20 October). Note especially the modesty of Cabrera's title: not "discovery of," but "first results from." The sting is that Cabrera's finding has never been repeated and is now largely disbelieved. But Cabrera's reputation remains intact because of the low-key way in which he presented the claim. This is why Cabrera is such a useful precedent for those who want to move cautiously.

Many of the themes were rehearsed again in a telecon on 6 December. Here are edited extracts from the minutes of the meeting; the "S#" identifier is meant only to recapture the ebb and flow of the argument and does not correspond with the tree names used in the rest of the book.

> S1: Title is bold. Does the CBC Group feel very strongly that this is a Direct Detection rather than strong evidence? . . .
>
> S2: It's about style, shall we be "measured caution" as a matter of style, like Cabrera was?
>
> S3: We're as confident as we'll ever be.

S4: We can never be sure, but we're as confident as we can imagine being.
S5: We'll be more confident when we have a dozen of these.
S1: There will be reluctance from some people to be bold in the title.
S2: We need to make the strongest statements we can about whether the detectors can fake this.
S6: This is the highlight of the Cabrera paper, and it is missing here.
S4: We can't rule it out, we've just never seen it yet. . . . There is no known instrumental mechanism to produce such a trigger. But we can't rule out unknown mechanisms.
S7: Can't make too strong a statement about one detector, but the confidence comes from coincidence. . . .
S8: . . . We've only been looking at these detectors for 1–3 years; don't know everything that can go wrong with this.
S4: [In conclusion:] However, we cannot completely rule out other possible causes, and look forward to more sensitive advanced detectors being brought online to observe rates and distributions of gravitational wave events.
S2: So can we say "Detection" or "Evidence"? . . .
S1: [Notes that the "bold" title "Direct Detection of Gravitational Waves from Compact Binary Coalescence" will bother some people; there may be a lot of difficulty getting it through the full LVC (LIGO-Virgo collaboration)]. How strongly are CBC folks against softening it to something like "Evidence for" or "Evidence for Direct Detection of" or something more cautious, in the spirit of the Cabrera paper?
[Minute taker's note:] *No one remaining on the call wanted to speak strongly against softening it. We'll bring it up again in tomorrow's CBC telecon.*

Note that, once more, the future is being invoked as a way of deciding what to do in the present.

Summing up the state of play as represented in a telecon in which took place on the following day, a scientist wrote to me:

> There is certainly discussion on the question of softening the claim. The question also came up in today's CBC telecom. The reasoning behind people who think this way seems to be independent of estimated FAR—rather, it is a general caution that a single event with no other corroborating observation (EM event, say) just shouldn't be the basis of a detection claim, period. This seems to be especially strongly held in Virgo, but as you know [Quince] is pushing this, too. [Ilex] resonates with this idea, too, I think. Meanwhile, I think that the majority of CBC would like to claim detection, but are per-

haps preparing to lose this argument in front of the full Collaborations. [Ilex is a generally cautious and senior member of the collaboration.]

Foxglove was the chief spokesmen for caution in the case of the Equinox Event, with Quince railing against the timidity of the collaboration, but they seemed to have exchanged positions, as exemplified by e-mails of 7 January.

> Ilex: and although I know there are many people who disagree, I think we should still say somewhere that it is impossible to definitely rule out a nonastrophysical cause.
>
> Quince: Absolutely! Nature can far cleverer than we are when it comes to putting artifacts into our data.

Foxglove responded:

> Please allow me to argue the other side of this question.
>
> Of course, it is true that it is impossible to definitely rule out a nonastrophysical cause. But that is equally true of any experiment or observation—you can never be sure that something you haven't thought of is spoofing you. And yet, you almost never see this written.
>
> I think it is tasteless to say this, not wrong.
>
> Presumably, we aren't going to keep saying this for every paper we write on into the future. At some point, everyone would believe we'd made a detection. But what will be the difference? Two similar events? But they could have the same nonastrophysical cause, fooling us twice. Or, are we going to require an EM counterpart, or require that detectors made by different designers give coincident detections, or that we see 100 events with a strength distribution that corresponds to a homogeneous spatial distribution . . .
>
> I personally think that this event is so good that we ought to claim it as a detection. Its FAR is remarkably small. The events really look like inspirals, visible to the eye in minimally processed data, and seen the same in two independent detectors. This is as close to golden as I ever hoped to see in Initial LIGO.

What is being argued here under the guise of physics, is, once more, philosophy and sociology. First the philosophy: it is always true that any apparent discovery could be an artifact. Now the sociology: the culture of physics is such that this philosophical truism is not mentioned in every paper. And now the sociology of gravitational-wave detection in particular: the truism *is* being mentioned in this paper. Gravitational-wave detection is a cautious science!

Like every internal state of an individual, the positions of Quince and Foxglove can be rendered consistent with the appropriate interpretation. Quince was angry that the Equinox Event was not taken seriously enough, but this does not commit him to saying that an event which is taken seriously has to be taken all the way to "discovery." Foxglove believed that the Equinox Event was too weak to be treated seriously as a candidate for a gravitational wave, but he believed that Big Dog was at the top end of the spectrum of discoveries. And Quince could have argued that even though the culture of physics is such that the possibility of being fooled by an artifact is not mentioned in every paper, it should be mentioned in first discovery papers. I invoke the "anti-forensic principle"; I am not a detective nor a psychoanalyst and the internal states of the actors in this story are not my business. My business is the envelope of possible internal states in the physics culture of a society like ours.[1]

More Papers Are Circulated

Around 11 January, a senior scientist from the Virgo group, Juniper, who was trying to build a foundation for the discovery, poured in another concrete mixer–load of physics tradition. Juniper prepared a "reading list" of papers, with their citation statistics and interpreted them as revealing the need for caution. The list of papers is given in figure 22. The Cabrera paper was included in the list and its details are repeated for completeness:

Quince circulated the list with the following comment:

> [Juniper] correctly points out that each situation is unique, and thus we can't follow any model exactly, but I think that seeing how other major discoveries have been announced may be extremely useful. In any event, these papers represent some of the most important physics discoveries of the past four decades, so reading them will never be a waste of time.

On 17 January, a senior scientist from the Virgo collaboration, [Kapok], expressed the caution that seemed to be characteristic of that group. Kapok wrote:

1. It was later explained to me by Quince that the role he was trying to fill was to "add capacitance to," or in my words, "act as a sheet anchor for" the collaboration, making sure it did not lurch too far in any direction,

1. **Discovery of the tau lepton**
   M. L. Perl et al., "Evidence for Anomalous Lepton Production in e+ - e- Annihilation," *PhysicalReviewLetters*35 (1975): 1489–92. *Cited 1,042 times.*
2. **Discovery of the W boson**
   G. Arnison et al. and UA1 Collaboration, "Experimental Observation of Isolated Large Transverse Energy Electrons with Associated Missing Energy at s**(1/2) = 540-GeV," *PhysicalLetters B* 122 (1983): 103–16. *Cited 1,420 times.*
3. **Monopole candidate**
   B. Cabrera, "First Results from a Superconductive Detector for Moving Magnetic Monopoles," *PhysicalReviewLetters*48 (1982): 1378–80. *Cited 205 times.*
4. **Discovery of atmospheric neutrino oscillations**
   Y. Ashie et al. and Super-Kamiokande Collaboration, "Evidence for an Oscillatory Signature in Atmospheric Neutrino Oscillation," *Physical Review Letters* 93 (2004): 101801. *Cited 538 times.*
5. **Discovery of parity violation in atomic transitions**
   M.A. Bouchiat, J. Guena, L. Hunter, and L. Pottier, "Observation of a Parity Violation in Cesium," *Physical Letters B* 117 (1982): 358. *Cited 88 times.*
6. **Discovery of the top quark**
   F. Abe et al. and CDF Collaboration, "Evidence for Top Quark Production in Anti-p p Collisions at s**(1/2) = 1.8-TeV," *Physical Review D* 50 (1994): 2966–3026. *Cited 788 times.*

Figure 22. Papers circulated by Juniper

It is a pleasure to see that the detection paper has reached a mature state. However there is still a weak point for me. I am indeed not comfortable at all with the title (and with the first sentence of the abstract as well). When looking at the past discovery papers provided by [Juniper] and [Quince], it appears that all these papers have had quite humble titles ("evidence" rather than "detection," or even more modest!).

I would then militate against "direct detection" in the title and in the abstract. Even if we are sure of us (and of course we can never be 100% confident), I sincerely think it is counterproductive to show too much certitude to the outside world (our readers), certitude which could be even perceived as some sort of arrogance. By the way I am not sure we have reached the famous "5-sigma" level I advocated in the recent past for the detection probability. Let's stay humble.

Notice the reference to the "famous '5-sigma' level." Kapok is saying, quite correctly, that the 5-sigma level has not been reached but is accepting it as the discovery standard.

A lively exchange followed, with Dogwood saying that the position being advocated by Juniper was misplaced since the papers being discussed were all tentative and the results were not confirmed until later papers were published that had better statistics and more confident claims.

Hornbeam responded:

> Regarding the title, I disagree with your summary of surveying the past "detection" papers. Those papers, taken as a group, have had more conservative titles than you seem to be saying.

But Dogwood pointed out that the results being claimed were only partial even if the statistics were strong:

> The 1975 tau paper doesn't say what they think the new particle is, because there wasn't any indication of its spin and very little information on its mass. Later papers with better experiments and analysis pinned this down to be a 3rd charged lepton.
>
> In the first W paper they could not measure the spin or the width or the branching fractions with only 5 events, so although the confidence in a real signal was very high, there was no clear proof it was W rather than some other kind of particle (which at the time was certainly possible). They claimed what they had at the time.
>
> The neutrino paper only has "3.4 sigma" confidence that the effect they saw was oscillation rather than decay or decoherence. I think this is why they claim "Evidence': the claim was specifically oscillation, rather than just neutrino disappearance, which was already established.
>
> The top quark PRL papers . . . have significances of 4.6sigma (1 in 2*10^6) and 4.8sigma (1 in 10^6) resp. and do claim a specific physical effect. Lyons [an authority on statistics] says that the top quark history is the model for "recent" particle physics practice.
>
> The second point was that, when you have strong enough evidence (e.g., 1 in 10^6) "Evidence" is no longer a suitable title. Nothing is 100% certain in physics, but after a certain point we stop drawing attention to the uncertainty. "Evidence" will imply not just skepticism, but significant doubt. What is the source of that doubt? OK, we only have one event, but it's a "gold-plated" one with coincidence and coherence between two independent detector outputs.

Simultaneously, however, Dogwood opined that having read the papers circulated by Juniper, a suitable title would be "*Observation of* a Gravitational-Wave Signal from Compact Binary Coalescence" (my emphasis). An immediate response came back from another scientist: "I like 'Observation of' (not as coy as 'Evidence for' or as crowing as 'Direct Detection of')."

On 17 January, [Linden], a youngish but active scientist, intimated that the right tradition to draw on might not be high-energy physics but astronomy and circulated some papers that suggested the astronomy community had less rigid standards.

> The "detection" papers we've been considering are all from the high-energy/particle-physics physics community. Since this paper (if published) will officially make us astronomers, it might be work considering practice in that area. . . . The exoplanet papers have absolutely no qualms about claiming detections, and don't even mention confidence levels, even though there were previous claims in the literature that turned out to be false (sound familiar?).
>
> I don't urge being as bullish as the exoplanet people, but I think that "Observation of" or "Observational Evidence for" would be perfectly reasonable in our case.

Unfortunately for Linden, Hornbeam argued that the papers actually had much higher significance than appeared on a quick reading so no new precedent was being set and that "Observational Evidence for" might be the right phrase. It was becoming clearer that it was not always easy to interpret the statistical significance of a published paper.

What Is the State of Play with Sigma Levels?

In fact, much of this argument was thoroughly out-of-date, since the half-dozen papers circulated by Juniper had mostly been published two or three decades earlier. The modern state of play for high-energy physics publications in *Physical Review Letters* has been researched by Allan Franklin.[2] Franklin is himself a physicist who now specializes in philosophy and history of physics and has been actively studying the history of discovery claims in physics; I am able to draw on his unpublished (at the time of writing) book manuscript "Experiment then and Now" (forthcoming,

2. Franklin sent me various personal communications throughout the exercise; I thank him for carrying out this survey at my suggestion.

University of Pittsburgh Press). Franklin documents the steady increase in the sigma level required for a discovery claim in physics as the standard became more formalized from the 1960s to the present. The present situation is, he says, well characterized by the phrase used by a physicist of his acquaintance, "5-sigma police."

Though Franklin initially concentrated on the discovery and observation papers, I asked if he would check recent "evidence for" papers. Eventually he completed an analysis of publications in *Physical Review Letters* during the period 2003–2010. He found that of 105 "observation of" papers, 90 explicitly stated that the observed effect had a statistical significance greater or equal to five standard deviations while eleven, though they made no explicit mention of standard deviations, reported effects that clearly satisfied the 5-sigma criterion. He says that only 2 "observation of" papers were published during this period that reported a statistical significance less than 5-sigma but they were published in 2003 and 2004. Finally, 2 papers made no reference to standard deviations and appeared not to satisfy the 5-sigma criterion but the last of these was published in 2005. Looking at the opposite side of the coin, 44 "evidence for" papers on high-energy physics were published in the same period, of which 40 reported effects with significance less than 5-sigma. Effects with a significance greater than 5-sigma were, however, reported in 3 papers, but the last was in 2006 while one 2005 paper makes no mention of standard deviations. These results seem to confirm the notion that in 2010 the convention has been established that "discovery" papers must report 5-sigma while, since 2006, all 5-sigma papers are reported as "discovery" or "observation" and not as "evidence for."

Thus, it seems that at the same time as 5-sigma has become the standard for a discovery claim, physicists have more and more sought the "discovery" standard and become less and less inclined to label a finding that reaches that standard anything other than a discovery. Metaphorically speaking, a new examining standard has been set and, consequent upon this, when physicists achieve an "A+" they are less inclined than they might have been in the past to say that they had achieved merely "a good A." A more recent set of papers than Juniper's would tell a different story and might have affected the title and the way the research was carried forward.

The physicists were unaware of this most recent trend and the state of play as seen by Acacia was summarized on 17 January:

> To forestall complaints from collaborators that "Direct Detection of" is too bold a claim (since it's just one event and one can never by 100% sure), and (from other collaborators) that "Evidence for" is too weak of a claim given

> our confidence in the signal, Tom has suggestion "Observation of" as a (perceived) intermediate-sounding word. I think that's a good compromise and propose to make that change in the title, abstract and in the main text. I'm sure the debate will continue. I'm glad we're not submitting this to *Science* or *Nature*, they would likely write the title for us.

Quince said on 17 January: "Personally, I can live with 'Observation of' . . . although I prefer . . . 'Observational Evidence of.' "

The "Observation" Solution

The title of the circulated paper changed from, on 18 January, "Direct Detection of Gravitational Waves from Compact Binary Coalescence" to, on 19 January, "Observation of Gravitational Waves from Compact Binary Coalescence." The trouble with this title that I see is that it does not distinguish what had been done from the Hulse-Taylor observations because the word "direct" has gone—a problem which one of the scientists would refer to in due course. "Direct" seems to have disappeared because everyone is preoccupied with the fact that "observation" seems to finesse the argument about "evidence for" versus "detection," at least for this subgroup.

At the same time (during an 18 January meeting of the Detection Committee), another very senior physicist from the Virgo group, [Mulberry], opined:

> We have to be very, very cautious. The young peoples' impulse toward big success in this enterprise is right but we, with our experience, have to be very cautious. We have seen too many things in the past that were really bad [as a result of] being too much publicized in a certain way.

It is remarkable how these arguments were repeated over and over again in different locations without seeming to come any nearer to a conclusion. For example, the term "observation" did not finesse the matter for Juniper. Juniper pointed out (quite correctly) during a Detection Committee meeting of 25 January that "observation" means the same as "direct detection" or "discovery" as far as high-energy physicists are concerned. Juniper said that "observation" was wrong:

> Because it implies that we can exclude all nonastrophysical causes for the event which I think is regrettable, and I think it would not hurt to maybe only quote "evidence for the detection of gravitational waves."

Juniper said that "big discovery papers mostly have modest titles." In response, Foxglove, though a consistent supporter of a strong title, intimated a readiness to concede to the Juniper argument:

> I surely do favor a strong claim in the paper. I think this is the golden event we were all hoping for. And my feeling is that no discovery can ever be made on the basis of having ruled out every possible, even implausible, false reason for seeing it. I'm willing to learn the lesson that I think we've learned from seeing all these famous discovery papers—that sometimes papers that read as discovery papers have very modest titles. So I'm going to step back from insisting that the title ought to be really strong but I do believe that the message of the paper ought to be that this is a real thing.

The sentence pointing out that it could not be ruled out that some other unknown physical effect was responsible—the "epistemological sentence" that was initially inserted into the conclusion by Quince—was still being discussed. Ilex was championing it but was willing to accept that it should go somewhere less prominent than the conclusion—but others were unhappy.

On 25 January, a very senior scientist from the LIGO group, [Sumac], came out very strongly against it:

> I don't like that at all, that's utter nonsense. That's like a lot of those Italian papers from years ago and I think it's disreputable science. I think the case is very straight and we have established it as best we can with what we know. We wish we had other things . . . I think a philosophical paragraph at the end would make me and would make a lot of other people in the LSC [LIGO Scientific Collaboration] feel a lot better.

What is being referred to, of course, is the "Italian" papers published out of Frascati and discussed above. The Frascati claims were always marginal and were presented in a very *evidentially collective* way. The Frascati group said only, "we've found something suggestive here which is just about worth publishing—here it is, you make what you can of it." If we go back to one of the most notorious of these papers, we find in the concluding paragraphs a sentence that is remarkably similar to the "epistemological sentence":

> We think it is unlikely that the observed coincidence excess is due to noise fluctuations, but we prefer to take a conservative position and wait for a stronger confirmation of our result before reaching any definite conclusion

> and claiming that gravitational waves have been observed. Furthermore, although we have excluded that the events are due to cosmic ray showers . . . we cannot completely rule out that they be due to some other exotic, still unknown, phenomenon. (Astone et al. 2002, 5465)

So we can see just what a bind the current gravitational-wave community finds itself in given the vituperative condemnation of the "Italians" that many members of the LIGO collaboration handed out for writing equivocal sentiments like this.

I cannot refrain from pointing out that it was a lot easier to condemn the way the Italians presented their results when the interferometers had no results of their own to report—in an era when everyone in interferometry could still imagine that *their* signal would come in crystal clear form. Now that results, as they are found in the real world of experiment, are making their presence felt, evidential collectivism does not look like such a bad thing after all!

Sumac went on to say that publishing the paper with a strong claim and not too much of a qualification would "give a certain honesty to the thing": "we have done our best and we claim this is a real event," even though we would like to have had multiple events, a correlated source and so on.

Ilex insisted, against Sumac, that they could not rule out the possibility of the event being an artifact since they had only a single event. Ilex wanted a modest title and modest remarks in the paper, agreeing that even in the case of a less golden event they would still have to publish: "We would have to show the numbers and let the readers decide what they want to think about that" (which is evidential collectivism).

But [Oleander], another very senior scientist more closely associated with the LIGO group than the Virgo group, agreed with Sumac that putting out a statement that you can't rule out the unknown is obvious:

> It's so obvious that it undermines . . . it suggests that we don't have as much confidence as we are saying . . . I think we have to say that if it is something unexpected it is beyond us and we have made strenuous efforts and we cannot find it. Otherwise we are undermining what we are saying about the detection.

Sumac returned to stress the point again:

> We can't suppress this because there are too many years before we see anything else to confirm it so this is actually impeding science if we don't

> publish something. . . . I would have much preferred to wait until we had more events . . . to me that's the most important lack—that we only have one event. But that's what we have.

At the next meeting of the Detection Committee (1 February), even Juniper, who had taken a very conservative view throughout, argued against Quince and Ilex's epistemological sentence, describing it as a truism.

There is a lot going on in these discussions. They can be analyzed in terms of the "past/present/future" scheme and in terms of "evidential culture-space." Let us start with the present: the evidence and its significance.

The Present

At the time of these debates the sigma level equivalent of the signal is between 4.3 and 4.8. This alone would indicate that the appropriate claim in the title would be "evidence for"—at least, if it was high-energy physics that was taken to be the precedent. But that is a question about traditions so let us leave it until we get back to the past. More noteworthy is that there has been no argument about whether the statistics alone allow a "discovery" claim. A significance level is not something fixed in stone; if what the collaboration wanted to do was get to 5-sigma they would, I believe, at least have been discussing if and how it could done. This is the kind of thing that high-energy physicists would have been discussing—"how much longer will it take us to get to 5-sigma?" They could have been asking whether they could squeeze in more time slides with a shorter unit delay and what difference it would make; they could be questioning what is counted as the length of the run in the calculation of FAP; and, as we will see, they could have been talking about extrapolations of the existing time slides. But people are simply not talking about these things—their interests are elsewhere.

Foxglove suggested to me that, once more, their interests were in the future. Given that a day of observation with AdLIGO, scanning 1,000 times the volume of the heavens (if it works), will be worth around three years of current LIGO's time, why agonize over raising the significance of this event to 5-sigma? Also, he, at one time a strong supporter of an unequivocal discovery claim, suggested that I may have missed a "literary" nicety: the scientists wanted to say "evidence for" and leave it others to tell them they had made a discovery.

Though no one discusses this explicitly, it is hard not to notice that the Virgo group scientists have very few, if any, among them who do not take a very conservative view. Juniper, Kapok, and Mulberry are very well established Virgo scientists, and they are all conservative in respect of the title. In contrast, the powerful LIGO scientists are split fairly evenly: Quince and Ilex are very conservative, whereas Foxglove and Sumac strongly prefer a positive strategy, with Acacia favoring a positive strategy but being judicious about marshaling opinion fairly, as his committee chair position requires.

For Virgo the problem might be that if this really is the first direct discovery of gravitational waves, it was discovered by LIGO. The paper as written will disguise this fact—it will be described as the finding of the LIGO-Virgo collaboration and quite a bit of energy has been expended in working out how to say this without implying that Virgo actually observed a signal. Thus the first sentence of the abstract of the 5 February version of the paper read: "A coincident signal was observed at 16 September 2010, 06:42:23 UTC in data from a joint science run of the LIGO and Virgo gravitational-wave detectors." But, however matters are expressed in the paper, the physics community will probably find out that Virgo did not make much of a contribution because, at the time, its sensitivity was too low.

There is an interesting subtheme about how much contribution Virgo can make. There is agreement that it can make a contribution toward eliminating certain areas of the sky as the source because it can be said that, if those areas were the source, Virgo would have seen more; but having seen little, even thought it narrows the direction of the source, as with ever upper limit result, is not a great contribution. There is also, beginning in the middle of January, a growing movement to make more of the vestigial signal that Virgo can be said to show. This was triggered by Virgo personnel who, in the main, are quite positive about it whereas the LIGO group are again split with some for and some against. In the Detection Committee meeting of 11 January, Juniper asked:

> What is the probability [of] having seen something that we have seen in the LIGO detectors, what is the probability, even though the SNR is low, to still find a feature [in Virgo] that is coincident in terms of the templates and masses that we have found?

Though the question was first asked by Virgo personnel, some LIGO people join in with enthusiasm. Though such a weak signal would never have been noticed in Virgo as a stand-alone detector, something can be reconstructed post hoc using the template that best corresponds to the signal in the Hanford LIGO detector. There is enormous danger of statistical "massage," or post hoc "tuning to a signal" here, so the scientists agree that the Virgo data cannot be fed into the statistics; what is being canvassed is that this post hoc reconstruction of a signal adds to the confidence in the LIGO findings in a qualitative way. In a Data Analysis Council meeting of 11 February, [Piñon], another important Virgo scientist, went so far as to say that even if the two LIGO detectors could be construed as having a 5-sigma result it may still not be good enough without Virgo:

> I have seen in my life some 5-sigma events that turn out to be fake. . . . I still think that if we mention Virgo, and the indication we see in Virgo, that would reinforce for the external public that would be an additional indication that it's a real signal.

This debate was to become quite heated, as this e-mail exchange of 15 February indicates.

> [Ginkgo:] We can also say that the Virgo data is not inconsistent with our evidence for a detection. My concern, however, is that we don't let our desire to find a signal in the Virgo data influence our usual rigorous [quantitative] approach to claiming evidence for such a signal.
>
> [Piñon:] I'm not one to say that we should throw "our usual rigorous approach" out the window, but I think we shouldn't be too righteous about it either. I've used that parallel before, but talking about "using ambiguous statements without numbers," I think that showing a background plot without little dogs in figure 2 is not exactly a model of virtue. We don't justify why we can do it, we don't derive any number out of it, we just hope the reader will get away with the notion that our event is even more significant than we claim it is and that we'll get away with murder (of the little dogs).

Piñon is saying that there is no justification for showing a value for the FAP without little dogs and is using this to justify the nonquantified discussion of the Virgo signal.

Piñon did make clear, however, on 16 February that it was not being suggested that the small signal in Virgo (V), if there was one, should be treated as ranking alongside the signal in Hanford (H) and Livingston (L):

> What I'm _not_ advocating is changing the executive summary at the beginning of the paper and turning the HL event into an HLV based on the hints we have for a small signal in Virgo.

Many in the LIGO group think that Virgo wants a place at Big Dog's table or, failing that, would prefer to wait for an event which will enable their own machine to make more of a contribution. This was certainly a widespread feeling among LIGO scientists in the case of the earlier Equinox Event, where Virgo scientists were equally massed on the side of conservatism. In that case, after talking to a Virgo spokesperson, I concluded that the case could not be clearly made and, once more, I invoke the antiforensic principle: I am merely reporting what interpretations are available within the social milieu in which this debate is taking place.

The Past

There are two separate elements of the past that impinge on the way the debate has unfolded. The first is the tradition of high-energy physics. But, as explained, the high-energy physics model seems to have been misunderstood. Everyone recognized that the discovery/evidence papers circulated by Juniper were not necessarily a representative set, but they did not realize that they did not represent physics as it is currently practiced. To repeat, that it was not representative has been shown by the analysis of publishable sigma levels carried out by Allan Franklin.

The other element of the past has to do, as with the Equinox Event, with the fear of making another of the false claims that characterize the history of the field. It is the Weber results and the Frascati result to which Mulberry and Sumac refer in their remarks above, but they read the influence of the past in two different ways and reach two different conclusions. Mulberry is saying that the group must be cautious because of all the false claims in the past. Sumac says that to act too cautiously is to endorse the reviled evidential collectivism of the Frascati group.

Incidentally, Mulberry, as we have seen, suggests that it is the young people who are enthusiastic for making a strong claim and that those with older heads and greater experience are better placed to know the dangers and understand the need for caution. On the other hand, a good option for young people is to make a weak claim; they have nothing to gain by being associated with a strong but incorrect claim and they can afford to wait for the era of gravitational-wave astronomy to come in another five years or so when they will have a more prominent place in the vanguard. It is, if anything, the most senior people in the LIGO group for whom a strong claim is important: they may not still be active by the time the AdLIGO is producing astronomy proper. Mulberry, though senior, is from Virgo.

Sumac is one of the senior LIGO scientists and, as we have seen, is not cautious. Sumac wants the "epistemological sentence" removed and a firm claim to be made. Does a moderate title represent proper modesty, as Foxglove is ready to accept, or is it equivocation? Unwittingly, when Ilex remarks in response to Sumac, "We would have to show the numbers and let the readers decide what they want to think about that" it is an endorsement of the evidential collectivism that, in the 2000s, Sumac adamantly and vocally rejected.

The Logical Atomism Problem

We might say that the "icon" of the argument between Ilex and Quince on the one hand, and, on the other, Sumac, Foxglove, and a few others is "the epistemological sentence." It is worth revisiting. Some of the dilemmas were well expressed in an e-mail of 22 January in which Dogwood referred back to a comment by Ginkgo supporting the initial remark of Quince:

> I was [initially?] persuaded by points [Quince] made. In my own words, this was that "Evidence for" is appropriate since we have only a single event, and even if it is gold-plated, it might not look so once we see what real events look like in a[d]LIGO. (Several years from now we don't what to be saying, why did we ever call that big dog event a detection?)
>
> I don't entirely understand this argument (at least, phrased like this). This event isn't any less "real" than the first few GW-like events [that will be] seen in aLIGO. . . . There will always be a first "real" detection candidate . . . and lots of glitches too, and we won't always know the difference. . . .
>
> So it probably won't get any easier at the start of aLIGO to claim detection. We will still be subject to the suspected uncertainties [Quince] referred to, or even more so because detector transients [noises] are worse understood at the start of a run.
>
> If there is, now, a suspicion of correlated, chirp-shaped environmental noise at two detectors separated by thousands of miles, I don't see how it will ever go away: it remains into aLIGO, or until we get a 3-detector coincidence over two continents, or a coincident EM [electromagnetic] signal. Even if we see two separate inspirals, there is still the same suspicion of two instances of correlated chirpy noise. If it happens once it can happen twice.
>
> If we don't want to claim an undiluted detection or observation now, and a precedent is set, there may be no future undiluted LIGO detections, either. It will just be evidence for a GW observatory. . . .

> OK, [Quince] knows these monitors a lot better than I do, but I would be interested to know what physics anyone could envisage to ever cause coherent, correlated noise, within a time lag of 10ms, without showing anything on the [environmental] monitors.
>
> The bottom line for me is that "evidence" will be interpreted as "doubt," but what is the doubt? Philosophically, everything in science is "evidence" and there is never 100% proof, but beyond a certain point the uncertainty stops being relevant.

Let us try to put this argument into a more philosophical frame. It is always possible to be certain of one's claims so long as one claims as little as possible. This can be made clearer by considering the history of the philosophical movement known as "logical positivism." The logical positivists defined meaning as "the means of verification." Nothing could be meaningful unless it could be verified. There were two ways of verifying claims. The first was that the claim was a definition-in-terms such as "a rainy day is a wet day." These "analytic" statements are useful for deductive purposes, but they tell you nothing about the world; they tell you only about how words or symbols are used. Thus one cannot imagine that a rainy day could be anything other than wet—if it was not wet it would not be raining.

The second way of verifying a claim was by empirical observation. The claim "a rainy day is a cold day" is not true by definition; it is not analytic. To claim, "a rainy day is a cold day" is to try to do something "synthetic," to try to say something informative about the way world is. Whether that thing is true or not is not a matter of definition but of empirical observation. Logical positivism gave rise to a wonderfully attractive, stripped-down version of the world. Try the claim "God exists." It is not a definition in terms but there is no empirical evidence that can be brought to bear on the matter so it is, according to logical positivism, simply meaningless. The same goes for moral imperatives and artistic judgment and, as portrayed in the marvelous television series about 1950s university life, *The Golden Prizes*, a girlfriend's argument that it would be wrong to have sex. The world of meaning is, then, exhausted by logic and science.

But things were to go wrong. To bring evidence to bear upon the matter of whether a rainy day is a cold day, "cold" has to be defined. Different people feel differently about cold so it turns out that the claim is not as open to verification by observation as had been thought. The *verifiable* claim is something like "a rainy day is 20 degrees or less." But then there is the matter of the meaning of "20 degrees," and this leads on to the question of the instruments and procedures used to measure temperature and how one knows that they are reliable and whether the measurement applies just to the moment the measurement was made or extends to all day on all rainy days. And so it goes—on and on and on. In the end it turns out that real unshakable certainty from empirical observation is to be had only in respect of minimally informative claims such as "cold here now." Certainty, it turned out was achievable only for atomistic and momentary sense experiences—a position known as "logical atomism." The lesson is that to achieve complete certainty for a synthetic proposition it must claim almost nothing.

Ilex and Quince want certainty, so they pull in the direction of logical atomism—they want to say less. Sumac thinks that it is the responsibility of scientists to say what they think and not to hide behind a shield of equivocation and so pulls toward saying more with a little less assurance. Of course, neither party would go near the extremes but the argument about the epistemological sentence remains a matter of truism with safety versus new knowledge with risk.[3]

The Future

The force of the future has a palpable aspect: that LIGO has been torn down to begin the process of building AdLIGO (Advanced LIGO) makes a real difference to what can be done in the way of affirming Big Dog. Less palpable but still powerful is the influence of the putative thousandfold increase in likelihood of seeing a GW source that AdLIGO should provide. To resort to the precarious method of historical research known as "the counterfactual," it seemed likely that things would have been different if AdLIGO were not on the horizon. In this case the counterfactual method is a little less precarious than usual, because so many of the scientists invoke the future in support of caution. Indeed, the force of the future is right there in the last phrase of Quince's epistemological sentence: "However, we cannot completely rule out other possible causes, and look forward to more sensitive detectors being brought online to observe rates and distributions of numerous gravitational wave events."

While exploring the territory of the counterfactual, how big a part in the argument was played by the possibility of Big Dog being a blind injection? I thought the effect was small; as the months wore on and the moment for the opening of the envelope approached it seemed to me that the argu-

3. In Collins 2009 ("We Cannot Live by Scepticism Alone"), I argue something similar in respect of social studies of science.

ments on both sides increased in ferocity and that suspension of disbelief became more rather than less complete. I didn't see anyone write anything like "it's only an injection so why the fuss?" Even at the beginning of the final conference in Arcadia in March (chapter 10) when a speaker talked about what they would do if the envelope was empty, there were no knowing grins or sniggers—everyone simply listened with interest about what would happen under those circumstances. But as will be seen in chapter 15 and the third part of the book, not everyone agreed.[4]

4. In the course of the final debate I began to discover that there was more disappointment about it being a blind injection than I had imagined. For example, one senior scientist told me he had been hovering around 50/50 when the envelope was opened.

13 Closing Arguments

As successive paper drafts evolved, the team maintained a long document that recorded their reasoning about what was in the paper. Here is an extract, from the version current on 10 February, which bears on title and abstract. It is essentially a summary of the science and some of the other decisions that have been discussed so far:

> The current title is "Observation of Gravitational Waves from Compact Binary Coalescence." This is a compromise between the bold "First Detection of" and the wimpy "Evidence for." Many members of the CBC Group are very confident that this event represents a real astrophysical signal (or a blind injection), not an instrumental artifact. As detailed below, the event is not only too loud to be an accidental coincidence of instrumental noise triggers . . . and not improbably loud, it also "looks, sounds and smells" like a CBC signal . . . At the time of the event, all three detectors were in the middle of long lock stretches, well behaved with relatively low glitch rate [the relevant analytic teams] have never seen an instrumental glitch that looks like this. . . .
>
> Other facts speak for caution in the title, softening it from "Detection of":
>
> (a) This is one event, with no EM counterpart. The optical followup is still in progress . . . but let's assume for now that they find nothing. . . .

(b) Estimating false alarm probabilities at the level of 10−6 is notoriously difficult. We use time slides to do this . . . and these make the implicit assumption that the noise characteristics are at least approximately time-invariant over the observation time used.

(c) We can never be sure that there does not exist some source of terrestrial disturbance causing correlated chirplike glitches in the data (e.g., magnetic storms or Schumann resonances in the atmosphere). It is "impossible to definitely rule out a nonastrophysical cause" ([Ilex]), and "Nature can far cleverer than we are when it comes to putting artifacts into our data." ([Quince]). However, others argued that this is so improbable as to be unreasonable, or that this is so obvious and implicit in all scientific discoveries to be unworthy of mention. For now, we have dropped the following sentence from the conclusions: "However, we cannot completely rule out other possible causes, and look forward to more sensitive detectors being brought online to observe rates and distributions of numerous gravitational wave events."

This is a "direct" detection, as opposed to the "indirect" detection by Hulse and Taylor. We dropped "Direct" from the title and abstract to make it shorter, and we hope it is nonetheless clear. The word "direct" is used in the introduction.

The title also refers to this event as a "Compact Binary Coalescence." This is jargon-y, but we have used it in four of our [previous] publications, and it has the advantage of being as precise as we want to be, here. We see only the inspiral phase of the coalescence; the merger and ringdown are in the high frequency noise; but this late stage of inspiral can be considered as the "coalescence." "Compact" refers to the fact that these are neutron stars or black holes. The evidence from parameter estimation . . . very strongly suggests that these are binary black holes: the astronomical community conventionally refers to compact objects with mass > 3M_ as black holes, and with mass between 1 and 3M_ as neutron stars . . . but they're both compact objects. But at present, the parameter estimation group cannot rule out the possibility that one of the components can have a mass as low as 2.6M_, so we can't (yet) call them Binary Black Holes.

By 11 February, opinion in the Detection Advisory Committee was hardening against the epistemological sentence. Acacia, running through the current draft of the paper step by step, was dismissive of this kind of sentiment:

> We left out other statement like "Oh, of course, we can't be a hundred percent sure after all, this only one event and there may be weird correlations between detectors that we don't understand and you can never really be sure"—we left all those things out. We also left out a statement saying, you know, "since it's only one event we look forward to more events."

But arguments are hard to close. On 14 February, Hornbeam wrote in an e-mail:

> I still wish the title and abstract of the paper could say "binary black hole coalescence" instead of "compact binary coalescence." I don't put much stock in the argument that our past papers say "compact binary coalescence"; this paper will probably be read by 50 times as many people as those past ones. I think the significance of this paper is not ONLY the detection of the gravitational waves, but also the fabulous astrophysical event that produced them. Note that the text currently in the discussion section, about merger rates, presumes that it was a binary black hole merger.

To which Dogwood replied:

> Yes, I'm sure a lot of people . . . wish we had sufficiently strong confidence in the measured masses to have "binary black hole coalescence" in the title. . . . The reason why we don't write that is that we don't have sufficiently strong confidence now. Wishing won't make it so.

I am puzzled. There is very little chance that the lower mass is less that 3 solar masses—only one model gives this answer and then it is 2.6. But, worse, I cannot understand why no one has suggested "binary coalescence involving at least one black hole."

On the same date another scientist revived the "direct detection" argument:

> About the title: when reading the content of the paper, I had not the feeling that "Observation of Gravitational Waves . . ." was an appropriate title. The result on PSR1913+16 precession [the observation of the binary star for which the Nobel Prize had been awarded to Hulse and Taylor] had no other interpretation than GW, so it was already an "observation of gravitational waves," like a top quark is considered observed even if an associated track in the detectors was never directly observed. So, I would replace the current

> title by "Direct evidence of Gravitational Waves from Compact Binary Coalescence."

Also on the same date, Quince, in the context of the argument about whether Virgo's vestigial signal could usefully help to establish the reality of Big Dog, wrote the following:

> Let's assume that we all accept that we have a significant coincidence between H and L. There are two possibilities I can think of for this—it might be a gravitational wave, and it might be a correlated glitch due to some undetected defect in the LIGO detectors. Because I am an experimenter and I think we did a very good job of designing the instruments, I think that there is less than a 1% chance of there being such a defect. On the other hand, we have astrophysical rate estimates for BH-BH systems, and that they say (just applying simple Poisson statistics) there is less than a 10% chance (and for the most likely astrophysical rate, less than 1%) of us seeing one BH-BH inspiral in the length of time we observed. In other words, I have an approximately flat prior with regard to these two possibilities. [I believe them to be equally likely.]

I found it hard to make sense of this logic and wrote to Quince:

> It has long been agreed that Initial LIGO [iLIGO, which for the purpose or this e-mail includes eLIGO but not AdLIGO] was not going to see anything unless it got lucky. Blind injection challenge aside, Big Dog is a case of iLIGO getting lucky. Now—the team has spent decades building an apparatus that is meant to be capable of detecting such a lucky event—i.e., it is carefully designed to avoid unknown correlated glitches. But if you are going to say that your prior in respect of an unknown correlated glitch is the same as your prior regarding a lucky event then you could never detect a lucky event. Surely the unknown correlated glitch prior must be much lower than the lucky event prior if the iLIGO project is to make any sense at all as a detector. Am I misunderstanding something?

And I wasn't only one. Later on the same day that I wrote, Dogwood replied to Quince:

> if the probability of such a thing [an unknown correlated glitch] across LIGO is really as high as 1% over a run, we shouldn't be writing an observation paper at all—we should be doing a much more detailed study of

possible environmental effects. I hope this number was just for the purpose of argument.

After a short exchange of e-mails with Quince, I thought I had come to understand that, actually, Quince had no confidence in any quantitative expression of either the prior for the rate of unknown correlated glitches or the prior for the likelihood of an event such as Big Dog that emerged from the astrophysicist's rate calculation so this "setting them equal" was really, as Dogwood put it, for "the purpose of argument." The very least this exchange shows, however, is how easy it is to slip into a way of talking that can make any signal go away.[1]

Indeed, I discovered that Quince had expressed the negative quantitative argument still more strongly in an earlier e-mail (1 February):

> There are things about the instrument we don't understand after all these years, and there are undoubtedly things about nature that we don't understand. It is very hard, nay impossible, to put a number on how likely it is that something we haven't thought of might have caused this, but it is my personal opinion that this probability now exceeds the FAP we have from chance coincidences in the data streams.
>
> So the "lack of confidence" due to such noise could possibly be larger than the estimated false rates from the CBC analysis. At least, you would not be entirely confident in ruling out correlated noise at the level of (say) 1 in 10^4 over the run. I'm just picking numbers out of the air here, please correct me. [This paragraph is quoting someone else.]
>
> Let me say "correlated effects" as opposed to "noise," but with that nuanced change, the answer is yes. It's hard to pick a number, but 10^{-4} is probably about where I place my personal confidence.
>
> I am reasonably certain that all of the experimenters have their own estimate for this number, and it is likely that mine is more pessimistic than many others. But then I have been around long enough to know how much easier it is to make a mistake than to do things right, from personal experience.

1. It may be that both this exchange and the earlier one with SB about the point at which you become sufficiently sure that the Big Dog was signal, not noise, that it is removed from the time-slide analysis are based on my failure to be sufficiently immersed in the discourse of this physics community. It may be that physicists sometimes express their loose remarks in terms of exact-looking quantities, knowing that other physicists will understand that nothing genuinely exact is intended. Dogwood's response to Quince (above), then, would have been rhetorical, whereas mine was genuine puzzlement borne out of overreading the exactness intended by the quotation of 1 percent for each prior.

On the face of it, this sentiment seems to rule out iLIGO as ever having been a confident detector of gravitational waves, at least in terms of statistical significance. But when I queried this e-mail, Quince responded in an essentially sociological way with two long e-mails (20 February). Quince claimed that statistical significance was not, in the end, what counted in making the community believe in a discovery. I have edited the e-mails quite heavily:

> I don't think that most physicists really require that standard of proof. I believe that there is an intuitive level of confidence that trumps the pure statistical measures in many cases . . . It is also equally clear to me that really requiring an error rate of 10^{-5} would not serve the best interest of science. Scientists should make their results known when they have reasonable confidence in them, not when they have absolute certainty that they are error free. We have the ability to judge things as likely but not completely certain, and to act on that knowledge, and to be able to accept that honest errors do happen when results are not confirmed. Was science harmed or helped by Cabrerra's monopole paper? I think helped, and he didn't suffer personally from it, because he was cautious about how far he pressed his claim. However, there is no doubt in my mind, that had later experiments confirmed the existence of monopoles at a rate that said Cabrerra had only a 1% chance of seeing one (i.e., he was very lucky!), he would still be hailed as the discoverer of the monopole.
>
> I am personally convinced that the reason physicists talk about 5-sigma results is because it is hard to do real false alarm rate estimates, and requiring "5-sigma" means that typically less than 1% of the results have significant errors.
>
> . . . To see something significant we have to see it at a very low false alarm rate, but whether 10^{-5} is really the standard that we (and the broader community) should set is far from clear in this case, at least to me.
>
> The statistical purists say yes, but I think that the average physicist will recognize something that shows up at the 10^{-4} level, or even at the 10^{-3} level as very convincing, providing we have really made an accurate estimate of the false alarm probability. The key point is that it is the community which makes the decision about how convincing our result is, not us. We need to put forth the case, make modest claims about it ("evidence for"), and let the community decide what they believe.

This last sentiment is, of course, evidential collectivism. As it turns out, Quince was to act according to these principles, agreeing that the event was "gold-plated" when it was later established that the, noise-based, little dog profiles looked quite unlike Big Dog's components, making Big Dog still less likely to have been a noise event. He, however, along with most of the other scientists, still wanted to keep "evidence for" in the title—which could, as has been pointed out, be construed as hedging one's bets.

Did Virgo See Anything Once More

A flurry of e-mails starting around the middle of February, which reached a crescendo around the twenty-fourth, once more concerned the contribution of Virgo. Vigorous arguments were mounted from both sides. Again, the point is "philosophical." Everyone agrees that there was nothing in Virgo that would have stood out above noise had attention not been directed to it by the events in LIGO. Once the best "template" that fitted the LIGO signal was applied to Virgo, however, something could be seen. With this template applied it could be said that there was only something in the region of one chance in a few hundred that the thing that could be seen was due to noise. "One in a few hundred" counts for nothing in this statistical context—as someone said, things that are likely to be due to chance one time in a few hundred are happening all the time. Therefore, it was also widely agreed that the "event" in Virgo could not be used to bolster the statistically expressed confidence in Big Dog—to do so would be retrospective statistical massage. As a scientist put it (16 February):

> What I think is making a number of us uncomfortable is the idea of "waveform shopping," searching through the data with nonstandard waveforms and paying more attention to the results that give more significance to the Virgo trigger. In trying to convert the upper limit on the FAR for the HL trigger into a number, the group considered more and more time slides, but kept the same trigger from the original . . . search.

And, from Dogwood on 24 February:

> It's also essentially different from the result in H1/L1 where almost *any* template family gives you a strong signal. In Virgo you only see something when you feed in all the information on the signal from the other detectors.

They are saying that in its calculation of the background the LIGO search restricted itself to using the waveform that first indicated that there was something unusual going on even though a more complex "spinning" template would have given the Big Dog signal more salience. More complex "tailored" templates were not used in this way because it would have been retrospective data manipulation. Yet some were stressing the importance of the Virgo signal by reference to these more complex templates.

On the other hand, it was argued that it did seem proper to ask astronomers if they had found something in the electromagnetic spectrum that occurred at the same time as Big Dog and this was a retrospective question.

> [From 24 February:] "Are we biased when we ask if Virgo sees something below the threshold?" I think the answer is "yes, as biased as when we ask whether there is any observed EM counterpart." I still think that's one of the legitimate a posteriori questions we need to answer in such a paper. Then, I can conceive that the choice of words to translate numbers into a judgment about this event is delicate as we all have different background and guts, as difficult as the choice of words for the title where we already could see very different opinions, but I can't see how we can avoid it.

In other words, if some vestigial flash of light or burst of neutrinos had been seen in the right part of the sky at the same time as the Big Dog signal, that would certainly be taken as an interesting corroboration even though it too was not going to be expressed statistically. Virgo could be seen as a similar case as was argued by another scientist on the same day:

> Imagine that LIGO and Virgo were separate collaborations. LIGO publishes a paper announcing a discovery and specifies the parameters of the template with the loudest signal.
>
> Virgo then carries out a straightforward search with that template and finds a FAP of 1/300 for its loudest trigger for that template in a time consistent with the LIGO observations. I think Virgo would then (correctly) publish a paper stating it has corroborating (or supporting) evidence. Virgo would, in fact, feel obligated to publish what it finds and to characterize its significance. I don't think "weak" is the word that would be used or should be used.
>
> I realize that many persons worry about blind vs. unblind analyses, biased conclusions, etc. But in that introductory paragraph we appear to be stating our scientific judgment as to what was seen in Virgo—all things considered. We should not feel we must behave like robots who can only follow preordained instructions in a statistics textbook. Although we do quote the robot numbers elsewhere in the paper, they are not all that we should present.

> Many persons have agonized over the wording of the title and the wording of other sentences in the paper because they represent qualitative scientific judgments that go beyond quoting a number. I think the qualitative characterization of the Virgo signal is in that category. We are obligated (I think) to state our judgment and not merely state that no trigger exceeded a predefined threshold.

It was Quince (a senior LIGO scientist!) who pushed hardest for the line that the Virgo signal was important. Quince, remember, had on the face of it argued at one point that the chance of Big Dog being an artifact—a correlated glitch of unknown origin in the LIGO detectors—was about equal to it being a real signal. From this position Quince was now able to say that the vestigial Virgo signal was important because it appeared on a remote detector of different design and made it far less likely that unfathomable correlated glitches had caused the effect. When the LIGO signal was put together with the Virgo signal, Quince was more comfortable about making a positive claim.

There could be something going on here that was to do with the politics of collaboration. It could be that some senior scientists' attitude to the inclusion of the Virgo signal had to do with maintaining good relationships across the whole gravitational-wave community. I do not know how far such considerations were playing a part in the attitude of the LIGO scientists.

The remaining question was how Virgo's evidence was to be described in the publication. "Weak evidence" was tried out and one or two other terms but about a week before the Arcadia meeting, "some evidence" seemed to become the acceptable compromise.

New Kinds of Evidence

At the Detection Committee meeting of 22 February, it appeared that some new ideas were coming forward and Big Dog was turning out to have fingerprints after all. Ilex chaired a subcommittee charged with deciding whether the event reached a satisfactory level of confidence to be reported as a finding. The subcommittee report, which was then included in the full Detection Committee report circulated to the entire collaboration, included two key arguments, two supporting arguments (such as the "weak" corroboration from Virgo), and three additional arguments that were mainly about why the signal was not incompatible with astrophysics and an absence of electromagnetic signals. Here I quote the two main arguments.

1. The probability of the CBC search producing a non-astrophysical accidental coincidence in S6 with the same ranking statistic or larger is 0.007% [172 days / 7000yrs].

The false alarm rate is measured with 4 million time-slides for H1L1 triggers . . . multiplied by a trials factor of 6 (due to considering double and triple coincidences, and three mass bins). Extrapolation to all of S6 (172 days of live time) yields a FAP of 7 × 10 – 5, corresponding to a 4-sigma signal if taken from a normal distribution. For comparison, note the single-trial FAP on S6D/VSR3 corresponds to a 4.5-sigma evidence.

2. There are no known instrumental or environmental causes that produce simultaneous transients in H1 and L1, or transients in the gravitational wave output with the same time frequency pattern as the candidate.

There are possible reasons that could cause simultaneous transients (like magnetic storms), but they would be seen in environmental monitors, and have not been seen yet in any science run so far. None of the previous coincident candidates that were followed up in earlier runs or in S6 revealed sources of common environmental or instrumental transients in H1 and L1. . . . None of the identified causes of transients in S6, including of both instrumental and environmental origins, have produced simultaneous transients in H1 and L1.

Note the reduction from 4.5-sigma to 4 sigma presented as being a consequence of the choice to count the FAP as being calculated over the whole of S6.[2] Still more interesting is the second key argument, which points in the opposite direction. What is being added here is the outcome of a considerable amount of work that had been taking place over the weeks. The components of Big Dog had each been examined very carefully and it had been shown that they did not look like anything that would be produced by a glitch. This finding was being reinforced by another analysis going on in the background, which showed that Big Dog was "coherent"—there was (give and take some "fishy" behavior), a match in the waveform of the components in H1 and L1—and the coherence, of course, was what had led Big Dog to be spotted by the Burst Group in the first place.[3] On the other hand, the little dogs were not coherent—they were just ugly splotches of

2. Collin Capano suggested that the difference was less marked than this.

3. The Burst Group uses coherence as a central feature of their analyses.

noise. For these reasons Big Dog stood right out from the noise in a qualitative way that was not captured in the calculations of false alarm rate and false alarm probability. Something more than simple statistics was feeding into the argument, and it seemed to sway the scientists, even conservative scientists such as Quince. Interestingly, we now had a signal that nearly everyone believed in but which was still below the 5-sigma threshold! Naturally, this was to lead to further arguments over the title of the paper.

At this point it seems hard not to go back to the little dogs theme. Now that it was known and accepted that the components of Big Dog looked nothing like the little dogs, wasn't there yet another argument for taking them out of the time slide analysis? To do the time slide with the Big Dog components included was to include in the analysis something that the community was now as certain as it could be was a signal; this was to do it wrong. The "time slide regress," then, had been pretty well resolved by examination of gravity's fingerprints. No one, however, responded in this way. Perhaps the reason was that it would involve too much retrospective tinkering but here we see the tension between following the rule about avoiding retrospection and following physicists' rule about correctly describing the truth of the matter, insofar as it can be known.[4]

The Title Revisited

On 25 February, there was a collaboration-wide telecon. For me the most notable remarks comprised a reprise of parts of the original argument over black holes. One scientist wanted "black hole" back in the title: "two Nobel Prizes rather than one," as this scientist put it. This was rebutted by scientists who said that there was nothing more direct about the detection of these black holes than their detection via the x-ray emission from accretion of orbiting material:

> Our model is simpler—we have a binary system instead of an accretion disk with radiative transfer and so on—so, in that sense, maybe it's a more secure identification of a black hole but I don't think we're actually seeing the black hole dynamics so able to claim that we actually identify the object.

In other words, in both cases something was being deduced about the masses of the objects (i.e., in this case, perhaps only the larger object) by the way thing around it behaved rather than it being seen "directly.'

4. With a resonance with the "airplane event," described in *Gravity's Ghost*.

Following the 25 February meeting there was a long e-mail interchange in which the title of the paper was reconsidered again. Quince initiated this by proposing a new title: "Evidence for the Direct Detection of Gravitational Waves from a Black Hole Binary Coalescence."

In this e-mail, Quince was saying that he thought that this was truly a signal (blind injection aside); it was clear from other exchanges and from what he later told me that his confidence was increased by the analysis of the little dog glitches which showed that they were nothing like a gravitational-wave signal—their wave forms did not cohere between the detectors in the way that the Big Dog waveforms did cohere. His confidence was also increased, as he explained to me, by the weak signal apparently seen in Virgo. But Quince still was not ready to give up on "evidence for"; he did not believe that they had enough to call the event "a detection." He wrote:

> I argue for "Evidence for" as an acknowledgement that a single event is far from an ideal way to make as important a claim as this is. I don't believe that we do ourselves any harm by adding those words of caution—I predict that almost everyone in the community will forget those two words almost as soon as they see our case, but it is better to let them decide on the strength of our claim than to possibly over-sell it. We have a very good case, about as good as I think one can do with a single event. However, if we really want to hold ourselves to the "5-sigma" standard, we fall short.

Quince also argued that the collaborations had held themselves to too strict a standard, by refusing to use the phrase "black hole binary" or "binary black hole," if only one member of the binary was a black hole. He pointed out to me that the usage was more complicated and variable than was apparent from the debate and he had found in looking at previous paper titles that the term "binary thing" or "thing binary" was often used when there was only one "thing." He pointed out to me that the object observed in the famous Hulse/Taylor "indirect" detection of gravitational waves was usually referred to as a "binary pulsar" even though there was only one pulsar.

Quince also put "direct" back into the title. This title, it seemed to me, should be attractive since it gave quite a bit back to the more positively-minded physicists—"direct" and "black hole"—while still satisfying the lobby who wanted a, modest, "evidence for" title.

As it proved, however, this was only the start and by the end of this burst of the debate, which faded out around the beginning of March, there had been about fifty e-mails on the topic with the "evidence for" versus "observation" theme still very much alive.

The authors of the majority of the first few e-mails wanted to go back to "observation" instead of "evidence for" with a couple on the other side—one urging caution and one suggesting that LIGO had not done any "observing" only "detecting." This, remark was trumped, however, by it being pointed out that the "O" in LIGO stood for "observatory" and that battle over this term had been fought and won against the astronomers a decade previously.

One of the correspondents reinforced the point made above about black holes:

> I like the suggestion for putting "black-hole binary" in the title: this only says that one of the components is a black hole.

Since from the very beginning everyone was sure that at least one of the components was a black hole it appeared, on this reasoning, that there had never been any reason to take "black hole" out in the first place so long as the order of the words was thought about! On the other hand, a second physicist thought it depended upon the tradition to which one was referring.

> There is some ambiguity surrounding this—a (numerical) relativist will think of a binary made of two black holes while an astrophysicist will think of a binary containing at least one black hole.

While a third physicist wrote:

> Anecdotally, numerical relativists who also work on relativistic hydro sim[ulation]s of BH-NS mergers use "black-hole binaries" to refer to both types of merger sources.

To which the response from the second physicist came back:

> At least Caltech-Cornell, Kyoto, AEI-LSU (and, I think, Rantsiou et al.) always specifically speak of BH-NS binaries when they mean black-hole–neutron-star systems. That aside—it's always better to be specific than ambiguous.

"Tradition" was now being invoked at the level of specific groups.

A fourth physicist pointed out:

> "Evidence for" appl[ies] not only to the detection but also to the black hole(s) claim. So, we should feel relaxed about the exact mass limit between NS and BH.

Going back to the question of confidence, one experienced physicist, [Nutmeg], invoked the future in arguing for a conservative title:

> We should use milder wording [because] everything is based on a single event. . . . an important test the first detection will have to pass before it is declared as such is the rate test. [This event] has to be statistically consistent with subsequent measurements . . . Being conservative at this point and not rushing to claim this (at this very moment) as "the first detection," is I think a wise choice.
>
> From the historical perspective 10 years down the road I do not think anyone will doubt Big Dog was the first direct detection of gravitational waves, if the event passes the rate test in the advanced detector regime. Clearly, this is independent of what we choose to put as a title in this paper! So, we will . . . save ourselves from an embarrassment in the remote chance [that] rates in the advanced detectors regime turn out to be inconsistent.

This, however, is exactly the kind of "hedging" that Sumac wanted to avoid in order to be consistent with the way the LIGO community had argued against the earlier Frascati claims. Dogwood wrote on 28 February:

> I have to continue arguing for the "Observation" claim which I think is well justified.
>
> From what I have seen, "evidence" seems to be the standard wording when the statistical confidence is not overwhelmingly strong and it is considered possible that the event or effect will "go away." Publishing with "evidence" is usually the rapidly made first step on the way to a stronger claim in particle physics.
>
> But for this event the Detection Committee hasn't identified a source of significant doubt. This need not have been the case. . . . But everything checks out. In fact, everything looks better than we would have expected.
>
> I am interested to know if . . . anyone has substantial doubts that this isn't a "gold-plated inspiral event." It is certainly possible to have doubts for various reasons, and we would want to do as much as possible to address them.
>
> Or [is it being said] that "Evidence" is an appropriate title even if there is no serious doubt? Will everyone understand that we are only pretending to be cautious? (Then why pretend . . .)

To which Hornbeam responded:

Fundamentally, yes, what I and a number of other people have said is that "Evidence" is an appropriate title even if there is no serious doubt.

The paper presents evidence, as well as our interpretation of it. It is very strong evidence, which will be clear to the reader when they read the abstract and/or the paper, but there is no need to challenge their healthy skepticism before they have even gotten that far. Any while it is very impressive for just a single event, it is NOT the strongest evidence for GWs we could hope for; having a few events, even if all of them were somewhat weaker than this one, would be more statistically significant and more convincing.

There is a tradition of choosing conservative titles, especially when the significance is less than the conventional 5-sigma threshold which we do NOT claim to surpass in this paper, so naturally we will be understood and respected if we do so too. There is no reason to lay this out as the "mission accomplished" paper. We're in this business to detect lots of signals, at which time their reality will truly be beyond doubt.

Hornbeam, as can be seen, also invoked the future. Dogwood replied:

I understand a conservative title as a style choice, but it seems a bit odd in context. If this is the real thing there will be a media circus and we will be unable to avoid saying things like "first detection" and "gold-plated inspiral event"—which is anyway the content of the paper.

Can we stop ourselves, or the media, saying that this is with high probability the first direct detection of GW? I doubt it. Once that is clear no one will consider the title to be meaningful either way. Sure, the ultimate goal is observation of many GW signals, but we shouldn't be underestimating what first detection means historically.

[Nutmeg] was saying—at least, as I read his argument—that just by altering the title we could get to back gracefully out of the claim, if and when future events don't appear consistent. But I think once a media circus has happened there is no backing out. If we want to make some sort of public claim about a signal but without really committing to it, we need to be consistently lukewarm, not only in the title but in everything we present. I personally think that's almost impossible.

Whatever we do with the title, the question that needs to be answered is "Is this the first detection or not?" If yes, we can't back out of it.

Dogwood added what, to me, is a very interesting "p.s."—interesting because it picks up the points made earlier (chapter 5), which are repeated in outline form above.

P.S.—oh, this "5-sigma" chestnut . . . If you look into the recent PHYSTAT conferences and Louis Lyons' talks, you will find "5-sigma" is taking a bit of a beating from experienced statisticians in the particle physics community. It was never more than a rule of thumb, because there was never a clear procedure to apply it to an experiment as a whole. With or without all possible systematic errors and biases? With or without considering all possible fluctuations that could produce an equally significant deviation—i.e., the trials factor / "look elsewhere effect"? No clear answer. The criterion was just to show that *something* is rare at the level of 1 in a million.

That's one reason for a very strong criterion: anything which was 5sig significant *before* accounting for systematics, bias and trials would remain very unlikely after doing so. If you are cynical, it was a way of being able to claim discovery without being very careful about systematics and bias and trials factor. (Which is difficult in most particle physics experiments, as they are way more complex than our detectors.)

Could be that our 1-in-14,000 total false alarm probability, which accounts for trials and we know to be free of tuning bias, is more solid than some "5-sigmas" in particle physics.

Summed up by Dogwood, we see that these physicists now understand how insecure a statistical confidence claim really is: it depends, as it is expressed in chapter 5, on so much that is hidden or unknown—willy-nilly, a statistical claim is a historical claim. The other side of the coin is that the confidence in the result is based on more than statistics—it is also based on the waveform of the signal which, in retrospect, has been shown to be nothing like the waveforms of the little dog noises.

The outcome of all this discussion was a decision made by Acacia, based on the sense of the "weight" of the e-mail thread, that the title should be changed to that suggested by Quince. That was to be the title of version 15 of the paper circulated on 7 March. It did not pass without negative reaction, including the remark that people who want change are always more salient than people who don't. In any case, the paper that went to the Arcadia meeting was Quince's: it had "evidence for" in the title, not "observation of," but it also had "direct detection" and "black holes." Arcadia was going to be fun!

14 Twenty-Five Philosophical Decisions

Looking back through the account of how the title has emerged, around twenty-five decisions that might be called "philosophical" had been made by this point. Remember, we are not talking about the interpretation of quantum theory or the anthropic principle—matters that are philosophical at the moment of conception—we are talking the decisions that lie behind the making of a single measurement; the kind of measurement upon which traditional philosophy of science based its theories of how science worked; we are looking at the meaning of "empirical."

Whether the number of decisions is twenty-five or a few more or a few less depends on exactly what is counted as a discrete problem. As will be seen, it would be very easy to get the number up to thirty by adding some subdivisions. I have chosen to count them as twenty-five and group them under nine headings, giving page references back to where they were first discussed.

Little Dogs

The most "classic" of the philosophical problems concerned whether the little dogs should be included in the time slide analysis. (1) The "time slide regress" arises because to know how to do the time slide analysis, you have to know whether or not to remove Big Dog (or equivalent); and to know this you have to know whether Big Dog was real but to know whether Big Dog was real you have to do the time slide analysis (p. 213).

One way forward is to do an initial analysis with putative signal left in and, according to whether this initial analysis makes it look more like a real analysis, you can then take it out and redo the analysis. (2) But this could leave a potential "kill your grandfather" problem if the initial decision is based on reaching a certain sigma level and the traditional sigma level changes over time. Something that counted as a signal even according to a more demanding level of significance that was applied at a later date, should not really have reached that level of significance in the first place according to the new significance level (p. 217).

And then there are other arguments: for whether Big Dog or equivalent should be removed for such a search. (3) I argued that removing the little dogs seemed right because only that way could staying on air longer produce a better sigma result. This, again, is "philosophical" argument (p. 222).

In my accounting system I have just dealt with three philosophical matters out of the twenty-five or so, and one heading out of nine.

How were these problems solved? They were solved by reaching a compromise that offered to the reader of the paper both sets of results—with and without little dogs—though this compromise was found only in a diagram. In a text the "little dogs in" result was quoted and this was the result of the power of certain conservatively minded individuals to win their way in the argument. Even though the number in the paper was the result of a calculation, there was no way to calculate the *right way* to do the calculation so the number could have come out differently. This pattern, the outcome of calculations depending on decisions that were not the outcome of calculations, was found to be repeated over and over again. Thus, not only were the decisions that went into the *title* not based on calculations, so were the decisions that went into the *calculations* that produced the sigma value. These are the seemingly incontrovertible measurements upon which traditional philosophy of science builds its models. The point will recur when other calculations are examined.[1]

Trials Factor, or "Look Elsewhere Effect"

Under this heading there are two problems. (4) The first is the general question about how to take account of trials factors or the "look elsewhere effect." This problem was discussed at length in chapter 5, as well as here (p. 223). (5) The second was the concrete question of whether to apply

1. One might see these elements of this chapter as extending Kuhn's brilliant 1962 articles.

a trials factor of 12 or 6, which in turn depended on whether there were counted to be 2 or 4 coincidence searches running in parallel (p. 000). The decision was that it should be 2 (a rare, nonconservative choice), because the signal was seen in a run where only 2 coincidences were running. But it could have been decided the other way.

What Is "the Same" when It Comes to the Instrument?

The two subheads here concern, respectively, the length of the run and the time slide analysis. (6) To calculate the false alarm *probability* it was necessary to multiply the false alarm *rate* by the length of the run (p. 225). But what counted as the length of the run depended on whether the "same" experiment was being done with the "same" instrument.[2] It was decided that the instrument running during S5 was not "the same" as the instrument running in S6 but that the instrument running in S6A, S6B and S6C was "the same" as the instrument running in S6D. It could have gone another way. It went the conservative way it did by default, without further discussion after the Detection Committee decided it could not reach consensus.

(7) When a time slide analysis is done a similar decision has to be made: one cannot do time slides which include stretches of data that are not "the same" in character as the period of the event being analyzed. In this case it was the whole of S6D that was counted as "the same but not the whole of S6 (p. 225).

Sigma

(8) What sigma level counts as a detection as opposed to evidence for? There is no way to know what the "right" level is (p. 227). In the Big Dog debate the resolution turned on the choice of a physics tradition to take as precedent—in this case high-energy physics—and then the argument about the proper interpretation of high-energy physics' changing tradition. Different conclusion could have been reached by interpreting the traditions in different ways but in the end the default position reached was the modest one.

2. Many philosophical problems turn on what it means to be "the same"—a question originally posed by Wittgenstein 1953.

False GWs and Trust

(9) The decision over what to take seriously as an imitator of gravitational waves is straightforwardly traditional or cultural—psychokinesis and strange versions of relativity has not been considered; malicious sabotage has been considered (p. 206). There are no calculations that can tell one what to consider (unless Bayesian priors are expressed in numbers so as to make a social choice look like a calculation). This is the "dam of disbelief" argument (note 5).

(10) The matter of sabotage was solved by a consideration of how hard it would be to fake the data and, it not being possible to eliminate this possibility entirely given the skills of insiders, it was decided that insiders had to be trusted (p. 207).

What Does "Direct Observation" Mean?

Here there are three subsections. (11) First, what does "direct" detection mean for the at least one black hole that was observed (p. 197). Was it observed directly since the gravitational waves were emitted by its own movements rather than the "indirect" observation of electromagnetic waves emitted by local matter? Or is it that this effect is no more direct that what has been seen before?

(12) Second, what is meant by "direct" observation of a gravitational wave and are the interferometers seeing gravitational wave more directly than the Hulse and Taylor observation or not (p. 197)? There are two viewpoints and no way to calculate which is right. It is a philosophical or, if you believe Damour and Taylor, a sociological/political debate.

(13) Finally, there is the question of whether anything is observed "directly" by modern instruments in which all that is seen is, as one of the scientists said, a "blip" on the instrument (p. 199). Or is all that is observed a string of numbers that, in this case, measure only the feedback signals sent to the mirrors to hold them still and show nothing without a huge train of statistical and mechanical inference? This is simply a philosophical debate that, by tradition or cultural habit, at best, rarely is discussed when a discovery is being discussed.

Have Two Black Holes Been Seen?

(14) The lower bound of the lighter star in the binary system was 2.7 solar masses (p. 195). No neutron star heavier than 2 solar masses has been seen

and no calculations allow for neutron stars as heavy as 2.7 solar masses. But the astrophysical tradition is that the description "black hole" is never used for objects of less than 3 solar masses. Here tradition trumps everything; there is no argument.

(15) Except there is an argument about whether the odds against the mass being as low as 2.7 solar are so great that this lower limit possibility should be ignored (p. 255). There is no way to know for sure.

(16) There is the question of whether the title should refer to a coalescing binary black hole even if there is only one (p. 264). Some traditions, such as numerical relativity, suggest this should be the case, other traditions, such astronomy and astrophysics, suggest it should not. A tradition to follow must be chosen!

Retrospective versus Prospective

(17) The question about what counts as acceptable changes to the data analysis method was discussed at great length in *Gravity's Ghost* (e.g., pp. 27–32). At the outset, the collaboration set itself very rigid standards: no changes could be made once the "box" had been opened on the main body of data (p. 26). As described under the heading of "the airplane event," this standard cannot be maintained. In that incident it was discovered retrospectively that an airplane was flying over the site at the time a certain signal was seen. The rules indicated that this could not be taken into account, as it had not been anticipated but the conclusion amongst most of the scientists was that this was madness—though one scientist was to resign from the collaboration in the light of the decision that was made. So though the group managed to extract itself from a "crazy" interpretation of the no post hoc changes rule it is still something that is very much present in people's minds. In my discussions about what could be done to enhance the signal to 5-sigma, I was told more than once that many people would not countenance any change that looked like retrospective readjustment of protocols.[3]

It remains, however, that what counts as retrospection is not a matter of unambivalent logic. (18) For example, it seems permissible to increase the number of time slides up to a maximum even if this maximum has not been set in advance and likewise for the invention of new methods for

3. The airplane event and the question as it appears again here both illustrate Wittgenstein's (1953) point that rules do not contain the rules of their own application.

doing time slide equivalents, such as the "find all coincidences" method (p. 212). Would it be retrospection to shorten the time slide interval so as to squeeze more in? Would it be retrospection to have kept the detector on line after 20 October so that more and longer time slides could be done? Would it be retrospection to take out the little dogs once an initial analysis had shown that Big Dog was likely to be real? Was it retrospection that what counted as the length of the run was only decided after the box was opened? Was it retrospective analysis when everyone agreed that Big Dog must really have been a signal after it was found out that the little dogs were unlike Big Dog in that they were not coherent and it turned out that Big Dog had fingerprints after all?

These questions were not discussed by the collaboration but settled by default. (19) The second point under this heading concerns Virgo and its observations (p. 245). Here there was a heated debate about whether the signal, extracted from Virgo only after the most favorable template had been applied to it, was something that had been generated by data manipulation. It seems that it could be so counted unless one imagined Virgo as belonging to some kind of alternative team who had been asked whether they had a corresponding signal just as one might ask the astronomers whether they had seen anything corresponding with the signal.

Logical Atomism, Epistemology and Conservatism

(20) Of course, a huge philosophical conundrum is the problem that, at its extreme, turns into logical atomism: how much doubt is it possible to express before one ceases to be a scientist (p. 250)? A scientist is someone who must take the risks needed to put forward "synthetic" claims about the world. (21) The "epistemological sentence" proposed by Quince was a symbol of this dilemma (p. 232). How much does one want certainty at the expense of risk? We have seen that mostly conservative choices were made.

(22) But conservatism, while it is better than recklessness, cannot always be the answer (p. 249). There comes a point when too great a combination of conservative choices leads to misreporting of what has been observed. One sees this in the frustrated comments of some scientists that if they are really certain that Big Dog is not noise they should say so whatever the sigma level. Too much conservatism amounts to deceit. At what point is integrity best served? Is it best served by scientists reporting their instincts, as Quince put it—their tacit understanding of what it is to make

1) First direct observation of gravitational waves and first direct observation of black holes in a *binary black hole* coalescence
2) Evidence for observation of gravitational waves from a compact binary coalescence
3) Evidence for the direct detection of gravitational waves from a *black hole binary* coalescence

Figure 23. Three possible titles for the paper

a discovery, as the sociologist might put it—and forgetting the false scientism of the statistics of significance tests?

Evidential Culture-Space and the Past, Present, and Future

(23–25) The context of all of this is, as explained in chapter 9, is the choices belonging to evidential culture-space and the way they combine with the forces coming from past, present, and future (p. 244 passim). The debate about evidential collectivism is intimately combined with the debate about risk versus certainty. The choice of evidential threshold is intimately connected with physics traditions coming from the past and present, factoring in the unlikelihood of seeing a signal as large as Big Dog in the first place. The choice of evidential significance is equally connected to past and present. And the future hangs most heavily on the shoulders of the present generation: one day of AdLIGO will be worth three years of current LIGO! How can one escape the tyranny of the future?

Consider the three titles for the paper presented in figure 23 and read them as signifying what was being claimed. Title 1 is the most that could have been claimed and notice the phrase "binary black hole"; Title 2 is the least that could have been claimed. The twenty-five philosophical choices fill the space between these two titles: the space was entirely filled with what could not be calculated. Title 3 was the compromise on the table on 14 March; "evidence for" is still there.

15 Arcadia
Opening the Envelope

Before the Meeting

The envelope is going to be opened on Monday, 14 March 2011, after a session in which the gathered community would make their final arguments about what the paper should contain. The idea was that following this discussion there would be a vote on whether the paper was ready to be submitted should the envelope prove to be empty and Big Dog real. At this point the community should have reached the point where they had a paper ready for immediate submission. I am writing this passage in my hotel room at 3:15 a.m. on the morning of 14 March and I am looking forward to the next few hours and will be quite glad when they are over, because the strain is intense.[1]

I have been in Arcadia for a couple of days already. The weekend has been spent working on this book—trying to clean up the manuscript as much as possible before the envelope is opened. I have attended a couple of meetings and talked a lot. There are many things you can get from talk that you cannot get even from thousands of e-mails and long-distance telecons. I can, however, confirm a number of impressions.

When I ask most of the physicists I bump into whether this is a blind injection, they tell me, "of course it is." Some also tell me

1. Working at 3:15 a.m. is not quite as heroic as it seems, since my body clock was on UK time.

that if it is *not* a blind injection there is going to be a lot of anger about why the detectors were not kept on air longer. On the other hand, discussion in the meetings and the corridors is continuing as though Big Dog is real; a thoroughgoing suspension of disbelief when it comes to making a scientific argument seems entirely compatible with whatever reservations are held in mind. It is just like being at the theater—complete suspension of disbelief unless someone nudges you and says "Is this a play or real life?" In the meetings the extra work needed to complete parameter estimation (see appendix 3) was discussed without sniggers or grins—it just went along like ordinary business. So it was truly the "nonlogic" of *discovery* that we have seen here, it is not a piece of playacting. I am as sure as I can be that it would not have been much, or any, different if the possibility of Big Dog being a blind injection had not been there in the background.[2]

Certain other points have been confirmed. At my request, his supervisor has asked Collin Capano to find out if it would be possible to push the limit without little dogs to the magic 5-sigma by extrapolating from the results of the time slides that actually have been completed—a reasonable procedure in physics. Collin says he will do this. In conversation with Collin and the supervisor, I ask why it is that no one has yet done it. That is, why, unlike what seems to be the case in high-energy physics, every sinew has not been stretched to reach 5-sigma? Why is it me, the outsider, who is triggering this calculation? The answer is consistent with what I have already heard: once it was agreed that the little dogs should be included for the purposes of the main report, 5-sigma was beyond reach, so there was no point in doing more and the no-dog extrapolation was forgotten. In any case, it seems more and more widely accepted that "evidence for" is good enough. People are saying that it is just a matter of modesty or "style." It is said that the rest of the scientific community will soon decide this is a discovery if they think it is and, in any case, AdLIGO will soon be here so why worry about it? I sense that I am beginning to irritate people with my continual nagging about 5-sigma. I take a little comfort later in the evening when, over drinks, a government agency representative remarks that it would not be good if they really had made a discovery but decided to report it as something less.

2. But see chapter 17 for an alternative view put by some of the scientists.

The Meeting

At last it is 8:30 a.m. and the meeting begins. Last discussions are held, last arguments mounted, and the envelope will be opened just before lunch at 12:30.

There are 350 people at the meeting, and they are all in the room. This is not a good meeting room. It is a long corridor made up by opening the partitions that normally separated a series of smaller rooms arranged side-by-side along one outer wall of the hotel. The podium is in the middle, with microphones and a half-dozen projectors and screens ranged along the front wall so that the long, shallow audience can see the slides. It is hard to maintain a sense of unity. Further, the lights are dimmed to make it easier to see the screens, but this meant people could not see each other and anyone not in the middle of the room is almost invisible. If there was to be a rebellion, this was the wrong room for it.

The LIGO Science Collaboration spokesperson, Dave Reitze, opens the proceedings and explains the timetable. My voice recorder, long an accepted part of the proceedings, is switched on. I am typing on my laptop, making notes on the proceedings, every now and again pressing the "index mark" on the recorder when something interesting is said. On my laptop I make notes of what point of interest corresponded to which index mark so I will be able to transcribe any important parts of the recording rapidly without listening to the whole thing again.

Dave Reitze says he wants a vote at the end of the meeting, just before the envelope is opened. This vote would decide whether the paper was ready for submission should the envelope be empty. He says he wants a "vote by acclamation" without the need for counting. It would take a tough individual to stand against the collective will.

Acacia opens proceedings with a very funny account of why the paper was written as it was. Acacia says, incidentally, that the question of whether Virgo had seen anything was still open. Going through the purposes of the meeting that was going to unfold he adds, to loud laughter, that the meeting was also meant to

> provide lots of grist for Harry Collins's next book—that's maybe the most important thing.

It is nice to be "part of the furniture."

Acacia goes through the title word-by-word, explaining the sequence of decisions that led to *this* term rather than *that* term being used. The title, by the way, was Quince's clever final compromise. For what it is worth, the only bit I don't like is "evidence for," but "direct" and "black hole" were in there.

Acacia acknowledges the Burst Group, who had first seen the event. I discover only at the meeting that the Burst analysis had found the event eight minutes after it occurred and that the signal had jumped all the hurdles for passing on to astronomers by the time forty-two minutes had elapsed.[3] Only forty-two minutes after the event a series of groups of astronomers were asked to point their telescopes in the rough direction of the source!

Acacia says:

> It's not surprising that in the CBC Group, where we've spent a lot of time on it . . . we prefer a bolder title like "First Detection of" or "Look what we did, isn't this great?"

The joke, however, is disarming, indicating that the group is not going to mount a serious argument for change at this stage.

Another thing I hadn't realized until this meeting was that all the little dogs were caused by combinations of the Hanford component of the Big Dog signal combining with noise glitches from the Livingston detector—there were none that came from Livingston Big Dog plus Hanford glitches. I simply had not noticed that before, though, in retrospect, I could easily have realized it from the e-mails. The salience of particular points is much easier to grasp in speech.

A poignant moment followed when Acacia reaches the reference list for the paper. Rai Weiss, the founder of interferometric gravitational-wave detection, spent a good part of the early development years of the big interferometers in bitter fights with Ron Drever (see *Gravity's Shadow*), often losing battles over the design of the devices. Drever has slowly faded from the scene starting when the big machines were being built, and, after spending many years in Caltech, had retired to Scotland. Weiss spoke up and said that Drever's name should be added to the reference list. This was generous, since Drever, who invented many of the features that make the

3. Entirely my fault—I discovered later that this had been much discussed, but it indicates how easy it is to miss things when one is simply sitting at home scanning e-mails.

big interferometers work, had made many enemies and could easily have been forgotten.

Acacia sits down and more detailed presentations are given by other speakers. I thought the crisis had been reached when [Redbud], an active junior scientist, is confronted by [Teak], a senior theorist, over the matter of removing the little dogs:

> Teak: You said that [the choice with little dogs in] was conservative thing to do. Can you convince me that that's the conservative thing to do not the *wrong* thing to do?
>
> Redbud: No, because I actually think it is the wrong thing to do. I'm not going to stand up and defend this.
>
> Teak: Can somebody else stand up and defend this?
>
> Quince: I will stand up and defend it.

Quince argues his usual line, that since you cannot be sure whether Big Dog was signal or noise one must assume it is noise and leave the little dogs in. Dogwood and Ilex also defend the position, but I cannot follow either of their arguments in detail. These defensive remarks are made forcefully and decisively. To fight on is not going to be easy, and in the dark room it is impossible for the widely separated people who may have held that view to spur each other on. The power and gravitas of the last three interventions seem to win the day.

A bit later, however, another scientist speaks up in a very quiet voice and says that the analysis was a mistake. It was severely biased by the little dog issue. This scientist says that in the advanced detector era we must have a process in place that makes sure we just don't make more mistakes like this. "We don't have to live with this imperfection on our analysis in the future."

But continuing discussion is headed off by one of the chairs:

> I think we could spend all day on this one. I'd like to move us on unless anyone's going to stand up and violently object and say "with this plot in the paper I refuse to let it go forward." I'd like to move us on . . . because we've got lunch coming at some stage.

Later in the meeting the title comes up yet again. Ilex says:

> I am one of the people who felt strongly that we don't need to be bold with the title. I don't think we can ever be sure with a single event we can never

> be 100 percent sure that this is not produced by something we didn't imagine but it's not a gravitational wave. Just as there are astrophysical events we can't imagine there are instrumental events and artifacts that we can't imagine so with just a single event I don't think we can ever be 100 percent sure so I don't think we should be bold about this and we don't need to. The evidence is strong enough so I think that with just a single event, caution is the right thing to do.

Teak, however, replies:

> I'm actually one of the people in favor of removing "evidence for" from the title and just saying "direct detection." And the reason why is because any kind of scientific work builds on a set of assumptions and that's normal in scientific work. If you look into any paper on any topic there are assumptions built in which could, of course, be wrong and the paper could in fact be wrong based on that. I think it's reasonable to temper a little bit the claims but I don't think the title is the right place to do that. I can think you can do that within the body of the paper and things like the assessment of false alarm rates and so on are far enough away from what we might expect that I don't think it's so reasonable to be putting doubt about the conclusions within the title.

I believe that the analysis of chapter 14 bears exactly upon Ilex's comment. Here, as Teak implies, Ilex was doing philosophy, not science.

At this point the two chairs step in:

> Chair 1: Dave keeps coming to tell me that we've got a lot to get through and we are likely to speak through our break so I'd like to just—
>
> Chair 2: Or you can say it another way—I'd like to open the envelope at the end of the morning session we have to get through all of the discussion ahead of that time.
>
> Chair 1: I think we can come back if there's time to stylistic issues and if we have agreed this we can discuss these other issues.

These issues were not to be discussed again.

Later, a scientist points to what are claimed to be contradictions in the Detection Committee report:

> I found it interesting that the Detection Confidence subcommittee said that if it were not an injection then it was definitely a gravitational wave but that

> the Content subcommittee said that we should be careful how we put the title so that we don't make too strong a statement. Can you comment on what seems to be somewhat of a disparity.

Quince replies:

> I strongly believe this is a gravitational wave. . . . On the other hand I still believe that making a very cautious title statement for it is the appropriate thing.

Ilex makes the "philosophical" point once more on behalf of the statistical confidence committee:

> Most of the people in the subcommittee said they were confident that this was an astrophysical event but we were not 100 percent sure and most of us supported the "evidence" title and not the "detection" title.

Not much later, Reitze says he will call for an "aye" vote. He has heard some people asking for a stronger title and thinks maybe there should be some more discussion of that, but says "we don't have a lot of time. . . . So maybe we can do this by acclamation."

I don't think there is anything too sinister about all this. It is, as a number of people tell me, "battle fatigue." This debate has been going on a long, long time. Any continuing argument about it would also go on for a long, long time. Everyone accepts that the "evidence for" lobby is too strong. So what is the point of dragging it out any further?

On the screen the substance of the vote is set out in two lines. The first line refers to the acceptance of the contents of the paper, the second the acceptance of its style. These, cleverly, are not to be voted on separately but together.

There is much laughter and a huge roar of "AYE." There is no sound when "no's" and "abstentions" are called for. Reitze says, "Congratulations, ladies and gentlemen." There is a loud and prolonged round of applause—the collaboration has written a paper and agreed on its contents!

Thanks and congratulations are issued all around, and each separate group who contributed is acknowledged with more applause.

Plastic champagne glasses are filled with much hilarity, signifying a huge relief of tension after an enormous effort. Jay Marx proposes the toast to the collaborations and all its detectors, GEO, Virgo, and LIGO. The toast is to the completion of the effort, not to the finding, the significance of

Figure 24. Fieldwork: computer, recorder, camera, champagne!

which we do not yet know. The champagne is quickly swallowed before the envelope is opened.

The Envelope Is Opened

Jay Marx plugs in a memory stick—"the envelope."—and says, "The first slide will tell you the answer: the Big Dog is a blind injection." There is applause. "So, Harry, you can publish your book in the next ten minutes." I say, "thank you."

But it turns out that, as with *Gravity's Ghost*, there were two signals injected, not one. One was the Big Dog and the other was twenty weeks of a continuous wave—a pulsar. This was not detected. But in this case nobody is too surprised by that: the detection was not made since the continuous wave group expects to look at periods much longer than twenty weeks in order to build a vestigial signal to the level of statistical significance. There is no shame that this second signal was missed.

What is a very big surprise, however, is that Big Dog comprised stars of 1 solar mass and 24 solar masses, not the roughly equal masses that were on the table. This revelation was greeted by loud, almost hysterical, laughter. A 1/23 scenario had been proposed early on in the game by the group tasked with the discovery but had been rejected by the more sophisticated analysts of the parameter estimation group and had long been forgotten!

As the day wears on, it turns out that although, in the main, the job had been done well, the detection exercise was always confounded because the very injection programs were faulty in various ways. This did not affect the main detection, but a postmortem was going to be needed to work out exactly what should have been discovered in the way of the parameters of the signal given that there was some mismatch between injection templates and detection templates.

The discussion ends right on time for lunch to another round of applause.

The Day or Two After in Arcadia

On 16 March I had new conversations that intimated that some of the heat was taken out of the debate about the paper's title because at least a couple of important actors were especially concerned to maintain good relations with Virgo. It also came to my attention that at least three important people, at least one of whom might have been expected to push hard for a more positive interpretation of the Big Dog, had, as a result of a mistake, come to know that the signal was an injection; they found this out some months into the debate. This cooled things a little at that final meeting. I also discovered that certain analysts whose voices were salient in the final couple of hours had convinced themselves that Big Dog was an injection (one told me his odds had been 1,000 to 1), because they had deduced that outdated injection templates had been used. By the time 14 March came round these scientists did not care much anymore what the title was and this certainly helped to quiet things on the day itself and may explain the lack of fierce argument, since one of these people was helping to chair the meeting.

On the other hand, some very senior people (including Quince) told me that on 14 March they had either convinced themselves that the odds were 50/50 or had good circumstantial evidence for thinking that the odds were something like 50/50.[4] Thus, by the day, there was a full range of views in respect of whether Big Dog was an injection but the way these views stacked up would have tended to damp down argument for a stronger title.

The spokesperson for the Virgo Group assured me that though Virgo personnel has been uniformly against a strong title, it had nothing to do with Virgo not playing much part in the detection and was solely to do with their traditional caution and experience and the fact that it was only one event. This is compatible with the Virgo view in respect of an anticipated future (see p. 165).[5] Another central member of Virgo said something similar and pointed out that the Virgo team's pressing for the inclusion of the vestigial Virgo signal was something that was directed at helping to establish the Big Dog, not the other way around.

Should Big Dog Have Been Seen?

After the meeting I asked [Willow], the scientist who had been partly responsible for the blind injection, what had been injected. Willow told me that the only two parameters he set were the size of the signal and the time. Of the size Willow said:

> It wasn't a no-brainer but my intention was that if the detectors were working right and our process was not broken it would be seen to be a signal.

Willow confirmed that, on the basis of what he had injected, my claim, that the interpretation of the signal had been too conservative, was supportable. Willow asked rhetorically, "How big a signal do they want?"

He suggested, cleverly, that the way to think about the matter was to ask whether, if the signal he had injected had been twice as big, it would have made a difference. The conclusion, of course, is that twice as big a signal would no more have resisted the essentially "philosophical" arguments of Quince and Ilex than did Big Dog. Even if the signal had been twice as big,

4. Quince had asked one of the senior astronomers if they minded pointing their telescope at the position of what might be a blind injection. The astronomer expressed a preference against this course of action. So Quince explained how to contact the injection team to find out the fact of the matter since he did not want to know the answer. When the telescope in question did not immediately point, Quince concluded it was a blind injection, but when it did point a couple of days later, it brought him round to the 50/50 level.

5. Note that the outturn of the Big Dog story does not correspond to the sentiment expressed by Jay Marx on page 165 (see the postscript of *Gravity's Ghost*) where he opines that 3 or 4 sigma should be good enough since gravitational waves are to be expected.

it remained just as much the case that one could not be sure that it was not caused by some unknown physical mechanism or by some cunning hacker, it would still have been a single event, and AdLIGO would still casting a shadow from the future.

Later I talked to three or four scientists, all of whom had been deeply involved in the analysis, in the sunshine outside the hotel. It was interesting that our conversation about the matter of the little dogs and the sigma level was very even-handed. In fact, I discovered that I had far more of the details to hand, including the numbers, than them. Partly, this is because I had been obsessed with 5-sigma and how to reach it and they had not; I had a lot of things at my fingertips that they did not and I could speak as an authority. I pointed out as they listened with interest that the limit on FAR without little dogs is 1 per 200,000 years, while with little dogs it is one per 40,000. It was me who pointed out that when physicists look for 5-sigma, it is before the trials factor is taken into account. It was me who pointed out that the choice of multiplying by the length of the whole of S6 or only the length of S6D when FAR is converted to FAP is somewhat arbitrary. It was me who pointed out that one might be able to get to 5-sigma by extrapolation from the nondog situation (Collin Capano had not yet completed his calculation). I realized that I had been studying these things much more closely than the physicists and they were interested in hearing what I had to say.

At the same time, however, one of them, Ginkgo, said something that made me reconsider. Ginkgo had always been of the opinion that the little dogs should be removed but that the remarks of Ilex in the meeting in favor of the other view had resulted in a change of mind. This is how conversation works: I had not "taken in" Ilex's remarks, but now they were made salient, I went back and listened again to something that went over my head when the meeting was in progress. This is the advantage of recording conversation. This is what Ilex said:

> Coincidence is the strongest argument we have for believing this is a detection. And the strength of coincidence lies in saying "what else can a chirp in one detector be coincident with just by accident?" . . . If you take that out of the game you are not giving your chances of accidental coincidences of something nice in one detector and you might as well do one-detector searches which we know we cannot do.

I can see why I did not make much of that at the time but, given what I had heard in the sunshine conversation, I decided that I ought to make sense of it and, thinking about it hard, I managed, perhaps, to achieve the right perspective.

There are *two* ways of thinking about the problem of the Big Dog and any other such signal, which are like the two perspectives that one can switch between when looking at the same drawing—for example, the Necker cube. One way to look at the problem is the way I had been thinking about it all along: *the unit of analysis* is *the coincidence*. Thus the Big Dog coincidence was either a signal, in which case it should be removed, or noise, in which case it should be left in. I believe this way of looking a thing was also Quince's way. (This does not mean that people who see it that way have to agree with Quince about what should be done about it.)

Ilex's way of seeing the problem, it seems to me, was interestingly different. It took *the unit of analysis* to be *the excursion in a single detector*. In that case you don't ask whether Big Dog was a signal or noise, you ask for the chance that a big excursion in Hanford, for example, could fall opposite a big glitch in Livingston. To answer this you take the big excursion in Hanford (that associated with Big Dog) and ask how often it falls opposite a big glitch in Livingston. Looked at this way the individual Big Dog components *have to be* used to ask the question—they cannot be excluded on the grounds that they comprise a signal.

Intriguingly, this way of looking at things also leads to two different answers because, in this case, all the little dog noise excursions were in Livingston. Thus if you start with the Hanford Big Dog excursion and ask the question Ilex's way you get a false alarm rate of 1 in 40,000 years. But if you start with Livingston and ask the question you get a false alarm rate of less than 1 in 200,000 years—that's because the Livingston Big Dog excursion never falls opposite a loud Hanford excursion (Big Dog aside). So, after all, the two results are the same as with the coincidence-as-basic-unit method with Big Dog in or out!

It seems to me, then, that the answer to the philosophical conundrum of whether to keep the little dogs in or to remove them has only one answer (keep them in) if you are thinking of the unit of analysis as signals in one detector but two answers (leave it in or take it out), if you think of the unit of analysis as the *coincidence*. I suspect there is something deep going on here but it is beyond me.

So, it turns out that I really learned something from that 14 March meeting. But I learned it only because of a casual remark in a conversation in the sun—Ginkgo said "I found Ilex's comment convincing," and that made me look back and discover something new. The second lesson is about how conversation works and how, as one might expect, it is still better than reading e-mails.[6]

15 March and Capano's Extrapolation of FAR

On 15 March, Collin Capano finished his calculation of what happens if you extrapolate the no-dog situation and sent it to me. He wrote:

> I get an extrapolated . . . FAR of 1 in 14 million years. Using the S6D livetime (0.12 years), this gives a FAP of 8.6e-9, or a sigma between 5.6 and 5.7. If you instead use all of S6 for the livetime (as we did in the detection paper), which is 0.23 years, this gives a FAP of 1.6e-8, or a sigma between 5.5 and 5.6.

In other words, it would have been easy to get to 5-sigma if the will had been there. I am not in a position to argue that this should have been done, but I am still surprised that the question was not asked until I asked it. I am surprised that it was as a result of my probing—me, the sociologist—that the calculation was carried out. I am surprised that no one else was interested. It suggests to me a conservative mindset.

Was Initial LIGO a Success?

Barring something like a strong correlation with a nongravitational signal from an astrophysical source, its seems hard to escape the conclusion that iLIGO would have found it very difficult to detect a gravitational wave and announced the fact unequivocally. Big Dog was constructed to be a detectable signal, yet even it was presented as only "evidence for." As Willow pointed out, if the signal had been twice as strong, it would not have affected the conservative arguments. On this analysis, LIGO was a tremendous success as a stepping stone to gravitational-wave astronomy because it led to the funding of AdLIGO. And it was an astounding success as a technological accomplishment. But it was not such a great success as a gravitational-wave detector. That it did not find any real gravitational

6. Since I was only able to look back because I had a recording, it might be thought that there is a third lesson here, but that is misleading. If I had had no recording I could have pressed Ginkgo to re-explain Ilex's remark to me or I could have sought out Ilex for a rerun. Incidentally, none of this would have occurred had I not spent so much time obsessed with 5-sigma as a result of going native.

waves was not the problem (the presence of detectable gravitational waves is in the hands of whoever it is that runs the heavens); the problem is that, on the evidence we have before us, LIGO was very unlikely to be able to make an unequivocal detection even if a real and detectable gravitational wave had been there to be detected. From the very outset of LIGO it was known that it could not make a discovery barring enormous luck. But Big Dog was (if it had been real) the piece of enormous luck and now we know that even under those circumstances LIGO was unlikely to announce a "discovery." So, though at the time it seemed an act of malice when the person who told the betting firm Ladbroke's that the right odds were 500 to 1 against there being an official report, before the end of 2010, that LIGO had detected gravitational waves, they were not far out. And the reason for this is not the technology but the sociology—the choices that were made in resolving the philosophical conundrums.

On the other hand, perhaps Blas Cabrera is the icon. Perhaps it is right that the physicists felt that they could never make a "discovery" from a single event because physics isn't history. Blas Cabrera's monopole is no longer physics, it is a piece of history—it is some strange, singular thing that happened in the past. That, perhaps, is why the physicists are always turning to the future. It is only the future that can turn a singular event, such as Big Dog, from history into physics. All being well, we now move on to the era of gravitational-wave astronomy and Big Dog, along with Initial LIGO, will soon be forgotten.

APPENDIX 3
Parameter Estimation

There were two stories in parallel going on here. Right from the beginning I decided I would concentrate on detection. The other story is about what kind of event was being detected—which is known as "parameter estimation." I thought this problem was too hard for me—the science is immensely complicated and I did not feel my expertise could cover it. To the extent that it has been discussed in this book, parameter estimation is what is at the heart of the debate of "one black hole or two": the conclusion was that the lower bound of the smaller star, 2.7 solar masses, and tradition, demanded that this could not be called a black hole. This argument aside, the parameter estimation debate was not a matter of discovery/nondiscovery but of agreeing on certain values none of which had any special sociological significance. Obviously there is a sociology here, but it is a very untidy sociology. It turns out that my choice was fortuitous.

Parameter estimation was worked on by separate groups and went slowly. Apart from the black hole question, it dealt with the mass of the other source, a function of their combined mass known as the "chirp mass," the direction of the source, its distance, the spin of the objects and whether the system was precessing. As we have seen, the question of direction is where Virgo came into its own. That there was no independently statistically significant signal on Virgo limited the region of the sky in which the source could have been located.

It turned out that parameter estimation proved very difficult in some respects. At the Arcadia meeting it was decided that there was something fishy going on because the exactly the same template wouldn't fit both the Hanford and Livingston signals. Note that when the envelope was opened it was discovered that the estimated masses were completely wrong. The injected masses were 1.7 and 24 solar masses; the extracted masses were around 5 each. An initial estimate by the detection groups had found 1 and 23, but these had soon faded from attention as the more sophisticated methods of the parameter estimation groups were brought to bear and the detection groups surrendered their rights (another problem was that 1 solar mass for the smaller object is "unphysical"; it has to be at least 1.4 to be a neutron star and no one was saying "I told you so," at least not after the initial gale of laughter).

But then it was discovered that the entire parameter estimation exercise had been confounded by problems with the injection exercise. I turns out that the injection templates were outdated and less accurate than the analysis templates, and, furthermore, a sign had been "flipped" that meant the direction of the putative source was entirely wrong. It should never have been called Big Dog in the first place!

None of this makes the slightest difference to the story told here, and it is of no importance to the physicists: they can congratulate themselves on noticing that there was something fishy going on and not pretending that it wasn't. These problems, let me stress again, have nothing to do with the question of "discovery" versus "evidence for." That did not concern the parameter estimation group.

For the record, table 3 shows the intended injected values of the parameters and the extracted values. Bold italics show injected parameters that fall outside the bounds of the extracted values.

Table 3. Intended injected values (right column) and extracted values.

Parameter	Lower	Upper	Injection
Chirp mass (M⊙)	4.38	5.18	4.96
Symmetric mass ratio η	0.16	0.25	***0.06***
Distance (Mpc)	7	57	24
Larger mass m1 (M⊙)	5.4	10.5	***24.8***
Smaller mass m2 (M⊙)	2.7	5.5	***1.7***
Spin on m1	0.67	1.0	***0.57***
Spin tilt on m1 (°)	84 (@40Hz)	111 (@40Hz)	***63*** (@30Hz)

GLOSSARY OF TREE PSEUDONYMS WITH DESCRIPTORS

Name	Descriptor
Acacia	Senior leader of CBC group
Balsa	Senior theorist
Quince	Very senior scientist who took a conservative view throughout and exercised great authority
Dogwood	Very active junior data analyst
Elderberry	Experienced astrophysicist/theorist
Foxglove	Very senior scientist with a more adventurous inclination
Gingko	An experienced analyst
Hornbeam	Active junior scientist
Ilex	Very senior scientist who took the most conservative view on everything throughout
Juniper	Virgo scientist who mostly leaned in a conservative direction
Kapok	Senior Virgo scientist
Linden	Active scientist
Mulberry	Very senior Virgo scientist
Sumac	Very senior and much respected LIGO scientist
Oleander	Very senior theorist
Piñon	Senior Virgo scientist
Nutmeg	Active scientist
Redbud	Active junior scientist
Teak	Very senior theorist
Willow	Senior scientist responsible for blind injections

III

THE TREES AND THE FOREST

Sociological and Methodological Reflection

16 The Sociology of Knowledge and Three Waves of Science Studies

For a philosophically inclined sociologist, the *sociology of knowledge* is the most exciting thing. It is a fact that the broad pattern of what you take to be true is nearly always a result of where you were born and brought up. Thus, if you were born and brought up as a member of a tribe living in the Amazon jungle, it is almost impossible that you could even consider whether any modern physical theory was true, since it would be impossible for you to hold the concepts in your head. The only chance for your ever coming to accept Western ideas of physics would be for you to be removed from the jungle while still fairly young and become immersed in the society of some country where such things were a normal part of life. And it doesn't have to be physics—it is also true for the ideas of a mortgage or an egg-beater. Of course, the same thing applies in reverse: you cannot really understand the tribesperson's deep belief in witches and their consequences for day-to-day life. There can be a *sociology* of knowledge because knowledge is drawn from society.

The sociology of knowledge becomes dizzying as soon as one applies it to one's own beliefs. Is my belief in the theory of relativity just as much an accident of birth as the Amazon tribesman's belief in witches and spirits? Worse, if I had been brought up in the right circumstances in 1930s Germany, would I have believed that Hitler was a brilliant leader and the Jews were an evil race who deserved destruction? Thoughts like this induce a kind of terror. They pose a desperate philosophical question: How can I find something solid on which to my base my knowledge? Moral philosophy grapples, largely unsuccessfully, with the Hitler-type question. Epistemology—the philosophical study of the grounds of knowledge—grapples, largely unsuccessfully, with matters of knowledge of the natural world.

Here our topic is knowledge of the natural world, but I am going to suggest that the way we live our moral lives still has something teach us. The problem of our knowledge of the material world cannot be solved by induction or deduction, or, at least, it has not been solved by these means in spite of the effort that has been expended. Setting out the problem clearly, however, might help to avoid the paralysis that the "epistemological terrors" can induce. If we can come to accept that many things that once seemed a matter of logically compelling truth are more a matter of choice it might make the choice easier to make; epistemological choices—our preferences for how we like knowledge to be made will be seen as like moral choices. Only the morally crippled are paralyzed by the intractability of the problems of moral philosophy. We know, without having to think too hard, that we want to live in a society free of mass murder and gratuitous torture. Thus, if the almost-inescapable theory of moral relativism does not make us doubt our moral choices a clear view of the problems of epistemology might help us reach a similar understanding for our choices about the way knowledge should be grounded. Choosing science as the preferred way of grounding knowledge is the basis of "elective modernism"—which will be discussed below.

The First Wave: Science as Self-Evidently the Finest Form of Knowledge

Epistemology has been subject to a new kind of trouble since the middle of the twentieth century, a trouble that grew out of the sociology of knowledge. The history, philosophy, and sociology of science are collectively known as "science studies," or "social studies of science." I am going to call the aspect of it that emerged from the new kind of trouble, the "second wave of science studies," or wave 2.[1] The second wave contrasts with a third wave, which I will introduce toward the end of this introduction, and a first wave, which preceded it.

1. For the first introduction of the three-wave scheme, see Collins and Evans 2002, 2007. Even social scientists, in spite of the sociology of knowledge, like to think of themselves as independent reasoners, and it is only fair to say that the three-wave scheme explained here has been heavily resisted within the traditional heartlands of science studies. It is nevertheless a huge success in the sense that the 2002 paper is now the second-most-cited in the forty-year history of the journal.

For a century or so before the second wave it seemed that science provided the fixed anvil of sound knowledge on which epistemology could be hammered out. It was taken that theory and experiment made science into an especially sound kind of knowledge so the knowledge puzzle could be solved by describing how science must work. To do this was the job of philosophy of science, with history and sociology making contributions where they could. This is the approach of the first wave. We might say that the first wave was trying to get some sense of the shape of the overall "forest," or "wood," of science.[2] How must science be in order that the trees of theory and experiment define a path that leads to sound knowledge? What prevents the process becoming distorted and our becoming lost in the forest?

An example of a solution that is fairly widely known to natural scientists is that associated with Karl Popper. He said that we can never be sure we have the truth but we can sometimes be sure that we are wrong; it follows that the path to sound knowledge is that of trying to falsify what you know and holding to be true what you can't falsify, or at least, holding it to be provisionally true. But Popper is just one philosopher among many who have tried and failed to find a quasilogical justification for science. Popper's falsifiability scheme, though it is a valuable guideline, was shown to be philosophically questionable by his pupil Imre Lakatos; he revealed through close analysis that it is just as hard to falsify something as to prove it in a positive way.[3] One might say that Lakatos discovered the problem by examining the actual process of falsification—its trees as it were—rather than thinking only about a general, forest-like scheme.

Robert Merton, with his "norms of science," is the best-known sociologist associated with wave 1; he said that science was the product of certain ways of acting including the sharing of knowledge, the organization of skeptical analysis of knowledge claims, and the acceptance of anyone's right to make a scientific claim irrespective of race or creed.[4] Of course there is a vast literature based on popularized versions of these philosophers' ideas, often written by scientists.

Richard Feynman, by the way, is famous, among other things, for being credited with the saying, "Philosophy of science is about as useful to scientists as ornithology is to birds." There is some truth in it—scientists

2. British people tend to contrast the "trees" and the "wood," but I have used "forest" for "wood" throughout, as this is less confusing to others.

3. See Popper 1959; Lakatos 1976, 1976.

4. Merton 1973.

know how to do science without having philosophers tell them. But it also follows that if the job is to analyze the nature of science then one would no more ask scientists than ask birds about ornithology. I will try to demonstrate that, nevertheless, it not philosophy on its own that leads to an understanding of science; it is a more sociological examination of the activities of science bound up with a philosophical curiosity.

Of course, it was not always obvious that it was science that produced the soundest knowledge. Shapin and Schaffer revealed that when, in the seventeenth century, Thomas Hobbes was insisting to Robert Boyle that Boyle's new air pump could not possibly work, he was arguing about far more than pumping air: Boyle and Hobbes were disagreeing about whether an individual could produce new knowledge that could stand against the power of the king and the people embodied in his Leviathan-like form. Hobbes feared that if just anyone was allowed to produce new knowledge then society would collapse into disorder; each person could compete with the traditional authority of kings. Boyle, knowingly or unknowingly, was as much trying to create a rival to kingly authority as he was trying to create a vacuum.[5]

To most minds, Boyle won the argument. In establishing the experimental method as the road to truth he established that individuals could stand against the might of kings, at least where knowledge of the natural world was concerned. These individuals would need some support: their claims would only be accepted if their procedures were witnessed by a body of gentlemen, or, at least, "virtually witnessed" via the carefully crafted written account.[6] Thus was born the scientific paper, its penumbra of peer reviewers, and the other social institutions of science. Still, the important thing was the individual in dialogue with nature.[7] Among educated peoples, when it came to the natural world, science became the authority, even over kings, popes, and emperors, and its heroes were individual discoverers and experimenters.

So it remained for centuries, for many coming to seem self evident. The role of philosophical and historical studies of science, therefore, was restricted to distinguishing between the pure and the pathological, while sociological studies were aimed at working out how society could best nur-

5. Shapin and Schaffer 1987.

6. Shapin 1994.

7. There is a "twist": since Boyle needed witnesses to affirm his observations, the facts were, to some extent, made in a collective way. This led Shapin and Schaffer to conclude that Hobbes was right. On the other hand, given the conclusion that most commentators drew from the events, Boyle was right when we consider the social construction of the matter.

ture an undistorted science. This, to repeat, is the first wave. In the first half of the twentieth century, even the spell of the sociology of knowledge was said to be broken by science and mathematics because they touched epistemological bedrock.[8] Science and mathematics, then, could free us from the stranglehold of our conceptual birthright. During this wave it was the job of the outside analyst of science to explain and nurture science's manifest success.

The Second Wave: Sociology of Knowledge Applied to Science

But the sociology of knowledge did not go away. One culture's self-evidence is another culture's myth—there is witchcraft here and there is science there, both equally impossible to disbelieve if the circumstance of birth is right. In the middle of the century, analysts finally began to ask if the sociology of knowledge could be applied even to science and technology.

It is rare for there to be a clean start or finish to any historical trend, and the sociology of knowledge's assault on science's epistemological privilege is no exception. An early ragged edge was Ludwik Fleck's book *Entstehung und Entwicklung einer wissenschaftlichen Tatsache: Einführung in die Lehre vom Denkstil und Denkkollektiv* (1935).[9] Fleck was a medical scientist, and the book is a philosophical or sociological analysis of his own work on the Wasserman reaction for the diagnosis of syphilis. The book introduced the term "thought style" (*Denkstil*), which is very similar to Thomas Kuhn's "paradigm" in its sociological sense. But the time was not right for Fleck, and his book was little known in those Western academic circles when, much later, the sociology of scientific knowledge was being invented. Fleck's book was not translated into English until 1979, under the title *Genesis and Development of a Scientific Fact*. The social circumstances that gave the sociology of knowledge breathing space in its confrontation with science were not to occur until the 1960s—that period when, for Western academics, anything seemed possible.

Thomas Kuhn's book, *The Structure of Scientific Revolutions*, which is said to be best-selling academic book of the twentieth century, was published in 1962.[10] Kuhn showed how the very language—the conceptual structure of

8. Karl Mannheim (1936), the founder of the sociology of knowledge, took this view.

9. Nye (2011) locates the start of the second wave much earlier than I do: she claims its origins are to be found in Fleck and Polanyi's generation.

10. Kuhn did acknowledge the influence of Fleck's work but it was to be a decade or more before the significant extent of that influence became clear to others.

science—changed in the course of what he called a "scientific revolution." When Einstein supplanted Newton, the conservation laws changed: mass was no longer mass, length was no longer length and, correspondingly, the interpretation and the craft of cutting-edge experiments changed. What was once an anomaly was now a discovery. As Kuhn put it, scientists on either side of the divide "lived in different worlds." We now had a science that could radically change its concepts and practices according to the academic subgroup in which the scientist thought and worked; this was an invitation to apply the sociology of knowledge. If it was possible for different groups of scientist to live in different conceptual worlds then anything in the way of belief was once more possible. Kuhn, then, or Fleck, or maybe it was Wittgenstein, or maybe the whole atmosphere of the 1960s led social analysts to start to try to apply the idea of the sociology of knowledge to science itself.[11] At least within a small circle, the enterprise was a great success. This was the beginning of the second wave of science studies.

A key principle underlying the new approach was first expressed by David Bloor—it was the notion of "symmetry."[12] Bloor argued that the social analysis of science should proceed in exactly the same way whether the scientists considered the findings to be true or false. The social analyst's question was how *this* came to be seen as true while *that* came to be seen as false. Such an investigation would be circular if the "actual" truth or falseness of the matter played a part in the investigation. To develop a sociology of knowledge, a very difficult cognitive revolution was required because the analyst had to assume that the actual truth and falseness of some matter had no bearing on how it came to be agreed to be true or false; at the time, such a thing involved a feat of mental gymnastics, and many still cannot manage it.

The most widely accepted philosophical position, which accords with common sense and which persists to this day in the minds of all but the most careful and assiduous analysts of science, is that things come to be perceived as true because they are true. The only kind of sociological explanation of knowledge—the only knowledge issue that it is even conceivable to try to explain sociologically—is why false things sometimes are taken to

11. The major influence on this author, and on David Bloor (see below), was the later philosophy of Ludwig Wittgenstein (1953). Wittgenstein was initially introduced to Collins via Peter Winch's *The Idea of a Social Science* (1958). I did not encounter Kuhn's book until some time later and then I saw it as an application of these Wittgensteinian ideas to science.

12. Bloor 1973, 1976.

be true. In other words, there could be no sociology of truth because truth was its own explanation; there could only be a "sociology of error."[13]

The social and cognitive strain associated with the initial breaking away from this dominant view was such that it required a kind of mental "boot camp." It was necessary to practice thinking as a relativist of one complexion or another with support and encouragement from one's academic friends. Relativism became a credo, and it was relativism that gave rise to the bitterness that characterized the period known as the "science wars."[14] But in the longer term, relativism was important only as a source of mental discipline and later, I would argue, as a methodology. It is said that philosophers must exercise their minds by believing three impossible things before breakfast. To establish the sociology of scientific knowledge—SSK, as it came to be known—required, at least for some, analysts like myself, that for a while it be believed that there really was no such thing as true and false; only that way could the study and analysis of the *establishment* of the true and the false be done with sufficient rigor. Only that way could one steer a passage away from the siren call of truth as its own explanation.

My first contribution to SSK, as I see it, was to show what you could actually do in the way of empirical studies with the idea of the sociology of scientific knowledge. It seemed to me, as someone who had done science in (high) school, that the chief obstacle that scientists would always place in the path of something like SSK was replicability: the truths of science could be shown to be universal rather than rooted in society because anyone, anywhere, could repeat an experiment and get the same result. As Boyle was believed to have shown, nature spoke equally clearly to anyone, anywhere, peasant or king. So I studied the process of replication of scientific results in great detail and it turned out that it was not and could not be the royal road to truth in seriously disputed science. This was because the disputants were always disagreeing about what counted as a competent replication. I showed, in a repeatable way (!), that in the case of scientific disputes which turned on experimental replication, experimental results melted in the heat of mutual accusations of experimental incompetence.[15]

13. Laudan 1983.

14. Gross and Levitt 1994, 1996; Wolpert 1992; Hendry et al. 1994. Labinger and Collins (2001) believe they pretty well put an end to the worst of this period by bringing together the less extreme elements of both sides for a civilized debate.

15. Though Bloor wrote down his idea in 1973 and my key publications came later, in 1974 and 1975, I was coming from somewhere different and discovered the coincidence of the two approaches only after I was well on my way. The most complete account of my approach is in my *Changing Order: Replication and Induction in Scientific Practice* ([1985] 1992).

This kind of exercise involved looking much more closely at the trees rather than the forest of science. Instead of saying science must be true because its results are replicable one looked at the actual trees of individual replications of experiment. One discovered that where serious dispute was involved the path out of the forest would not reveal itself via replication alone. For example, you cannot know for sure whether a replication has been done properly; if it has been done badly it tells you nothing at all. Thus, while the "forest" of replication remains a valuable, indeed vital, idea, and is an essential part of the process of science, the way it works in practice is far more complicated than it looks from the distanced perspective (I first tried to show this by examining the dispute over Joe Weber's claims to have seen gravitational waves in the late 1960s).[16]

If you concentrate on the trees alone—what scientists actually do in the heat of a dispute—you cannot see any pattern to the forest and stories about its shape begin to seem more and more like comforting myths. One might characterize the difference between wave 1 and wave 2 as a preference for arguing about what the shape of the forest must be, on the one hand, and close examination of the pattern of trees on the other.

Though practiced only by a small group, all this examination of the trees was fruitful. For example by the time the dispute over cold-fusion came along the small group of SSKers knew exactly what was going to happen and so it turned out: Pons and Fleischman, the discoverers of the effect, saw cold fusion over and over while their critics did not. The critics said Pons and Fleischman were incompetent, while Pons and Fleischman said the critics were incompetent. The debate over cold fusion continues to this day, carried forward by Pons and Fleischman's successors, but it is mostly hidden from view—no one talks about it in public any more.

The relativists referred to the moment when continuing debate was forced underground as "closure." For all but those very close to the inside of the matter—those actually engaged in the work—that moment looks like the moment of decisive refutation. And for all practical purposes it is. After a couple of years the scientific argument over cold fusion was dead, even if it continued to have a subterranean existence. In terms of my own 1975 metaphor, closure was the moment when the glue dried on the ship in the bottle and the strings that had pulled the masts upright were cut so no one could see how the artifact had been put into place. Like the ship, the nonexistence of cold fusion looked as though it has always been there—it was the truth. Only close examination of the process of putting the ship

16. Collins 1975, 2004.

in the bottle could reveal that the truth, like the ship, had been "made," or "socially constructed." In that respect, "closure" is the right term because it implies that visible debate has ceased rather than that some quasi-mechanical ratchet has slipped over. The hiding of debate from view to all but a tiny residue of still active experimenters is a social process—by looking close enough and using one's relativist microscope could one witness "the social construction of reality."

Of course, philosophers of a realist persuasion and scientists pushed back against these ideas. From their point of view it was rather that cold fusion had been shown to be wrong and that was all there was too it. This is the right and necessary point of view for scientists, but certain philosophers' angry responses were badly misplaced and very "unphilosophical." There is no need to settle this argument if what is at stake is good historical and sociological analysis of science; what counts is what produces interesting insights. In science of every kind we never know if we have the metaphysics right; what we should argue about is what can be learned given the way the subject is approached. I believe that the concentration on the trees that came with the second wave has produced new insights that are valuable irrespective of philosophical persuasion. Some of these will be listed below.

My early research along these lines, as I have mentioned, was based on looking at Joe Weber's claims to have detected high fluxes of gravitational waves. A group of rivals repeated his observations with similar apparatus but found nothing. In a 1975 publication, I argued that the methods of science alone were incapable of working out which set of disputants had done the experiments in a competent way. I subsequently referred to this as "the experimenter's regress."[17] The experimenter's regress turned on the fact that experimentation is a skilful activity based heavily in the "tacit knowledge" of the experimenter.[18]

Relativism, as I have intimated, can have a variety of complexions. Philosophical relativism takes it that there is no truth of the matter. I held this view, and for the reasons explained—mental boot camp—it was important for me to hold it at the beginning of my work in the sociology of scientific knowledge; I held the view from around 1972 until 1981.[19] *Methodological*

17. Collins (1985) 1992.

18. For more on tacit knowledge, see Collins 2010.

19. A symmetrical treatment of really "crazy" claims, such that plants have emotions or metal can be bent by paranormal means, is a really tough and interesting exercise. After a bit of practice one becomes sensitive to the fact that nearly all skeptics of the paranormal are quite unconscious of the extent to which their firmly held views are founded on nothing but their orthodox scientific

relativism, on the other hand, proposes that whether there is a truth of the matter or not, the social analyst of science must *act as though* there is no truth if the work of studying the establishment of truth is to be done properly.[20] I have held this view from 1981 to the present and expect to continue to hold it.[21]

In the early days, few people knew how to look at science from SSK's point of view, and to those who did not understand the idea it seemed insane—people denying the self-evident. One of the criticisms thrown at the relativists was the so-called *tu quoque* argument. I said above: "I showed, in *repeatable* way (!), that . . . experimental results melted in the heat of mutual accusations of experimental incompetence." The *tu quoque* argument goes: "If you say truth is socially constructed, what you say yourself must be socially constructed, so why should we believe you?" Or in this more narrow case: "If you have shown that replication does not prove anything how can *you* claim to have proved something in a repeatable way?"

The *tu quoque* argument should be ignored. Either things are socially constructed or they are not, and a logical trick is not going to prove it either way. In any case, it does not apply to methodological relativism, which does not claim to know how what the world is made of, only how to investigate it best. It may be that some analyst of social science could look at my work on replicability and show that it too is subject to the experimenter's

---

socialization, just as the sociology of scientific knowledge would lead one to expect; it is quite depressing to discover how unaware of the source of their arguments even brilliant scientists can be in this regard. Note, I have said nothing here about the validity of paranormal effects or otherwise, I have spoken only about how we come to believe or disbelieve in them and the reasons that scientists give for disbelieving them. This is a dangerous thing for me to write, leaving me vulnerable to accusation of being a crank or heretic (I use the word advisedly), so it is perhaps prudent to say that nowadays I do not believe it scientifically worthwhile to research the paranormal further. The reasons are to do with the topic already having had a long enough "run for its money" without demonstrating good control over its signal-to-noise ratio, considerations I would apply to all heterodox scientific claims of a certain vintage.

20. Around 1981, I realized that case studies of the trees of science could not prove that the world was socially constructed, only that when scientific disputes were examined at their height, the force of nature was invisible—they were settled by social agreement. Whether there was some "hidden hand" at work, which showed itself in the long term by creating paths in the forest but could not be known by me or anyone else. I therefore shifted to methodological relativism. In my view, the approach of methodological relativism is vital for anyone who wants to study the creation of scientific knowledge; it would be the right approach even for a natural scientist who wanted to do such work in their "time off" from science proper.

21. In contrast, David Bloor and his colleagues believed there was a true and a false and that social processes were the midwife for their emergence. Their view is put forward in contrast to methodological relativism in Barnes, Bloor, and Henry 1996. I can make no sense whatsoever of their criticism of methodological relativism as it seems to me to be a perfect fit with Bloor's symmetry.

regress, but I do not have to be that analyst. We live, and we have to live, in a compartmentalized world; all we can do is our best from within our own compartment. In the compartment the social analyst occupies, physical truth is socially constructed, whereas social truth is demonstrable by replication; for physicists, as physicists—as opposed to what they do in their time off—it has to be the other way round or they would be doing our work not theirs.[22] The productive question is, "Is there something we can learn from each other and what is it?," not "Who has the ultimate truth?"

Wave 2 produced a host of new activities, new case studies, new understandings, new approaches, and noteworthy publications. To sum this up would need a large book. I will try to provide some of the flavor of the changes brought about by the new movement. I'll do this with science's way of describing change in mind, as opposed to the approach more typical of the humanities. The humanities would talk of great works, new movements, and changed attitudes, but I'll list a few of the new things we found out in the early days about the way the world of science works. Other commentators would, no doubt, produce different, if overlapping, lists. Included among what we now know about science and technology that we did not know before, or that we did not know before in quite such a self-conscious way, are the following.

1) *Tacit knowledge.* Michael Polanyi originally coined the term "tacit knowledge": "we can know more than we can say." Under wave 2 it was shown how this affects the process of copying and repeating someone else's experiment since they cannot explain all that needs to be explained to ensure it is conducted with competence. Tacit knowledge was also shown to bear on expert systems and artificial intelligence.[23]
2) *The experimenter's regress.* Because of the problem of tacit knowledge if nothing else, replication cannot settle a serious scientific dispute on its own. Neither can theory plus experiment. This is because one does not know if an experiment has been done properly until one knows what it is supposed to produce. Thus scientists are forced to use "nonscientific" means—such as assessment of the reputation of the experimenters and theorists or their institutions—to decide what they should count as belonging to the set of proper replication attempts. The decision about whether an experiment has been replicated is coextensive with the decision about which set of the experiments have been competently performed

22. Collins 1982.
23. Polanyi 1958; Collins 1974, 1990, 2011; Collins and Kusch 1998.

and which set has not. The normal way of deciding whether an experiment has been done properly—inspection of the outcome—does not work where what counts as the correct outcome is in dispute. This means that the endings of scientific disputes are best thought of as "closure of debates," or "agreements to agree" rather than the automatic outcomes of sets of definable procedures. Other analysts have found similar regresses in other domains of scientific activity. For example, Kennefick analyses the "theoretician's regress."[24]

3) *The core set/distance lends enchantment/the trough of certainty.* The work of science happens among what is often a very small "core set" of active scientists who do the experiments or make the decisive theoretical contributions while the bulk of the scientific community merely comments. To understand science, one must investigate this core set and its relationship with the wider world of science. Popular accounts of some scientific dispute are misleading because "distance lends enchantment." The activity of the core set is intense and the bandwidth of the communication to the outer community is too narrow to carry all the information about what is going on so a simpler picture emerges. Mackenzie renamed the distance lends enchantment idea "the trough of certainty," because he noticed that policymakers and funders often have a good idea of what is going on in the core so there is one outer ring where the uncertainty of the core is reproduced.[25]

4) *Ships in bottles / black boxing.* In 1975, I used the metaphor of the ship in the bottle and later Latour used the term "black boxing" to describe what happens after closure. Debate is closed so the scientific "fact" that emerged from it is treated as a black box as far as the future of science is concerned. But this differs from science to science.[26]

5) *Interests.* Because scientific disputes are part based on things other than strictly scientific procedures, there is an interaction between wider interests and scientific results. A striking historical finding is that Pearson's correlation coefficient dominated Yule's proposal, because it fit better with eugenic ideas. Latour argued that Pasteur succeeded with the germ theory of disease because he drew the interests of farmers into the work.[27]

24. See Collins (1985) 1992, for the experimenter's regress, and Kennefick 2000, for the theoretician's regress.

25. Collins (1985) 1992 is the origin of distance lends enchantment. It was modified to trough of certainty by MacKenzie 1998.

26. Collins 1975, (1985) 1992; Latour 1987.

27. Barnes 1974 first developed the interests theory of science and this was taken up by MacKenzie 1981. Latour 1983 widened and generalized the point.

6) *Statistics.* That the results of statistical analysis are expressed as numbers makes them appear more "objective" than they are. The history of the correlation coefficient just mentioned is an illustration and chapter 5 is intended as another. Many of those whose work feeds into the new understanding would not, however, describe themselves as belonging to wave 2.[28]

7) *Modalities.* One can trace the firming up of a new scientific fact and its move from the laboratory to the outside world by studying the way claims are expressed in print. Initial claims will be of the form "on such and such a date, at such and such a place, X found Y, signifying Z." Gradually, however, all these "modalities" are stripped away and Z has an independent existence in the literature associated with no dates, places, or persons. This relates to the notion that there is a "constitutive forum" in science where the results of experimental and theoretical results are used to constitute knowledge and a "contingent forum" in which knowledge is constituted by talking of persons and places. Collins and Pinch argue that when fringe sciences like parapsychology are discussed, the boundaries of these forums are violated. For example, journalists, who normally remain in the contingent forum will enter the constitutive forum . Mulkay, Potter, and Yearley adapted this idea (making it less powerful in my view) by talking not of forums but of repertoires—scientists just choose to talk formally sometimes and informally on others.[29]

8) *Evidential significance/evidential cultures.* Pinch showed that scientists have to choose the extent to which they accord evidential significance and risk to their findings. Ray Davies could have said he'd found a few atoms of argon in his tank of chlorine-based cleaning fluid or discovered a certain flux of solar neutrinos with all the inferential steps included. I have added two extra dimensions of choice: the level of statistical certainty

28. Thus, where I say on page 94 of *Gravity's Ghost*: "That is to say, numbers appear to arise as a result of well-defined states of affairs in the world. It is possible to argue about whether this is a big pile of apples or a small pile of apples, but if there are fifty-six apples, then 'that's it.'" I could well have cited Porter's or Desrosières's studies of the way this state of affairs came about. Other historical studies of the emergence of statistics include those by Gigerenzer and his collaborators, and Hacking, and Stigler. We could also include Gerald Holton's reexamination of Robert Millikan's notebooks at the time he was trying to establish the discrete nature of the charge on the electron as a study of the use of numbers in physical science, or Krige's look at the way that Carlo Rubbia used his and others' statistics to win the Nobel Prize for the discovery of the W-boson. See Porter 1996; Desrosières 1998; Gigerenzer et al. 1990; Hacking 1975, 1990; Stigler 1999; Holton 1978; and Krige 2001.

29. Latour and Woolgar 1979 first set out the idea of modalities. Collins and Pinch 1979 are responsible for the idea of forums while Mulkay, Potter, and Yearley 1983 shifted to repertoires.

required to count a result as a "discovery" and the point at which a result is announced to the wider world for comment: "evidential individualists" keep results to themselves until they are sure, "evidential collectivists" announce uncertain results early and invite wide criticism. The three dimensions taken together comprise a "theory of evidential cultures."[30]

9) *Sociology of technology.* The evolution of the bicycle toward the "safety bicycle" was not a matter of technological inevitability; it depended on what people wanted bikes for. The dangerous penny-farthing (high wheeler) was good for "showing off" because it was dangerous. In the case of technology, consumers have a much greater role in deciding what counts as successful closure than in the case of science.[31]
10) *Epistemic cultures.* Different sciences work in different ways. When a physics experiment fails, the physicists struggle "forever" to find out why. When a biological preparation fails to do what is wanted of it, the biologists simply throw it away and try again. In my view the exploration of the differences between the sciences and their consequences is still a shamefully underexploited area of wave 2.[32]
11) *Critique of deficit model.* Given the emphasis on dispute under wave 2, it becomes obvious that some official bodies' view that if only the populace were better educated in science (their science "deficit" was remedied), everyone would agree about science policies. Since the most well-trained and experienced scientists disagree all the time, this cannot be true.[33]
12) *Computer proof.* A mathematical proof completed by computer is more reliable than a human proof, because it does not fall prey to the fallibility of the human mind; a mathematical proof that cannot be grasped by the human mind is unreliable, because all the bugs cannot be guaranteed to have been removed from complicated computer programs. The argument cannot be settled.[34]

Some case studies that illustrate some of these themes are gathered together in a series of easily readable books under the generic title "The Golem."[35]

30. Pinch 1985, 1986 is responsible for evidential significance, with Collins 1998, 2004 adding dimensions to make evidential cultures.

31. Bijker, Hughes, and Pinch 1987 is the iconic source for the sociology of technology.

32. Epistemic cultures are discussed by Knorr-Cetina 1999.

33. It is impossible to pin down the bibliographic source of the deficit model and the discussion is diffuse.

34. The analysis is that of MacKenzie 2001.

35. Collins and Pinch 1998a, 1998b, 2005. I thank Trevor Pinch for helping me assemble the above list. The following could easily have been added. *Boundary work*: scientists spend time estab-

This account begins with a historical account of the Boyle's air pump and its role in his political dispute with Hobbes. Shapin and Schaffer's analysis, published in 1987, depended on the already-existing wave 2 analysis. It was that analysis that showed how there could be continuing disagreement about the working of Boyle's pump in the face of what has been historically thought of as the indisputable facts about the vacuum that it produced. Their account turns on Hobbes taking apart the most intimate details of the construction of the pump; Shapin and Schaffer were in a position to do this kind of analysis only because Wave 2 had shown us how to see the trees of science instead of the forest—disenchantment came with learning how to look at the details of procedures.

After the initial excitement of the discovery of a new way of looking at science the theoretical ideas underpinning wave 2 broke apart; the aims of the field are now much less unified. The field sometimes appears to be in danger of turning into a general critique of science driven by changing political fashions. One theoretical tradition was founded by Bruno Latour and for a decade or so it was dominant, perhaps because Latour's approach was aligned with semiotics and literary criticism. This enabled literary theorists and other humanities scholars who knew little about the content of science, to feel they could analyze it in the same way as they would analyze any other cultural enterprise.[36] They could examine the way the scientific literature worked and show its relationship to other kinds of literature, *this* kind of phrase giving power to this actor in the story, *that* kind of phrase reducing that actor's power. Science became identified with its literature so the way the literature assigned power was all one needed to know to understand science.

For good or ill, a huge expansion in the activity and salience of science studies was brought about by the literature-based approach. The good,

---

lishing an defending the boundaries of their domains (Gieryn 1999); the idea that scientists talking across disciplines and world views must establish *trading zones* (Galison 1999; Collins, Evans, and Gorman 2007); *forensic evidence* is vulnerable to the view of the trees rather than the forest and that is why, in courtrooms, prosecution and defense can draw opposite conclusions from the same evidence (Smith and Wynne 1988; Lynch et al. 2011); in public enquiries responsibility for the *safety of technology* is passed between scientific, legal and parliamentary spheres (Wynne 1982); *testing* a new technology depends on projecting contexts of use (Pinch 1993); tests have to be carefully distinguished from *demonstrations* if demonstrations are not to mislead (Collins 1988); for scientists, everything changes, including how ambitious they are with their claims, once *funding* has been achieved (Pinch 1986). As intimated, this list could be much longer and a better scholar than I would blend it with much more work in recent history and philosophy of science.

36. In 2005, Latour published a summary of his "actor-network approach." For critiques, of which there are not many, see, for example, Collins and Yearley 1992; Collins 2012; and Bloor 1999.

for those of us in science studies, was that suddenly we were much more important within the wide world of academia, in particular that most articulate part of the world—the writers of books who tend to come from the humanities rather than the sciences. The ill, if one thinks it is ill, was that science studies became associated with that large, vague, antiscience movement known as postmodernism.[37] Worse, Latour's analysis, when it is not identifying scientific power with literary usage, stresses science's continuity with politics—Machiavelli being a role model—and this may have also have encouraged the increasing politicization of wave 2. It seems to me, and here I merely express the views of someone who has participated in this movement for forty years and occupied many of its major institutional roles, that taken together with the latent antiscience tendency in the humanities side of what in 1953 C. P. Snow called "the two cultures," a political critique of science was the most likely outcome, especially when combined with the prejudice for the "natural," as opposed to the technological, which came with the environmentalist movement. To be able to criticize science without any investment in its methods and practices is a very attractive prospect for many, though it counts as scandalous as far as most anthropologists and many other kinds of social scientists are concerned. Anyone who wishes to immerse themselves more deeply into this literature can easily locate sources additional to those mentioned here.

The Third Wave: Reconstructing Science after Wave 2

However fruitful wave 2 was, it does not offer reasoned guidance for action. Many of those coming from wave 2, nevertheless, have taken it to have implications for policy where science, technology and democracy come into contact. They argue that since scientific argument has been shown to be less different to ordinary argument than was thought under wave 1, then others, from the humanists who study science, through to the general public, have more rights to contribute to decisions affecting the technical heartlands of science and technology than they once had. But without being able to show how far these rights extend—how far into the heartlands they go—this is a recipe for reducing science to politics with no special place being reserved for scientists and other kinds of experts.

Wave 2 provides the sociological and philosophical tools for showing how each side of a scientific argument has its own rationality and could be

37. The other big influence on the salience of SSK were the attacks by the "science warriors," as mentioned above, which put this rather local and specialized activity onto a world stage.

sustained indefinitely—and would be sustained indefinitely if the logistics of social life allowed it and the champions of the heterodox stopped going to their graves. Wave 2, then, gives strength to any opinion on any matter expressed by anyone and simply slows down the mechanisms that make for closure. In some cases this would be a good thing, but it cannot be a good thing in all cases, for example, where lifesaving interventions are urgent.

The problem put me in a stark form by two anthropologists who worked in South African townships. They commended the fascinating work being done by wave 2 but asked how it could help them decide what to say to a man who assured them that AIDS could be cured by having sex with a virgin. Wave 2 can explain how the man came by such views and can reconstruct the logic of the man's position in a convincing way but it cannot provide a reason to tell the man to act differently.[38]

Wave 3, a position closely associated with me and my colleagues, is meant to provide such reasons without going all the way back to wave 1. Wave 3 tries to find a way of valuing science without abandoning the trees for the mythical forest and accepting that scientific claims are absolute in the way associated with wave 1. The trick of wave 3 is to concentrate on expertise rather than the truth of the matter.[39] It says that irrespective of who is right and who is wrong, more weight should be given to the views of those who literally know what they are taking about—those who know, through experience, the things of which they speak.[40] The technical branch of wave 3, therefore, is known as "studies of expertise and experience." The answer that wave 3 gives to the man from the township is along the lines: "I cannot prove with absolute certainty that your view of the world is wrong, but I can tell you that the people most experienced in studying these things are Western scientists and the opinion of nearly all of them is that sex with a virgin will not cure AIDS but is likely to transmit AIDS to the virgin. You should take their advice because they know what they are talking about and you have not studied the matter in anything like such depth: they are experts and you are not."[41]

38. For an anthropological account of the rationale of such a man's view, see Leclerc-Madlala 2002.

39. For the origins of Wave 3, see Collins and Evans 2002. For the first book length treatment, see Collins and Evans 2007.

40. See Collins and Evans 2007 for discussion of how to deal with heterodox expertises and the many other intricacies of this approach.

41. Of course, this could be used as a recipe for scientific imperialism, but nothing that is argued here is meant to justify the many, many pathologies of science. What is being justified here

Remember, one of the things that philosophers, historians, and sociologists are trying to do in both wave 1 and wave 3 (but not wave 2) of science studies is to tell a coherent story about why one should prefer scientific knowledge to other kinds of knowledge. There is a large body of thinkers who believe that scientific thinking should be accorded a central role in society, but according to wave 3, this cannot be a matter of going back to the unquestioned authority accorded to science under wave 1 nor to the failed models of science that underpinned it. Wave 2 has shown why such an approach is insecure. Furthermore, to say that we should value science because it works better is to give a hostage to fortune. In disputed science no one knows what works better, and in complex sciences like weather forecasting, climate change, or econometric modeling—the sciences that are of real concern to the citizen—there is no evidence that they work particularly well. Even medical science gets it wrong most of the time—we all still die!

In terms of the headline metaphor, wave 3 is trying to recapture some shape to the forest while not losing sight of the trees. Under this model the forest is no longer going to have sharp and clear outlines—it is going to be rough and ragged but not completely without form. As intimated, most of the general form was captured in the first-wave philosophers' and sociologists' "myths"; under wave 3 this form is retained but reconceptualized as a set of guidelines for how to walk among the trees rather than a clear path. Since there are streams, bogs, and thickets, there is nothing sure and reliable but there are hints that walking *this* way rather than *that* is what we do in science. The most top level guideline under wave 3 is "follow the expertise." The lower level rules are the old myths. To sum up wave 3 in a

is good science done with integrity—not bad science done with carelessness and driven by self-interest. Our ordinary moral sensibilities are sufficient to show why these pathologies are wrong; we do not need to destroy the whole of the idea of science just because some scientists are morally frail. What is assumed here is that any analyst invoking the experience of Western scientists would do it with the most careful eye on the value of the other's point of view. As is explained in Collins and Evans (2002, 2007), experts do not have to have formal training or scientific credentials; expertise can be experience-based. Nevertheless, this does not mean, and must not be taken to mean, that anyone's opinion as good as anyone else's. In the case of the townships it must not be taken to mean that the man's claim that sex with a virgin is a good cure for AIDS is experience-based, even though there is evidence that, for example, that "native cures" based on experience with herbal treatments and the like can make valuable contributions to our understanding of the world. None of this is intended to go against research that reveals the way some kinds of Western science has been dominated and biased by the gender and race prejudices of the societies in which it is embedded. It is, however, in conflict with the idea that science is merely a matter of one's standpoint. Expertise is real and our ambition must be to wrench it away from the biases of its practitioners.

single slogan, it is a shift from studying the making of truth—the topic of wave 1 and wave 2—to studying the nature of expertise. Expertise is exercised among the trees but since, in any sane society, the experienced are going to be treated as more expert than the inexperienced, we know who should be advising us if we want to follow our society's preferred way of finding a path out of the forest. And if the experts disagree, as they often do, we at least know who to include and who to exclude from the technical argument.

There is a nice paradox about philosophical justification—the more robust and lasting one wants one's philosophical story to be, the less powerful it should be. To provide a neat strong story about the nature of science—it is better at observing, proving, being objective, predicting, curing, falsifying—is to offer hostages to fortune; the exceptions will multiply from the outset. Tell a weak story, such as "scientists are the more experienced at studying this or that area of the natural world, and the claim is harder to deny.

Actually, the basis of what I am arguing here is still weaker. Why prefer experts? One can say only that it is a choice. Try to find a stronger foundation and clever arguers will find exceptions and counterarguments, but ask how one prefers to live and set the different models side-by-side and there is a good chance that most people who think about it seriously would find societies in which the view of the experienced expert on a technical matter is counted as worth no more than the view of the person in the street, abhorrent.[42] That is the foundation of the third wave.

To be consistent, this view implies that experts must be valued even where it is known that their expertise is not worth much. For example, we know that econometricians are pretty useless when it comes to predicting the future of a nation's economy—next year's interest rate, unemployment rate, and so forth—but on this model we must still value econometrics and, given that no one can do any better, we still have to rate the advice of econometricians on the economy a little more heavily than the advice of soothsayers, astrologers, or those whose views are solely the product of their politics or their standpoint.[43]

42. As is made clear in Collins and Evans 2007, an expert is not to be defined by their credentials, or the institution in which they work, but by their experience in working with the object or domain under discussion—that is why they are said, in a literal sense, to know what they are talking about. Of course, experts often get things wrong—we know that from the fact that they continually disagree. To repeat, the question being asked here is who should be included or excluded from the argument.

43. For the case of econometricians, see Evans 1997, 1999. One might add that in making policy decisions there is always a balancing act between the technical side of the equation and the

The name for the position under discussion is elective modernism, because one "elects for," that is, chooses, scientific expertise as the best expertise in respect of technical matters. It is a weak position. Start with this weak/strong foundation, however, and an interesting building can be erected. Wave 3 leads naturally to the careful study of the nature of expertise and has resulted in a new "Periodic Table of Expertises" with some new categories and a new experimental technique for investigating them.[44]

A new category that has emerged is "interactional expertise."[45] The category arose out of my puzzlement about how, after a decade of intense immersion in the field of gravitational-wave physics, I could talk gravitational-wave physics with physicists even though I could not *do* gravitational-wave physics in the sense of coauthoring papers or designing or building bits of apparatus. Interactional expertise is what is gained by prolonged immersion in the spoken discourse of a specialist community without engaging in its practices. The new investigative technique, the imitation game, starts with what was the inspiration for Alan Turing's "Turing test"—a parlor game involving humans alone (in Turing's case, men and women). In the modern incarnation of the imitation game, putative experts are asked questions by a "judge," who compares answers with those from a full-blown expert and has to guess who is who. Failure to identify the putative expert implies that he or she is as skilled as the full-blown expert at linguistic interchange concerning the expert specialism; this is to possess interactional expertise. At one time, the author passed a gravitational-wave imitation game test in which he was compared with another gravitational-wave physicist and found that other gravitational-wave physicists could not work out who was who. Further analysis of the notion of interactional expertise suggests it also fulfils a large and crucial and role within science. For example, gravitational-wave physics includes many specialists who cannot and do not engage in each others' practices but must understand them if they are to coordinate their actions; the me-

political side of the equation. We are saying only that technical experts have to be valued where the technical side of the equation is concerned. In the case of experts who are acknowledged to be weak, such as econometricians, the political side of the equation is likely to be given a much heavier weight. For more of these subtleties the growing literature on the Third Wave should be consulted. A guide can be found at the Expertise website, at http://www.cf.ac.uk/SOCSI/expertise.

44. Collins and Evans 2007.

45. For the growing literature on interactional expertise, see Collins 2004b, 2011b; see also the Expertise website, at http://www.cf.ac.uk/SOCSI/expertise.

dium is interactional expertise. The same applies to the interaction of gravitational-wave physicists and astronomers. And so on.[46]

What is going on in *Big Dog*, in those passages where I actually try to do some physics in a serious way, can be seen as a further exploration of the boundaries of physics expertise and its relationship to interactional expertise in (a narrow branch of) physics. In retrospect, it is an experiment that grows out of the third wave approach; it is a study within the domain of studies of expertise and experience and, this being a new domain, it is going to seem a bit odd at first. These matters are discussed further in the methodology chapter that follows.

Conclusion on Sociology: *Gravity's Ghost* and *Big Dog*

What are *Gravity's Ghost* and *Big Dog* in terms of the wave schema? Mostly, they are straightforward wave 2 science studies turning on the principle of methodological relativism. Each section looks at how the fact it deals with was made. The sections show the facts being made through debate about the meaning of certain measurements. These are to be made flesh in the form of the exact terminology of a text that the scientists had to produce for potential submission to the journal *Physical Review Letters*. The climax of this wave 2 element is, perhaps, chapter14, where it is argued that even the numbers that represent the outcome of calculations are based on philosophical judgments. The facts discussed in these books are, it has to be admitted, a little strange, in that they might have been fakes. I do not think much would be different had they not been fakes but, as we will see in the next chapter, it is possible to disagree.

I believe that readers will see from the wave 2 aspect of these books how the sociology of scientific knowledge works and how the world works. I believe the analysis shows that the old schemas that emerged from wave 1—the broad descriptions of the way theories are supported by or questioned by experimental facts, at how scientists are engaged in the falsification of theories, and so on, are as impoverished as myths. Like other wave 2 work, they will show that the world of science as it is lived has a richness that was scarcely ever glimpsed under the first wave. Science, as we can see, is as nuanced as the plot of any novel. It is not only a prolonged

46. The story about the author passing as a gravitational wave physicist can be found in Giles 2006; I doubt that I could pass nowadays though this does not mean I possess no interactional expertise. For more on interactional expertise as a central feature of science and of social life in general, see Collins 2011b.

adventure story, it is a tale of choices based on conscience, power, principle and tradition. What counts as truth emerges from human interactions as subtle and complex as those found in any sphere of human endeavor.

I use the word "myth" in the previous paragraph but it is a technical usage that has positive as well as negative connotations. The myths of wave 1 science are not without value. On the contrary, once thought of not as quasilogical descriptions of scientific practice, but as guidelines for living a good life in science, they are immensely valuable. As has been said, one element of wave 3 of science studies is to recapture the old wave 1 stories about science but recast them as guidelines or hints; they are no longer the "the logic of scientific discovery" but guides for action in the good scientific life to be lived under elective modernism. The envoi to *Gravity's Ghost* includes an attempt to capture something of this. The main wave 3 work is done in *Big Dog*, however. *Big Dog* will be discussed at greater length in the next chapter, where expertise is explored and conclusions drawn about the relationship between science and technology and the public.

17 Methodological Reflections

Big Dog presented some old methodological puzzles in a new and pressing form. As intimated, they have to do with the extent to which I became involved, or tried to become involved, in the physics itself. I did set not set out to become quite so engaged—it was not a deliberate methodological strategy—but Big Dog was exciting and I was sucked in. In this chapter I try to draw out the methodological lessons of the Big Dog experience. Part A concerns the details of my fieldwork and my subsequent interactions with the scientists. The big question is whether I went too far in trying to become like "the natives." Part B is primarily for social scientists; it is unrelated to gravitational-wave physics but was *inspired* by the Big Dog experience. In that section I draw up a general scheme describing the methods of the social sciences driven by the extent to which they attempt to base themselves in actors' categories.

A: The Author and Big Dog—On Going Native

In this kind of research it necessary to be sucked into the science. First, to elicit revealing reactions one has to be caught up. One has to argue and to confront; one has to be interested in the world of one's respondents if one expects one's respondents to engage with you. Respondents must become something closer to colleagues—people who are happy to argue physics, not just react as though filling out a questionnaire. Second, if one adopts the approach known as "participant comprehension," which is a philosophically worked-out version of what is more widely known as "participant observation," the object will be to become as like the actors one is studying as possible; the idea is to come to understand the society under investigation from the inside. The aim is to come to see the world through the same sets of categories as the actors one is investigating.

In my case there were limits. I am not—and did not become—a physicist in the sense that I will never contribute to formal publications, never help with building the apparatus, never do a real calculation or piece of theorization (barring the little dogs exception, discussed below), never make the sacrifices needed to tend the machines or analyze their output, never write the computer programs or develop the underlying algorithms, and never live the social life pertaining to a physics department or the commitment to a big science. The gaps, as I will argue below, are filled to a large extent by "interactional expertise." The aspiration that goes with the participatory method may help to explain why the main body of *Big Dog* is full of my opinions on the physics and why the text records a number of instances where I argue physics with physicists. These passages, I believe, represent good field craft whether or not they represent good physics.

The kind of sociology being done here, however, does not stop with participant comprehension. I now see that my paper on participant comprehension, "Concepts and Methods of Participatory Fieldwork" (1984), does not tell the whole story. It claims that the end point of participant comprehension is the acquisition of native competence, but the social analyst has to do more. The social analyst has to analyze sociologically, and this sometimes means analyzing in ways that are not natural to the members of the society being analyzed. Sometimes the social analyst has to disagree with the native members. For example, whenever the analyst is avoiding what I have called, above, "the siren call of truth as its own explanation" he or she is likely to find that scientists disagree. So disagreement with the native members is not only acceptable, but may be essential. The big problem is not whether I was doing my best to become like a native (that is integral with this kind of exercise) and not whether I sometimes disagreed with my respondents (this too is an essential part of the exercise) but whether I sometimes went too far in basing my analysis of Big Dog in my view of the *physics* rather than my perspective as a sociologist. If that is what I did then I certainly went native in a damaging way. Was my analysis sometimes *driven by* the actors' concerns rather than sociologists' concerns? In retrospect, sometimes it was.

I became involved as a "native" in at least three ways that might be construed as going beyond the rigors of the fieldcraft demanded by participant comprehension. First and worst, from the very beginning I made my own judgment about whether Big Dog was a blind injection and quite of lot of my initial analysis was based on that judgment. This was definitely a mistake. Second, I tried to do some physics and tried to persuade the physicists that I had a better idea than them about how to handle the question of whether the little dogs should stay in the time slide analysis. The physicists did not take this well, nor did one reviewer of the submitted manuscript. I tried to do the physics because I really thought I was on to something and because I was following my instinct about how to act adventurously as a sociologist. In retrospect, I think the instinct was sound even though not everyone will agree. Trying to do some physics was both an interesting "breaching experiment" and a good way to explore the nature and boundaries of expertise. The third way I became involved was in assessing the overall success of Initial LIGO. As the closing remarks to *Big Dog* show, I judged iLIGO not to have been a great success as a detector/discoverer of gravitational waves because the team failed to announce Big Dog as an unequivocal detection. I still think this was not unreasonable.

In what follows I am going discuss each of these aspects of my involvement in turn, admitting my fault in respect of the first one and trying to defend the other two. I will do my best to be fair to the scientists by quoting their criticisms, retracting my claims where I think they have been shown to be wrong or mistaken, but not entirely surrendering the analyst's perspective for the sake of good social relations. In what is now a longish career I don't think I have ever written *anything* of significance about science or artificial intelligence that some scientist or science-defending philosopher, or gang of scientists and science-defending philosophers has not told me is rubbish.[1] One has to become hardened to criticism if one is to survive. The countervailing danger is that one becomes so hardened that one stops being able to hear any kind of criticism even when it is justified and constructive. Somehow the analyst of scientific knowledge has to find a passage between these two extremes. One thing that helps is extensive quotation of one's critics' words and, as will be seen, in this

1. Given the duty to adopt the sociological perspective one can see how this is bound to have been the case. Historians who work with archival material might like to reflect on how their work relates to a real time analysis where one's interpretations are open to critical review by those one is interpreting.

section of the book I have again relied heavily on what the scientists said and wrote as a counterbalance to my interpretation of events. The lessons to be drawn from this exercise are about the insider/outsider dynamic—what happens when you poke your nose into a beehive—the risks, and benefits, associated with "going native," and something more about the nature of expertise.

This published draft of *Big Dog* differs significantly from the "circulated draft" that I sent to a number of the physicists for comment. The circulated draft was written in real time and only later was I advised that I had been getting some things wrong. Initially I tried to turn this into a virtue—"My real-time reflections are worth something as a historical document even though I had to qualify them later"—but a longer period of reflection has led me to change my mind. I have adjusted the main manuscript to eliminate the worst of the mistakes and I have shifted the physicist-critics' remarks and my responses to this newly written, third section of the book. Drawing everything together in a section which explicitly discusses the meaning and methodology of *Gravity's Ghost* and *Big Dog* has made it easier to admit the mistakes and reveal the lessons they provide while, making the main section of *Big Dog* crisper, cleaner, and, I hope, truer.

Was Big Dog Real or a Blind Injection?

The big mistake I made was to think I had an especially acute insight into the question of whether Big Dog was a blind injection. From almost the first moments of the Kraków meeting I was sure that Big Dog was not real and the earlier drafts included about eleven manuscript pages (nearly four thousand words), explaining why I was so sure. It was only six months later that I discovered that some of the scientists were much more ready to believe in Big Dog than I, so, in relying so much on my own view, I was being less than even handed in my description of events.[2] Nearly the whole of the eleven pages has now been deleted and the published version is much more open on the question of the reality of Big Dog.

Worst of all was my violating the antiforensic principle, which holds that the sociologist should not try to do detective work by exploring the motives of individuals but just record the range of available motives available to members of the group being investigated.[3] For the range of avail-

2. That some physicists had a much more favorable opinion of the reality of Big Dog than I did is revealed toward the end of chapter 15.

3. See p. 50 of this volume and pp. 412–13 of *Gravity's Shadow*.

able motives I had substituted my own internal state—Big Dog was an injection—and, still worse, tried to base my view on motives imputed to Jay Marx, the project's director. During the Kraków meeting I had asked Marx if he had looked inside the envelope and knew, therefore, whether Big Dog was an injection or not. He confirmed that he had peeped though I neither asked nor did he tell me what he saw. I then wrote to Marx on 2 September 2010:

> One prediction! If the envelope is empty it is inconceivable that you would allow S6 to be shut down. Hence the latest date for effective envelope opening is Oct 20 [2010].

Marx made no response except to acknowledge the e-mail.

To explain, iLIGO may have been able to detect a relative change in the length of its arms in the region of 1/1,000 of the diameter of a proton but this is only good enough to detect the very occasional gravitational wave—one every few years at best—and there was never any certainty that one of these would show up during the machine's lifetime. The upgrade to "AdLIGO," with its tenfold increase in sensitivity and thousandfold increase in the likelihood of seeing an event, had been scheduled; to begin building on time the existing LIGO had to be decommissioned on 20 October. But it seemed to me (and most of the physicists I was talking to at the time) that if the analysis of Big Dog had not been completed by 20 October, and the Dog was real, it would be irresponsible to take down the machine. There were two reasons. One was that it might have been necessary to run iLIGO for longer to establish more of a noise "background" for the signal as measured by time slides (this is explained on p. 9). The second reason, and the one that was seen as more pressing by the scientists, was that an element in the analysis of the reality of a signal is the exploration of any conceivable artifact in the machine that could mimic it. To explore such possibilities exhaustively requires that the machine be up and running. Therefore I concluded that if the device was shut down on 20 October, then Marx must know that the envelope contained the details of a deliberately injected signal and Big Dog was not the real thing. That is why I concluded that the effective date of envelope opening—the day we would know whether Big Dog was real—was not 14 March 2011 but 22 October 2010. In the event, the LIGO detectors were shut down on 20 October and I was 99 percent sure, thereafter, that origin of Big Dog was not in the heavens. Note the extent to which I drew on interactional expertise in gravitational-wave physics to reach that part of the conclusion that turned on technicalities and on some rough common sense to try to deduce Marx's motives.

But when I circulated the draft based on my certainty that Big Dog was an injection I was upbraided by some of the scientists. I still thought I had an ace up my sleeve in that these other scientists did not know that Marx had peeped inside the envelope. Therefore I wrote to him and asked him for clarification of the significance of his knowledge that he knew there was no real signal. He replied (15 April 2011):

> I would have shut down the run even if the event was real. Why? Because in my mind the most important priority was to get Advanced LIGO on the air as soon as possible and that shutdown date was determined by the readiness of Advanced LIGO to begin the installation phase. I've always taken very seriously the thought that 1 day of Advanced LIGO science running at design sensitivity is worth several years of science running with initial LIGO. As an aside, I'd also say that if the event was real, the chances of another one occurring in, say a few months of additional running of Initial LIGO after Oct 20 would have been, in my view, very small and so to try to capture a second such event would have entailed an extremely long delay in Advanced LIGO.

So, assuming that, in the absence of the kind of forensic investigation reported in courtrooms or psychiatrist's consulting rooms, the preeminent authority on a person's internal state is that person, we must take it that my inference from what happened on 20 October 2010 was insecure.

I say "insecure" rather than plain "wrong," because on 20 April I sent Marx another e-mail:

> Think back to Kraków: you have looked in the envelope and found it empty. The Big Dog is the best thing ever. Will you still keep the lid on the fact that the envelope was empty just for the sake of maintaining confidentiality when you know that you are going to get more assiduous work out of at least a subset of people if they know they are analyzing a real event?

Marx refused to speculate about what he would have done so there is an element of uncertainty.

Thus, I am not particularly ashamed about having drawn the poor inference if such it was. These are the kind of inferences and the kind of judgments we have to make about other peoples' states of mind if we are to live our lives. But, as a sociologist, I am ashamed of writing a manuscript that turned on individuals' internal states—mine and Marx's. That was bad field craft. I had gone native in a damaging way. Feeling I was one of the insiders I acted as though I was entitled to make the kind of judgment insiders make. I even concluded that, with my sociologist's insights, I was better placed to know whether Big Dog was real than the scientists themselves and, in consequence, I failed to explore what more than a subset of them was thinking and allowed my analysis to be driven by supposed insider's views. Seductive as it is, sociologists should avoid the trap. Use this incident as an object lesson for how easy it is for even an experienced fieldworker to fall into it. Always test a fieldwork style against the anti-forensic principle!

My Attempt to Contribute to the Physics of Little Dogs

Little dogs, remember, are spurious signals generated when one of the components of the Big Dog signal falls opposite a sufficiently energetic excursion in the opposite ribbon in the time slides. This is the way a noise background is generated and it is against this noise background that the statistical significance of any putative signal, such as Big Dog, can be estimated. The debate about little dogs concerned whether Big Dog, as an apparently real signal, ought to have been removed before the time slide analysis was executed. Leaving in a real signal risks overestimating the background noise because elements of a real signal are being treated as noise excursions and this will result in an underestimate of the statistical significance of the signal. Taking out an apparently real signal before doing the time slide analysis risks underestimating the noise if what was thought to be real turns out not to be real. There is a paradox: to know whether to take the Big Dog out before doing the time slide you have to know whether it is real but to know whether it is real you have to do the time slide analysis. I call this the time slide regress and page 213ff discuss the problem in greater detail.

It seemed to me that in respect of the little dogs paradox, physicists' normal way of thinking about things did not give them a special advantage because the problem was (a) unprecedented so they had no body of experience they could call on, and (b) the problem seems as much "philosophical" as technical; this meant that I did not have to move so far into the physics as the physics was, it were, coming to me and my professional world. The physics and the arguments surrounding it are quite complicated and detailed and so I have moved most of the material to appendix 4 retaining only the broader lessons for this main text.

In what is represented in appendix 4, I was breaking the rules; I knew I should not be trying to do physics as consequential as this. Sociologists will know of the famous "breaching experiments" invented by Harold Garfinkel.[4] Garfinkel asked his students to go home and act as guests in their family homes rather than as members of the family—they should be exceptionally polite and so forth. As expected, after a short time tempers were raised because unspoken social rules had been violated.[5] I knew I was breaking an unspoken rule in trying to do some consequential physics and I was not expecting too much sympathy. I knew I had long stretched the patience of the gravitational-wave community and the unspoken rule was, "we'll let you observe and talk with us as much as you like in an informal way but don't interfere in our official business" (on at least one previous occasion, early on in the fieldwork, this rule had been spoken out loud). The physicists work exceptionally hard dealing with intractable problems, using stunning mathematical skills learned after long apprenticeship, and they do not need an amateur nonmathematician who has not paid his apprenticeship dues wasting their time. Having learned my lesson early on, when it comes to really serious interchanges, I usually keep my mouth tight shut and my e-mails unsent and restrict the visibility of my attempts to do physics to one or two particular friends and acquaintances who are ready to tolerate my ideas and gently put me right. There are exceptions and when the stakes are not high, in private conversations, some of my physics is tolerated and even appreciated; there are illustrations in the main body of the book. Whatever, in the case of my analysis of the little dogs, the lesson of the breaching experiment was pretty clear; I had violated the outsider's etiquette and had to work very hard to get anyone to listen. That is the first lesson, and it is about how hard it is going to be to change widely accepted notions of what counts as expertise if that is what a wave 3 analysis of expertise, with its notions of experience-based and interactional expertise, is looking to do; the historical stream of expert activity has cut a deep channel in the terrain of social life and it is going to be dangerous to dive in.

A second point is more difficult to establish because it depends on the reader accepting that my analysis of the little dogs had some scientific

4. Garfinkel 1967.

5. An easier breaching experiment is to take items from others' carts in the supermarket—after all, no one owns the goods until after the checkout has been negotiated. One will find, nevertheless, that an unspoken rule is being violated. Where has it been spoken? Where are the notices forbidding it?

value. Note that it does not have to be the best possible analysis, or even be right, to be useful. As can be seen in these pages, and as anyone who has sat on a committee will know, not everything that everyone says has to be optimum, or even right, to be of value. All, or nearly all, of the physicists whose remarks are reported in this book said things that were wrong or offered solutions to problems that were not to survive the debate, but all these things helped form the eventual conclusion. What is without value is not the wrong but the crazy; it is the crazy that wastes everyone's time. As appendix 4 shows, I worked hard to establish that the physics/philosophy I tried to do was not crazy even if it was not optimum. At the moment of writing (February 2012), I'm still happy that no physics/philosophy claim I have made in this book is crazy. So I will leave myself hanging by the rope of physics argument round my neck, to strangle or have my weight supported by a passer-by or two.

If I am right about this, it is important to understand how it is that someone like me—a nonphysicist—can make potential contributions to physics, less than optimal though they might be. The question is central to the analysis of the nature of interactional expertise and its relationship to frontier physics and, therefore, to frontier science in general. To make proper sense of what is going on we have to separate out what is special to this case and what can be generalized to physics as a whole, to science and technology as a whole, and to the interaction of science, technology and society.

First, physics is a mathematical science, but it does not follow from this that every contribution to physics has to come from a competent mathematician. It is probably true that most research contributions to physics will turn on a mathematical presentation but we know that not all great research physicists were great mathematicians. It is said, for example, that Nils Bohr's brother Harald did the mathematics that formalized Nils's contributions, while Nils himself was not a good mathematician. Closer to home, Ron Drever is probably the most inventive of all the physicists associated with LIGO but is, by his own admission—which squares with the testimony of others—not a good mathematician.[6]

One of the reasons that physicists can contribute to physics without being good mathematicians, and this may well apply to the case of Bohr if not Drever, is that, as has been argued and illustrated several times in this book, physics sometimes comes close to the edge of philosophy. It was the

6. For the case of Bohr, see Beller 1999. For the case of Drever, see *Gravity's Ghost*, p. 561. For a general discussion of the role of mathematics in physics which includes a survey of the way physicists use mathematics, see Collins 2007.

philosophically paradoxical nature of the little dog problem that encouraged me to approach it. It is also the case, however, that (nearly fifty years ago, and not much exercised since), I learned some mathematics, achieving British "A-level" qualifications (high school final exam) in physics and math; these are sometimes said to be equivalent to American first-year undergraduate-level physics. So it would be misleading to say I have no formal mathematics and physics at all, though my arguments concerning interactional expertise and the like would be cleaner had I none. I have to admit that the remnants of those half-century-old, vestigial abilities may have helped in the few calculations, or quasicalculations, I have done in the book (or helped to lead me astray). I also have to think about statistics in a serious way to do my recent large scale research on imitation games though the area of statistical expertise is narrow.[7]

Above all, I have been talking physics with physicists for about forty years and I think that has kept me up to speed on physical concepts and, perhaps, helped me to think like a physicist. This is what I claim gives me interactional expertise in physics or, at least, in gravitational-wave physics—the interactional expertise that was at one time demonstrated by my passing an imitation game. It is these things combined that have put me in a position to make noncrazy arguments, if such they are, amounting to potential contributions, relating to the heart of the Big Dog debate.[8]

It is also important to understand that only a small bit of what goes on in respect of decision making in physics and all other sciences is technical in the sense of drawing directly on the mathematical and technical accomplishments of physicists. Attend a decision-making meeting of LIGO—even a technical decision-making meeting—and one does not

7. For those who know the "Periodic Table of Expertises" (Collins and Evans 2007) one could call this statistical knowledge "referred expertise" and the philosophical thinking that I do in my day-to-day sociological work serves the same purpose in respect of the Big Dog debate. Such knowledge of physics as I have from (high) school is a more direct kind of expertise and this mixes with my interactional expertise.

8. One anonymous reviewer wrote in respect of the initially submitted manuscript: "Collins does not have anything like the requisite expertise to intervene [in the physics]." In the normal way this might be fair, but experience is a good guide to expertise only if it is experience of doing something similar to what is at stake (Collins and Evans 2007). In this case there was no precedent so there was a gap between the problem as faced and problems that had been faced in the past. It may be that the very different judgments that the physicists themselves were making is an indicator of this divorce from the past and created a little more space for an outsider to think about the issues. In respect of my ability to write about this entire complex area of physics, one of the physicists was kind enough to tell me (also after the 14 March denouement), that I needed to invent a new term that goes beyond "interactional expertise" since in my conversations with physicists about the Big Dog analysis I was making a contribution not just evoking the arguments.

see mathematics being done. One sees arguments of the kind that have been explained and illustrated in this book. That is what most of the talk is like—you, reader, have seen it for yourself. That is one of the reasons that interactional expertise is sufficient to make potentially useful contributions—and the point stands whether or not I personally succeeded in making useful contributions. Therefore, if the argument is correct, we should be ready to spread the basis of input into technical debates so that it covers those with interactional expertise.

None of this is an attempt to get my feet, or those of people like me, under the table of research physics or any other frontier science. It is an attempt to demonstrate a point about the boundaries of technical expertise through the "hard case" of a technical research science. The policy implications do not lie with physics and the like, they lie in the courtrooms, policy-making committees, and public inquiries, and in medical and environmental debates, and any other place where science and technology are forced into contact with the life of ordinary citizens. It is how the boundary should work in those circumstances that is being explored by experimenting with the boundary in this much more demanding case. The point I am trying to make is that technical understanding can be found beyond the body of the formally qualified. The occasional interactional expert and the, less rare, experience-based expert, may have something to offer even where the substance of the debate is technically esoteric.

Equally important, or perhaps more so, is the demonstration of just how hard it is to acquire expertise, including interactional expertise that is good enough to make a contribution that escapes the crazy. The jury may still be out in respect of the significance of my contributions but even to get to the point where it seems not totally unreasonable for me to think I had a chance of making a contribution has taken decades of close contact and deep discussion with the relevant physics community. In recent years it has become fashionable, at least in certain social science circles, to talk of "lay expertise." This book demonstrates that there is no lay expertise. Nonqualified and nonaccredited persons may have expertise but they must have gained it either as a result of lengthy specialized experience or through occupying a rare social role such as mine. The lesson is that a moderate widening of the base of expertise to include experience-based and interactional experts is possible, and may be desirable, but the widening process cannot go too far if the notion of "expert" is to be preserved. The idea of cohorts of ordinary people, or people from the arts and humanities, joining expert committees as full members, and contributing to technical decisions, is not utopian but dystopian. Insofar as such processes have been endorsed by governments it can only be to sooth a public which feels impotent in the face of science and technology; it cannot be a matter of improving technical decisions.

The Scientists Criticize My Account of iLIGO

Near the beginning of *Big Dog* I make the claim that the large number of verbatim quotations anchors my account and prevents my obsessions and interpretations from floating completely free. Inevitably, as I point out, I am still the editor—the person occupying the "bully pulpit" (see page 129)—and I decide what gets quoted. As a further counterbalance to my power I am going to quote still more extensively from three senior scientists who were very critical of certain of the claims found in *Big Dog*.[9]

The scientists who were kind enough to put the work into a line-by-line reading of the circulated draft were Peter Saulson, Gabriella Gonzalez, and Stan Whitcomb. A few years back, Saulson held the position of spokesperson of the LIGO Scientific Collaboration for two terms, Gabriella Gonzalez is newly elected to the same important role, and Stan Whitcomb is one of the most senior and most authoritative scientists in the LSC and has held the posts of deputy director or interim director at various times. The roles that these three occupy, or have occupied, fall just below the directors of LIGO and Virgo and can be thought of as equivalent to deputy directors. LIGO personnel, who will have had no trouble identifying them when they were represented by pseudonyms in the main text, will realize that these three scientists span the range of opinions expressed during the debate and that Saulson is a close friend as well as a "physicist," so it is a specially serious matter when they are united in telling me that I am wrong.

iLIGO as detector of gravitational waves.

The most heartfelt and unanimous agreement concerned my claim that iLIGO was a not a great success as a gravitational-wave detector because it did not claim "discovery" even for a signal as strong as Big Dog.[10] I cannot

9. In fact, in this last chapter I can deal with only a small number of the remarks made by the three scientists. I have chosen those that would require the most significant changes to the argument had I incorporated them in the text. The many other remarks that touch on less significant points have either been incorporated in the text, sometimes with a precautionary footnote, or I have responded directly to the scientists in question.

10. In the circulated text I said it was "a failure," but the more moderate tone of this text is a more accurate way of expressing the matter. (Thanks to Trevor Pinch for pointing out to me that when my own research does not go exactly to plan I do not call it "a failure.")

go back on this, however hostile the physicists' reactions. Aside from the matter of analytic integrity, I have twice "nailed my colors to the mast" on this point—on pages 717 and 731–732 of *Gravity's Shadow* (note that the crucial interviews and analysis reported there took place in the early 2000s, long before LIGO was "on air"), and pages 134–35 of *Gravity's Ghost*; I have to take responsibility for those opinions.

There are three parts to the disagreement with the scientists. First, they claim that the Big Dog story leads to a conclusion opposite to the one I draw; they say the incident shows that LIGO *did* detect the event and this proves it was a *successful* detector. Second, they say that detecting gravitational waves was not an expectation for iLIGO so that not detecting them would not indicate any kind of failure. Third, there is the question of whether the possibility of Big Dog being a blind injection renders the incident a fair test of LIGO's capabilities in the first place. I now discuss these in turn.

Was Big Dog detected?

It is strange to find that after a book-length discussion it is still possible to disagree about whether Big Dog was detected or not, but possible it is! Here are the most uncompromising remarks put by the scientists:

> **A:** What I took from [Juniper's] papers was the idea that a paper can be (and often is!) considered a discovery paper even when the title says "Evidence for . . ." And we were stronger than that, making a real discovery claim in the first sentence of the abstract. I wouldn't even call it "evidential collectivism," but rather a literary strategy to subvert the 5-sigma police. We were all thinking of this as a discovery paper. Why do you feel it is justified to hang your whole take on the episode on the wording of the title? As your own literary strategy to focus your book, that is fine. But to draw such sweeping conclusions about success/failure, whether iLIGO should have had only one site, etc., after deliberately blinding yourself to any information presented outside of the title, or any examination of how physicists have learned to read paper titles, seems to me to be bad craftsmanship on your part.

> **B:** Your arguments using the coincidence numbers to demonstrate that LIGO couldn't "detect" anything ignore other features in the data. If the Big Dog were real, and only 10% closer (only 30% less likely than the actual Big Dog, then both the L1 and H1 signals would have been 10% larger, and in fact the L1 Big Dog would be its loudest event (as was the H1 Big Dog) and there would be no little dogs as big as the Big Dog.

> [But later, B wrote:] . . . I have advocated pushing for the strongest statement that I thought justified. However, with a single event, I find it hard to go beyond the "strong evidence" level.
>
> C: I disagree with this very strongly—"claiming" discovery and "discovering" are two different things. If the Big Dog had been real, and advanced detectors detected several more similar events, nobody would doubt the Big Dog was real and Sept. 16 would have been the date of the first detection. If, however, Advanced LIGO run for 10 years at 200 Mpc [megaparsecs] and saw nothing, nobody would believe in the Big Dog—just like Cabrera's monopole. It wasn't about whether the event looked real or not, it was about claiming it as real from the outset that we disagreed on.

Starting with B and C, the disagreement turns on what can be made of a single event. I quote Willow (p. 289), surmising that even if Big Dog had been twice as loud the group would still not have presented a discovery claim and that this is true seems clear in B and C's remarks in respect of the evidential fragility of single events. Furthermore, the Blas Cabrera episode is mentioned by all three scientists as the appropriate precedent. On reflection, however, I am not so sure that the monopole is the right comparator and this is brought out by what C says.

If Cabrera's claim had been confirmed by the later appearance of further monopoles, however long it took them to appear, then he would have been retrospectively, and quite properly, credited with having made the discovery in the early 1980s. This case is different, however. Advanced LIGO is not meant to confirm the existence of gravitational waves, it is meant to do gravitational astronomy—and the only good reason for spending a billion dollars on detecting gravitational waves with ground-based instruments is the promise of gravitational astronomy; the existence of gravitational waves is not seriously in question and has already been demonstrated by the Hulse/Taylor binary-star results and similar. When gravitational astronomy comes along it is true that, had Big Dog been real, it would have been confirmed retrospectively, but who would care? The initial sighting with a terrestrial detector becomes redundant if gravitational astronomy is already in place—all the interest would be in the astronomy. The crucial point is that Initial LIGO has not, in its lifetime, independently proved the existence of gravitational waves via their interaction with a terrestrial detector, and that would still be true. This is what I mean when I say that iLIGO *could not* unequivocally discover gravitational waves.

At the time LIGO was being funded and built, the claim that it plausibly could detect gravitational waves was based on the possible occurrence of something like Big Dog—something extremely unlikely. The claim was not based on the simultaneous appearance of an electromagnetic source or the appearance of multiple sources—one, lucky, purely gravitational event was at best all that could be reasonably expected. The Big Dog incident has shown that such a thing was unlikely to be forthrightly claimed to be a discovery.

To make the issue concrete, a Big Dog–like event was not going to justify the article in *New Scientist*, which formed the criterion for the Ladbroke's bet (p. 140). Also, it was not going to justify, *on purely scientific grounds*, the building of two detector sites rather than one (see *Gravity's Shadow*, pp. 733–34ff); we could have waited for Advanced LIGO before building a second site if there was going to be no confirmed discovery until then. So using either of those two criteria, if the envelope had been empty, Big Dog would not have counted as a discovery.

Of course, there are other criteria for what counts as a discovery. I cannot argue with A. I accept that many of the scientists believed it was a discovery and were prepared to talk of it in this way in their conversations with each other. Furthermore, just as A points out, the first line of the final version of the abstract did say "observation," which amounts to "discovery":

> We report the observation of a gravitational-wave signal in data from a joint science run of the LIGO, Virgo and GEO 600 detectors.

It was indeed bad field craft on my part not to have reported this sentence from the abstract in the main body of the book; the reason was my obsession with the title and the only excuse I can offer is that I am still practicing my craft here and now—interaction with scientists being a vital part of the craft—so it has been reported at the last. It is clear, however, even from those few questions that were still being asked on 14 March, that it was not thought by all scientists that the abstract had fully discharged the community's duties in respect of discovery. What the strong abstract along with the "evidence for" title would have meant for the way Big Dog was received in the wider community in respect of the two criteria I have set out above—the Ladbroke's bet and justification for building two detectors—we cannot know for sure. So scientist A's reading is not ruled out by my analysis, but not ruled in either, and perhaps the most interesting conclusion we can draw is that the notion of discovery is fraught with difficulties. As has been pointed out, we still don't know if the draft paper was a discovery claim or not. And for those who want to claim it was a discovery paper, what is to be made of the repeated insistence by some of those most deeply involved in the debate that a single sighting cannot comprise a discovery? This is a good question for future analysis by philosophers, historians and sociologists alike: under what circumstances can a single sighting of a novel event count as a discovery and what does this mean for the process of detection?

Was Initial LIGO expected to detect gravitational waves?

One of the scientists active at the time LIGO was being proposed and built wrote as follows:

> (i) There is a theme running through the manuscript that somehow Initial LIGO was approved and built as a detection instrument while you can demonstrate that it can't "detect" anything. Aside from the issue of what it means to "detect" something . . . it was not known at the time that LIGO was approved that it could not detect anything and, indeed, it is not true that it couldn't detect anything. At the time LIGO was approved and built, there were many aspects of the problem that were unknown—rates for events had larger error bars than they do currently, the glitch rate for the detectors was completely unknown, and so on. So it was approved and built knowing that it might not detect anything, but that it was plausible that it could.

I feel more confident about my response to this charge because the circumstances of LIGO's approval are set out at great length in *Gravity's Shadow*. Memory is unreliable and, in this case, even archived documents may be unreliable, but the advantages of contemporaneous analysis is that at least some of what people were talking about at the time LIGO was funded has been captured.

Everything that scientist (i) says is correct. No one could have been sure about the glitch rate. Worse, I report in *Gravity's Shadow* that many scientists distant from the detector community were sure it would not work—the vacuum system was too ambitious and so on—and I report that on the eve of the switch-on it was even doubted by some well-placed persons that the device would ever be made to lock. But if any these things had turned out to be serious problems then Initial LIGO would have been an unambiguous failure and we can be fairly sure that there would have been no Advanced LIGO. The only acceptable reason for failure at the time LIGO was being funded and built was unknown rates for events. It has turned out that there were no detectable events in the lifetime of iLIGO and no one who understands the history of the device can consider it a failure for this reason (though they might consider that it was overhyped). The question still remains, however, whether the approval (e.g., of two sites), was based on the idea that it was *plausible* that it could make a detection given the astrophysical luck that would be needed. As I argue in the main text, Big Dog was an artificial version of that astrophysical luck (and was put together with just that in mind). Yet, aside from scientist A's argument (above), the outcome suggests that it was not as *plausible* a detector as it was thought to be when it was being funded and built. That change in plausibility has been brought about, in part, by the way Big Dog and the Equinox Event were analyzed and interpreted. The contemporary record at the time LIGO was funded and constructed, at least as set out in *Gravity's Shadow*, says nothing about single events not being counted as detections—that wasn't part of the plan. And it would be surprising if it did say such a thing, because a single event was the best that could be hoped for. And that such a thing was thought plausible by the majority of scientists even up until 2004 can be seen in the fevered response to the Ladbroke's bet (p. 140).[11]

It seems to me that the argument over this point also has something to do with the different roles of social analyst and physicist. When Initial LIGO was funded "a ratchet clicked over"—the money had been won and it became the job of the community to do the best science they could with it without worrying too much about what had caused the ratchet to click. When Advanced LIGO was funded the wheel moved on another notch and attention was once more redirected. This is how funding works—there is a huge contrast between the ordered world of the grant application and the disorderly world of the actual research. In the world of the grant application everything is still taking place in the imaginations of the scientists and the imagination is a weak when compared to the mischievous ingenuity of the real world. As the real research unfolds expectations are inevitably adjusted and these altered expectations feed back into memory. These processes will only become more marked as the funding situation gets tougher and applicants have to promise more and more even to get started.[12]

11. It can also be found in the many promises made by such gravitational wave detection enthusiasts as Kip Thorne.

12. The mismatch between grant proposals and actual research is, of course, what made it so easy for LIGO scientists to tell "the Italians" how to handle their data at a time when they had no data of their own; it why things are so much more difficult now that LIGO scientists do have data coming in.

There is nothing wrong with this redirection of attention; now that gravitational astronomy is "around the corner" it is only proper that attention has changed from the past to the future. Indeed, this is a justification for Marx's shutting down of the detector on 20 October irrespective of what was in the envelope. Without a change of research focus the money would be misspent.

The social analyst, however, has to ignore the click of the ratchet. One reason is that the social analyst wants to provide materials to make possible broad comparisons between the few long term case studies of the sciences that we have. If we are going to understand how science and science-funding relate to each other we have to know how initial promises compare with outcomes even if those initial promises have lost their purely scientific significance; that's why I have to maintain the coherence of my story and make the current account match with what was said in *Gravity's Shadow*.[13]

Was Big Dog *a fair test?*

The main body of *Big Dog* was written in accordance with the conviction that scientists were arguing very much as they would have done if Big Dog was real. I noted that the arguments were fierce and I noted that after the end of the debate I found that some scientists were very disappointed that it was not real. But one of the scientists responded to the circulated draft as follows:

> (1) You seem to forget, in the conclusion, the possibility that you discussed earlier in the book, that we didn't really fight this thing hard enough because we were play-acting in a BIC [blind injection challenge] rather than believing in our hearts that this was likely real. Why is it OK to neglect this? We were stuck on a hard (i.e., statistically confusing) point of what was legitimate in data analysis—whether little dogs belonged in or out. (Note that we didn't have outside resources to fall back on here, since our "peer experiments" don't make use of coincidence in anything like this way.) . . . Given that . . . none of us saw the stakes as high enough to strain friendships (personal friendships within the collaboration, LSV-Virgo friendship

13. Thus, the relationship of initial promise and ultimate expectations of LIGO follow closely the relationship found in the case of solar neutrino detection when the number of neutrinos promised dropped steadily as the instrument came closer and closer to going on air (Pinch 1986). Furthermore, in that case too, the scientific significance of the eventual finding was quite different to what was promised in the application.

at the political level), why turn the question of stopping-at-4-sigma vs. going-for-5-sigma into the life or death matter that you make it out to be; I'm referring specifically to your using it to argue that iLIGO could never have made a discovery, which I don't accept.

I cannot deny this point either. As explained in the main text, I also discovered shortly after the envelope was opened that at least some of the scientists had decided not to make as many argumentative waves as they might have done because certain contingencies had led them to the knowledge that the putative gravitational wave was a blind injection. I have to accept that scientist (1) is telling me that the knowledge of the possibility of it being a blind injection was sufficient to have made at least that scientist press less hard than would have been the case under other circumstances. So the reader should bear this in mind—maybe the title would, finally, have boiled up to full-blown "discovery" had the blind injection possibility not been the coolant. The reader can only take each argument found in the book on its merits—it seems to me that there are some arguments that are vulnerable to scientist (1)'s point and many that just aren't—for example, the epistemological sentence argument, the vulnerability of a single observation argument and Willow's point that doubling the signal would have made no difference.

Was Initial LIGO a success?

Note that I do not claim that iLIGO was not a success. I think it has been an astonishing success in terms of organization and construction—to be associated with the building of a machine that can detect a change in the length of a four-kilometer arm of a thousandth of the diameter of a proton has been one of the most exciting, elevating and formative experiences of my life; I have been lucky to have been able to spend so long in this world. Furthermore, as I say in *Gravity's Shadow*, "If Advanced LIGO is funded, then LIGO I will have done its job of building a platform of political, technological, and scientific credibility from which the leap to the next level of sensitivity can be made" (p. 737; anticipating Marx's 20 October shutdown). Finally, one of the scientist respondents pointed out to me that LIGO (so long as you believe the upper limit results of a machine that has not yet shown it can detect a real gravitational wave), has provided a lot of evidence about what is not in the heavens. I am making only one minor claim about what is probably the least important feature of iLIGO: it was not a great success as a detector of gravitational waves. It was not a success as something that could lead, in its own lifetime, to what would count as an unambiguous direct detection of a gravitational wave—at least, not if we take the unfolding of Big Dog as a good example of what iLIGO could do. Setting aside some reservations that I will add below, this is what the blind injection exercises have shown.

Too Much Going Native?

The method of participant comprehension is especially open to the danger of going native. It is whether the right balance between participation and distancing was achieved in *Big Dog* that has been discussed in this methodological section. I have said that I had the balance wrong in my initial treatment of whether Big Dog was a blind injection that was found in the circulated manuscript, but I hope I have it closer to right in the published version of the book. Some of the scientists thought I had the balance wrong in my attempt to do the physics of the little dogs (appendix 4), but I claim that I had it about right even though I went much further than usual in this unusual case; it led to my learning a lot about both the science and the nature of expertise. The scientists nearly all think I had the balance wrong in the case of my discussion of Initial LIGO's ability to detect gravitational waves. I say LIGO was not an unequivocal success at detecting gravitational waves. I think I have it right with the following reservations: (a) if it had not been the case that some scientists knew Big Dog was a blind injection, and others were highly conscious that it might be, the demand for "discovery" might have been more fierce: (b) likewise if Advanced LIGO had not been pushing back on Big Dog from the future. Whatever, the fact is I decided to step in and make my own assessment of the success of the program from my analysts' perspective and my assessment does not match the actors' assessment. I might be wrong for reasons *a* and *b*, but I do not think the problem arises out of my stepping back too far from my respondents' view or going too native in wanting to contribute to them. In this case it is sociology that is being done, not anthropology or ethnography, and in this case the sociological analyst has to preserve the right to make an independent judgment. I made an independent judgment in the case of whether Big Dog was a blind injection—that was a wrong thing to do and led me to do bad social science. I made an independent judgment of whether little dogs should be kept or thrown out and, though resisted, it seems to have been useful in drawing out features of the world that would otherwise have remained hidden. I made an independent judgment of whether Initial LIGO was a success as a terrestrial detector of gravity waves—that was the right thing to do even if my judgment is wrong.

Does *Big Dog* Contradict *Gravity's Ghost*?

Finally, by going too native I may also have created a problem for my own analysis rather than my relationship with my respondents. As can be seen, during the six months of the Big Dog debate I found myself impatient with the scientists' conservatism; I wanted them to be ready to announce a discovery if the envelope was empty. On the face of it, there seems to be some contradiction between this and what is written in the envoi to *Gravity's Ghost*. There I advocated caution in science and only a gradual moving toward what counts as truth. Is this a *volte-face*?

I think it is more a consequence of the difference between The Equinox Event and Big Dog. The Equinox Event was a marginal signal and the argument set out in the envoi was that scientists should be ready to talk in public about how they work with marginal signals. I argued that how scientists work with disputable evidence could be the most important thing we can learn from science. Big Dog, on the other hand, was about as good a purely gravitational-wave signal as Initial LIGO was going to see. The position put forward in *Big Dog* is that, given such a strong signal, *philosophical* arguments about the irreducible uncertainty of scientific knowledge should not have been invoked. A public display of uncertainty may have been appropriate in the case of the Equinox Event but inappropriate in the case of Big Dog.

Science and philosophy (or sociology), are different. In the envoi it is argued that to hide every scientific proto-finding that cannot reach the level demanded by evidential individualism is to lose what might be science's most valuable contribution to life in the twenty-first century—the demonstration of how to apply expertise in conditions of uncertainty. Under other circumstances, however, science must remain a producer of a kind of certainty that exceeds what is possible in most other cultural endeavors and comes only a little way behind the strength of universal moral imperatives. For science to go down the road of logical positivism (p. 249), always stressing its weakness, would amount to a transmutation of the social elements—science would become philosophy. It is the job of science to take risks so as to generate interesting *synthetic* propositions. Without risk, nothing meaningful can be said about the world. Skeptical philosophy will always trump science, but skepticism does not tell us how to live our lives.

There is already a sociology/philosophy of science which insists on the uncertainty at the core of any scientific finding—this is wave 2. It is the job of this sociology/philosophy to work out what follows for technological decisions in the public domain—to develop an appropriate "citizen's view" of science—this is wave 3. But science itself has to continue to make the best discoveries that can be made. The standard for discovery is set within the evidential cultures of the sciences; the evidential cultures of sociology/philosophy work according to different rules. Science would no longer be science if scientists were to begin to state, on a regular basis, "no discovery is ever certain so scientific knowledge has no special warrant."

Of course, scientists are citizens too, and from time to time will reflexively examine their own work through the citizen's lens. But the citizen's lens can only blur the vision needed in the laboratory. The paradox is expressed in the Preface of *Gravity's Shadow*:

> What we see described here is the mundane but it is, nevertheless, an example of the very best that humans can do. There is one more irony—to do the mundane as well as we can do it we must act as though we are divine. Here, as in so much of life, the "is" of how things are done, does not provide a guide to the "ought" of how we should try to do them. . . . I want [this] to be a book that safeguards science against its critics by setting more reasonable expectations for what science can do even while maintaining our aspirations for how scientists do it.[14]

There are a new set of problems facing science in an age when the overriding urge of many of its critics is to democratize every technological decision. The arguments that underpin the critics' position come from philosophical skepticism, the sociology of knowledge and, more recently, literary criticism, and that is where they belong. Without science as our certainty maker and, by that fact, risk taker, we would no longer have the world we know and value.

14. *Gravity's Shadow*, p xvii. The sentiment, incidentally, can be found one of my papers which was written me as early as 1982: "it is no part of the natural scientist's job to consider the social construction of scientific knowledge" (Collins 1982, p. 140). In that paper I go on to say the same applies to social scientists when they are doing social science. The strong view expressed here in respect of the physicists actions would be a *volte face* only if I had *not* followed my own prescription; in fact, wherever they could bear the weight, I have always expressed my sociological findings in a forthright way without reservation and argued against those, such as the "reflexivists" who insist on a reservation and a backward look at every turn.

In sum, *Gravity's Ghost* and *Big Dog* are contemporaneous descriptions of closely observed science, they are analyses of the making of scientific facts under the model of the second and third waves of science studies, they are detective stories, and they are explorations of how good lives are to be lived.

B: The Methods of the Social Sciences—Analysts, Actors, and Expertise

As already mentioned, even setting aside the strong version known as participant comprehension, a long-standing tradition in the philosophy and methodology of social science takes it that sound analysis of social things can be accomplished only by working through the perspective of the actors. The way social beings interact depends on what they experience as the objects belonging to their world. There is no equivalent in the natural or life sciences. In these sciences the objects that are found in the world—quarks, neutrinos, cells, mitochondria, insects, and at least the lower animals—are defined, demarcated, and classified by the analysts and the analysts alone. Mortgages, witches, and myriad other such objects as order the lives of groups that exhibit cultural differences are, in contrast, defined, demarcated, and classified by the actors whose ways of living created them. Natural and life scientists have only one set of classifications to contend with—that established within their own scientific communities; social scientists have two sets to contend with—their own and those belonging to those they study. In the accepted jargon, for natural and life scientists, there are only *analysts' categories*, for social scientists, analysts' categories have to be based in *actors' categories* for it is actors who construct the world that is to be studied.

There are, however, academics who work in the area known as social studies of science who seem to have found ways to study science that involve no deep engagement with the science itself. They do what is almost the opposite of participant comprehension. This kind of unengaged analysis is accomplished by restricting the gaze to what I will call a cross-section of the activities of scientists. One of the most revealing and successful versions is the literary or semiotic approach. It begins from the insight that much of the activity scientists is recording findings and drawing together those "inscriptions" to create published articles. The semioticians of science feel they can do their work by analyzing only the written output; there is no need to get any closer to the world of the actors.[7]

Gravity's Ghost and *Big Dog* both turn on the writing of a paper for publication. Much of *Big Dog* is about the exact wording of the title of a proto-publication. Does this not show that science is, at its heart, a matter of literary production and that science studies should be, at its heart, about literary activity and the study of signs? Is it not, then, the humanities that hold the key to science? And yet, the scientists, the actors who construct the world that is being analyzed, do not think that the act of science in which they are engaged is the act of writing papers even though their activities are targeted on writing papers. The mistake made by the literary analysts is to describe the transformations of text associated with scientific creativity *as* the science rather than as an aspect of science. If the science of Big Dog *was* the production of the paper the scientists could simply write it with whatever title they preferred: Big Dog would be a "discovery" or not according to literary preference. The resolution is, of course, that though there is intense talk among the scientists of literary style and content, and though the paper was the end point of the activity, most of the work of constructing the paper consists of building machines, endlessly discussing the meaning of their output, spending months having computers produce time slides, working out the time slides' significance, deciding whether the little dogs should survive, and so on. The act of scientific creativity is multidimensional, whereas a determined semiotic approach treats it as one dimensional—as though the carapace of words that is the visible part of the science *is* the science.

Look at it this way: the publication of the *Physical Review* also involves print and paper. One cannot have the publications and the scientific findings (at least, not before the internet) without making paper. Paper making is, therefore, part of science and part of the proto-discovery of Big Dog. But this does not mean science is paper making. Ironically, the external view—the cross-sectional view—is the view of typical of the natural or life scientist with no responsibility for what the natives think. Only a *natural scientist* of science could legitimately concentrate on paper production (in whichever sense), ignoring its meaning for the actors.[15] I am arguing that though the literary and semiotic approach is insightful and has led to many interesting findings (I will explain below), to understand science properly one has to do everything one can to acquire the actors' perspective and this would not allow science to be described *as* either paper-making or the shuffling of inscriptions.

As has already been intimated, though social science of the kind that is done in these books starts with the actors' perspective, it does not finish

15. If one was an entomologist one might well consider scientists to be one of a class of paper makers; and if one was a bibliometrician one might think the same in the other sense of "paper."

with it. *Sociological* analysis, even in the case of interpretative sociology, requires a stepping back—"distancing," as it is called—so as to gain an analytical perspective. Distancing, however, comes only *after* every attempt has been made to acquire the actors' perspective.

Interpretative sociology requires, then, a balancing of the painfully acquired actors' perspective and the creative analysts' perspective.[16] One implication of what is called "going native" is that the analyst becomes so absorbed in the actors' perspective that it becomes impossible to step back. The problem of going native, which has been discussed at length above, is the opposite of the problem of the external, cross-sectional, analysis.

Following this path and the reflections on the analysis of Big Dog I want to suggest that the methodologies of the human sciences can be understood in terms of the "conceptual trajectory" of the research in respect of actors' versus analysts' perspectives. By "conceptual trajectory" I mean the way the analyst moves between actors and analyst's categories over the course of the entire project culminating in the final conceptual synthesis. The conceptual trajectory may not be congruent with the temporal unfolding of the work though, as intimated, in interpretative sociology a fair degree of acquisition of the actors' perspective has to come first.

There is, alas, a complication—one that does not seem to have been discussed before.[17] It is pointed out above that disagreement with the actors is inevitable in a subject such as sociology of scientific knowledge. It follows that the analyst has to say that the actors do not always know, in a self-conscious way, what their categories are even though it is these that organize their world. For example, most scientists do not accept that their world is ordered in the way sociologists of scientific knowledge say it is (at least, not without a lot of persuading). Consider the following extract from an e-mail of 13 March 2001 sent to me by one of Joe Weber's early antagonists; he is referring to my early analysis of his controversy with Joe Weber:

> I do not consider you "a trained observer of human behavior," so far as concerns the gravity wave field. Science and technology move ahead through advances in instrumentation and publication of results. Not through gossip

16. In the words of Peter Berger (1963), it is a matter of "alternating" between the two. In the early days of the "science wars" hostile scientists claimed that social scientists could not analyze science unless they were full-blown scientists themselves. But the history of the social analysis of science shows that such a level of understanding is not necessary; the concept of interactional expertise probably shows why.

17. I noticed it only around 2007 when putting together a paper for a conference (Collins 2008).

or "science wars" or deep introspection about what the other guy is thinking or what one is thinking oneself.

This, then, is the actors' perspective as far as some, or most, scientists are concerned. An analyst such as myself somehow has to claim that analyst's categories are superior to actor's categories even though they are based upon them. I think the way the position is to be rescued is with the, in principle, claim that, given enough time, one could reveal to the actors that though they believed their world is exhausted by their view of it, there is a deeper and more complete explanation. Crucially, that deeper explanation would have to be compatible with the actors' world as they formerly understood it. To say to a scientist actor, "your experimental results are actually determined by the machinations of witches," would not compatible; to say, "what you count as an experimental result is a matter of what your group comes to agree is the set of the experiments that were competently performed and this depends, in part, on your judgments about the scientists not just your judgments about the science," has a chance of being accepted after a period of reflection by the actors on the world they know. I have to believe that I could get even the fierce critic quoted above to accept this so that I can still say I am basing my analysis on actors' categories. That, perhaps, explains why I feel it vital to argue with my scientific respondents and explain myself so carefully in the earlier part of the book. Whether the effort has been a success is another matter.[18]

Notice another implication of this model. There is a clear separation between analysts' and actors' categories only if we think in terms of what the various parties are able to evince in a self-conscious way. As I initially concluded my first pieces of research on the sociology of gravitational waves, I could explain what the actors thought and I could explain what was different about what I thought—for example, that there was an experimenter's regress. My analyst's category, the experimenter's regress, was built upon the actors' world but was not a self-conscious part of the actors' world. But my hope has to be that as time goes on the idea of the experimenter's regress will become part of the actors' world. One could say that an analyst's category will have changed the actors' world in a small way even if it is only

18. Once more, this issue seems worth consideration by historians, who cannot go back and argue with those they are analyzing.

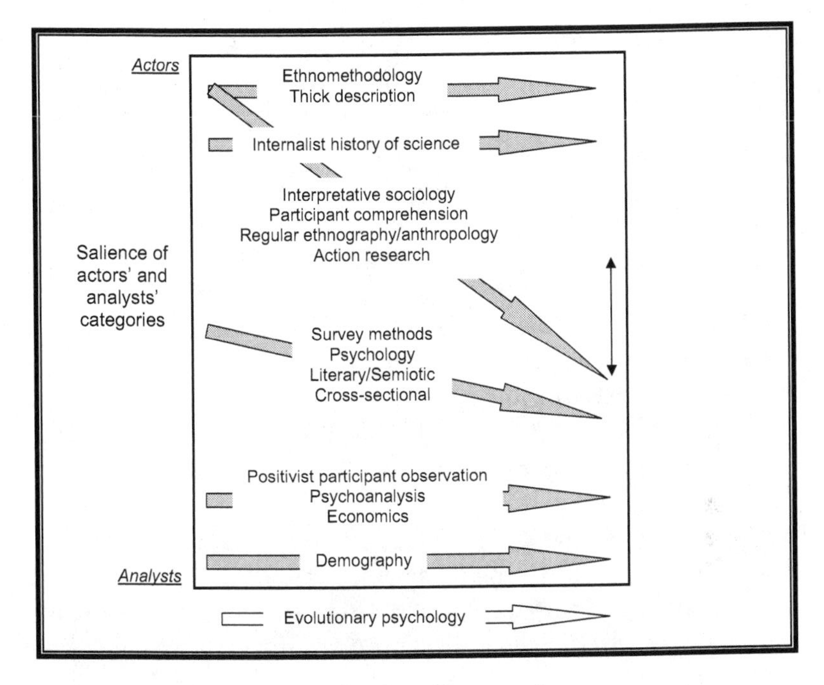

Figure 25. Conceptual trajectories of studies of human action

as a consequence of actors better recognizing what they already do.[19] The borderline between actors' categories and those of analysts is continually shifting, or so the analyst must hope.

Figure 25 shows the conceptual trajectories of a variety of methods thought about in these terms; there are some surprising bedfellows. The figure, unfortunately, is more complicated than it looks because, as we have begun to see, the notion of actors' categories is more complicated than it looks. Two versions have already been mentioned: there is the set of which actors are aware and the set, initially belonging to analysts, that actors might be willing to accept as describing their world after some discussion and reflection. But there is also the matter of what analysts declare they are doing. This is often set out as a methodological doctrine and does not always correspond with what they actually are doing. Inevitably, that distinction is tied up with the viewpoint of yet another analyst—myself.

19. Shapin (2001), points out that a number of natural scientists achieved a reflexive awareness of the nature of science not dissimilar to that found in this volume without the benefit of sociology of scientific knowledge.

What the arrows in the figure represent depends on which notion of actor's category is in play; sometimes the arrows represent methodological doctrine, sometimes the true position of the analyst, and sometimes my version of what is going on. I will make clear what each arrow is doing as the figure is explained.

To describe what happens in these kinds of researches it is also useful to think of four related factors that combine as a research trajectory unfolds:

1) The physical location of the analyst in the actors' world
2) The social location of the analyst in the actors' world
3) The degree of intervention by the analyst in the actors' world
4) In the case of science studies only, the extent to which the analyst must understand the technicalities of the science

In respect of item 4, five kinds of understanding of science can be taken from the Periodic Table of Expertises.[20] Starting from the bottom there are three kinds that are better thought of as assemblages of knowledge rather than true expertise since they can all be gained from reading rather than from immersion in the social life of the scientific actors. The first of these three is *beer-mat knowledge*. This, as the name implies, is the kind of thing one might learn from a coaster in a bar—a sentence or two that might be useful in the game of Trivial Pursuit but cannot be used in any kind of practice or nontrivial discourse. *Popular understanding* is gained from popular science books and the like, while *primary-source knowledge* is obtained from the technical journals or the internet in the absence of contact with the social world of the domain specialists. Because the material in primary sources is technical it gives the impression that it conveys understanding but deeper consideration shows that in the absence of the means to assess what is a sound technical contribution and what is not, primary source knowledge can be very misleading. Internet antivaccination campaigns and the like illustrate the dangers but even the physics journals publish many papers that the expert community does not take seriously. Only contact with the oral community enables one to filter the technical literature in the way that the specialist community filters it. Taking the actors' perspective seriously means acquiring one of the upper two categories of expertise. As explained above, *interactional expertise* is acquired by being embedded in the spoken discourse of the community, even if expert

20. Collins and Evans 2007, chapters 1 and 2 explain the Periodic Table. The five kinds of expertise are found in the "specialist expertise" line, starting on the left.

practices are not shared. *Contributory expertise* is the full-blown expertise of the practitioners.

Enough is now in place to explain the elements of figure 25. Starting from the top, there are approaches that claim to aim solely at immersion in the actors' perspective and to eschew the analysts' perspective entirely. These are represented with a horizontal arrow, which represents the analyst's doctrine in that the trajectories are said to remain constantly within the actors' world. Unsurprisingly, the researchers who work in these ways colocate themselves both physically and socially with the actors.

The very topic of *ethnomethodology* is the way the actors make their world.[21] Anything that is not an actor's category is not supposed to be part of ethnomethodology, so there should be no analysts' categories. Ethnomethodology's descriptions of the actors' world are, however, often very difficult to access. It is hard to imagine actors reading and understanding ethnomethodologists' accounts of their world. In spite of the fact that analysts' categories are eschewed, on the face of it the terms of the analysis seem to belong to the analysts. Furthermore, ethnomethodologists sometimes claim to have intervened successfully in actors' worlds as a result of their studies of human-computer interaction, DNA "fingerprinting," air-traffic control, and the like. But it is not clear how such successful interventions as there might be could have emerged purely from actors' categories without the superimposition of the researcher's analysis of the world. Perhaps, then, the ethnomethodology trajectory should slope downward. Ethnomethodologists do take it as their duty to acquire a good technical understanding of the science under investigation; they would not be satisfied with anything less than interactional expertise.

Anthropologists and ethnographers locate themselves physically and socially with the actors. Some of these, the cases represented by the horizontal arrow, restrict their object to evincing of the actors' world in as complete a way as possible; this is sometimes, sometimes known as *thick description*.[22] There is, of course, a puzzle about how the actors' world can be evinced for those who do not share it so this arrow should, perhaps, droop downward too. As far as I understand, to intervene in the actors' world would go against what most contemporary anthropologists consider the norms of their profession. One would expect ethnographers and anthropologists to demand good technical understanding of the science, at

21. The founder is Harold Garfinkel (1967).

22. The term originates with Clifford Geertz (1973).

least at the level of interactional expertise, but, surprisingly, this is not always the case.

Stepping down to the next horizontal arrow in the figure, *internalist history* of science is an interesting case when considered in these terms because of the way it differs from sociology of scientific knowledge. The internalist historian reconstructs the way theories and experiments were put together as passages of science unfolded. The internalist draws only on the accounts found in scientists' notebooks, drafts, and publications. Thus the history is intended to reconstruct the scientific world of the actors as it was lived. Internalist history of science follows (the doctrine of) ethnomethodology and thick description in basing itself in the world of the actors but it provides an analysis that squares with the world of the actors as it is unreflectively understood so to this extent it is more pure. On the other hand, as illustrated in this volume, researchers such as myself would claim that this version of science does not, in fact, capture day-to-day scientific life. What it captures is the commonsense philosophy of science which happens to inform the actors' lives in a self-conscious way.[23] Internalist historians develop excellent technical understanding of the sciences they describe, though it is difficult to acquire interactional expertise as there are no actors with whom to engage in linguistic interchange. Our modern understanding of science, therefore, has to stand in for a contemporaneous understanding. Interestingly, some historians have tried to acquire contributory expertise by reproducing old experiments but whether this can truly be done remains a problem which is fascinating in itself.[24]

The first downward sloping arrow represents approaches which claim to start with actors' categories but finish by building analysts' categories on top of them. These approaches include *interpretative sociology* in general, with its *participant comprehension* variant, those versions of *ethnography* and *anthropology*, which try to draw general conclusions or make comparisons between cases, and *action research*. In all of these the researchers will, again, consider it necessary to immerse themselves with the actors both physically and socially. There is not much to be said about the first three of these approaches in addition to what has been said already, except that it is important to avoid political preconceptions when building analysts' cat-

23. There are, of course, many other kinds of history, but they are not discussed here.
24. See, for example, Sibum 1995.

egories upon those of actors.[25] In general, proponents of these approaches do not intervene in actors lives; my attempt to intervene in the physics is untypical to the point of heresy. Once more, these approaches usually imply a high level of technical understanding of the science which, I would argue, should be at least to the level of interactional expertise.

Action research is different in that *intervention is integral to the method.* Action researchers attempt to understand the world by *intervening* in the world of the actors, often in policy-significant ways. For this reason, perhaps action research is even closer to the actors than the other approaches represented by this arrow; the action researcher has to be a contributory expert as well as an interactional expert in the domain being studied. I know of no action research on the topic of science though the generous-minded action researcher might include my little dogs intervention as a candidate.

Survey-based sociology stands in contrast to interpretative sociology in that it makes no self-conscious attempt to touch the world of the actors. Ideally, the research is done at a distance, often employing market research firms to ask questions of respondents. The analysis of the responses is quantitative. For surveys to be useful, however, the questions have to make sense to the respondents and so they have to be asked in terms of actors' categories. In the case of surveys this happens by default, as it were, rather than as a salient feature of the method. The circle is squared because surveys are generally designed and conducted by native members of the society being examined so they take the actors' perspective without noticing that it is an important feature of what they do. There is, then, less distinction between surveys and interpretative sociology than the starting point of the arrows representing the trajectories seems to imply. On the other hand, interpretative sociology and its variants generally look at groups which are initially socially distant from the researcher so that the acquisition of actors' categories is a deliberate and often difficult process. Surveys can go wrong when their regular methodology is applied to esoteric groups, because the default position no longer corresponds with the world of the actors. In such cases the survey can examine only a cross-section

25. In Collins 2008, it is noted that there is a systematic difference in the ways that different researchers impose analyst's categories on top of actors' according to their political standpoint in respect of technological controversies. For example, social analysts tend to side with actors in the case of vaccine scares, arguing that scientists' claims that controversies are closed are premature, whereas they side with scientists in the case of, say, global warming, arguing that critics who say that the scientific debate is still open are biased.

of the group it is studying. For example, in the case of science, it can look at citation behavior. For such cases it is right to show the start of the trajectory at a relatively low point on the vertical axis. Surveys are often used as the basis of intervention in actors' worlds though this is not an integral part of the method. Surveys used for the analysis of science and are not taken to require any technical understanding of the science beyond that required to recognize the boundaries of the topics being investigated—popular understanding is the most that is needed to look at a cross-section of science.

Psychology is similar to survey research: it purports to be an analysts' science but usually works through the perspective of actors, notoriously, first-year psychology students. Intervention is also not an integral part of the method. The psychology of science seems to depend on no more than popular understanding of science.

In the terms of this analysis, it should now be clear that the *literary/semiotic approach* is in many ways similar to the survey method in that both look at cross-sections of the activity of those they are studying. It is less obvious in the case of the semiotic approach because the first (I believe) analysis of the literature of this kind was carried out by Bruno Latour who based it on his several-month sojourn at the Salk Institute, where he elected to work as a technician.[26] Latour's participation is not indicative, however, because in spite of his being a philosophically inclined anthropologist, he made it clear that this he made no attempt to understand the science.[27] Latour immersed himself physically with the actors but maintained his distance in respect of social immersion. Something like beer-mat knowledge was proclaimed to be sufficient. Indeed, he stressed his determinedly distanced perspective in the most dramatic way by including in the published account of his study photographs of the chimneys on the laboratory roof, implying that, for him, they were as meaningful as the inscriptions produced by the scientists.

Once more, if the actors' world is organized via the actors' perspective, how can it be that an approach that begins and ends with the analysts' perspective can shed any light on the actors' world at all, assuming that its aim is not, say, to count scientists, after the fashion of demography? The photographs of the chimneys do indeed show how futile it can be not to share actors' categories for they offer no insights on the way science works.

26. The first of these ideas has been described under the label modalities on p. 309, above.

27. Latour and Woolgar 1979. Humanists, who analyze science, generally do not spend lengthy periods working in scientific laboratories, as the cross-section approach does not require it.

The enigma is resolved as it was for social surveys: the default position of the analyst corresponds with that aspect of the world of the actors under analysis. There is an overlap between the scientific literature and other kinds of literature. Scientists base their specialist practice languages on natural language and use the same sentence forms to express strong claims and weak claims. Thus, Latour (and Woolgar, his coanalyst) inhabit the actors' literary domain from the outset and this cross-section of scientific life is what their research is concerned with.

Positivist participant observation stresses physical colocation with the actors. Positivist participant observation also eschews social immersion, in its case so as to avoid "contaminating" the situation being studied. The book *When Prophecy Fails* is a classic "participant observer" case study of this kind: investigators infiltrated a small millennial cult pretending to be believers while carefully remaining passive and uninvolved.[28] Their position was essentially that of the fly on the wall even though to achieve it they had to interact with the cult members to the extent of passing themselves off as believers in the millennial prophecy, going to all the meetings, and engaging in the physical indignities associated with the belief. In spite of this degree of mixing with the actors, it appears that the researchers maintained their analytic distance; they *succeeded in failing* to grasp the participants' world view in any deep sense. It was this *failure* that allowed them to describe the nonfulfillment of the millennial prophecy as *generating cognitive* dissonance for the believers—the analyst's category—whereas, when the prophesied catastrophe failed to materialize, the believers took it that they had saved the world through their vigilance and willingness to make a sacrifice—as in the Old Testament story of Abraham and Isaac. From the actors' perspective, this provided cognitive reinforcement rather than being an instance of cognitive dissonance. Here, then, is a clear conflict between actor and analyst conclusions even though the building blocks of the analysis are taken from the actors' world. Of course, the researchers could infiltrate the cult in the first place only because of the overlap between their world and that of those they researched, but they could only support their major hypothesis by studiously not understanding the central categories of the cultists' world.

28. Festinger et al. 1957. For a much more detailed analysis in the context of the contrast between the Festinger study and participant comprehension, and an explanation of why it is described as positivist in philosophical spirit as opposed to interpretivist, see Collins 1984.

Psychoanalysis does aim to intervene in the world of the actor, its interventions being based on an analyst's version of the actor's world. Just as in the case of sociology of scientific knowledge, this perspective is not usually understood by the psychoanalyst's "respondents," at least, not at the outset. In the case of psychoanalysis, actors are meant to learn to accept the alternative version of their world provided by the analyst and find it beneficial in changing their behavior—explaining the analyst's perspective to the actor and persuading them of its validity comprises an intervention in itself. Psychoanalysts' versions of reality are far less well connected to the world of the actors than those of sociologists of scientific knowledge, however; critics would say they are much more likely to have originated solely in the world of the analyst. It may be, then, that psychoanalysis is meant to change the world view of the actor rather than represent it a way that could have occurred to them had they had more reflective ability of their own. I will assume this critical perspective on psychoanalysis from here on.

Economics is similar to this model of psychoanalysis: it intervenes in the world of actors while basing itself on a stripped-down and disastrously incorrect version of that world invented by the analysts. It is much closer to a purely analyst-based science than its self-description, which takes it to be based on the world of actors, would intimate.[29]

If my accounts are correct, the self-descriptions of *psychologists* and *social surveyors* on the one hand and *economists* and *psychoanalysts* on the other are equal and opposite. Psychologists and social surveyors do not expressly take the actors' world into account but their methods are based upon it by default; economists and psychoanalysts claim to describe the actors' world, but they work within their own world.

One has to go to *demography* to find a human science that is purely and properly based on analyst's categories from beginning to end. Even here, however, there might be a need to access the actors' viewpoint when demography tries to *explain* migration patterns.

Sciences that try to explain human behavior as an evolving biological system, such as *evolutionary psychology*, fall outside the diagram since they can take no account of the way human behavior varies from place to place—for such an approach there are no actors' specialist skills or abilities to be acquired.

29. The economists' actors' world supposedly has perfect flows of information, actions motivated by a rational calculus of pure self-interest, and so forth.

I believe that the intended research trajectories of analysts' methods should be set out as a matter of course in the introductory section of research reports. An inventory of the level of actors' expertise that has actually been acquired at various stages of the research should come at the end. All social analysis involves compromise, however. The important thing is not to succeed in gaining the highest levels of expertise even when the methodology demands it; the important thing is to understand the aspiration that goes along with the chosen method. It is important to understand where compromises have been made and what their implications are. Good research is not vitiated by compromise, it is vitiated by the pretence that there is no compromise.

APPENDIX 4
A Sociologist Tries to Do Some Physics

The technical physics/philosophical arguments that supported my attempt to intervene in the little dogs debate went as follows.[1] Forgetting about gravitational waves having "fingerprints," no one can know by means other than statistical inference whether a zero-delay coincidence is a real signal or a lucky chance. Therefore, one should redescribe the search for gravitational waves as a

1. Some more minor arguments of my own are found in the main text. For instance, I work out that sigma is unaffected by increasing the run length unless the little dogs are removed and this, I argue, suggests, in a philosophical sort of way, that they ought to be removed leaving it possible, under certain assumptions, to improve the limit on the FAP.

I had an additional argument that, for economy's sake, I relegate to this footnote: In the case of something like the Big Dog, however the time-slide analysis in handled, it is going to be concluded that there is very little chance that the signal was anything other than a gravitational wave. Let us say that at the very worst it is concluded that there is something like one chance in a hundred that it is not a gravitational wave. Yet if it is a gravitational wave then it ought to be removed from the data stream before the time-slide analysis is done. In terms of the tradition of physics, one in a hundred is not good enough to license a discovery announcement but it indicates the extent to which the components of the coincidence ought to be discounted when the time-slide analysis is done—the effect of any little dogs ought to be divided by 100. In effect, this means throwing them out. This is not the most conservative assumption but it isn't bad. It might be that this last argument is not so different from this remark, made by Dogwood in an e-mail of 26 October 2010, but I am not really sure.

A really correct treatment would use Bayesian statistics to evaluate the odds ratio of any given coincidence being signal vs. background, but we don't have the mechanics to do that. Bayesian statistics would take into account the estimated prior probability of the Big Dog being a signal.

search for "zero-delay coincidences." The scientists, though they describe themselves as searching for gravitational waves, have no choice but to search for *coincidences* because, at best, that is, what they see. That they are searching for coincidences is not affected by the fact that under the right circumstances, they will later describe zero-delay coincidences as "observations of" or "evidence for" gravitational waves.

Each zero-delay coincidence has, as I saw it, a certain "scientific value" made up of two components. The first component has to do with the quality of the coincidence as gauged by whether the devices are in a good, quiet state when the coincidence is found, whether the environmental monitors have picked up anything that would cause the setting of a warning flag, whether the waveform is plausible, and so on. The second component relates to the offset coincidences. Scientific value is reduced every time an offset coincidence is found in the time slides. Find a real time coincidence along with few or no offset coincidences and confidence is much higher than if lots of offset, and therefore noise-generated, coincidences of similar size are found in the background generated by time slides. Under this philosophy one never asks if *this* zero-delay coincidence *is* a gravitational wave because one accepts that one cannot know, one simply asks what is the "scientific value" of "this" zero-delay coincidence.[2]

Imagine that there are lots of zero-delay coincidences in the data streams and hardly any noise so that, in the absence of data-quality flags or other reasons for vetoing the coincidences, they would have high scientific value. Such a situation can be represented schematically as in the top half of figure 26, which shows "ribbons" from two detectors, A and B, each showing excursions. Here there are five zero-delay coincidences and a couple of noise excursions that are not opposite each other.

To make it easier to visualize and talk about, the bottom half of the figure shows a second version of A and B with the zero-delay coincidences marked as exclamation marks. It must be understood that, barring "fingerprints," which did not feature at this stage of the argument, there is no knowable difference in substance between an exclamation mark and a solid bar. In other words, the only thing an exclamation mark indicates is

2. I originally wrote up these ideas in pseudoscientific notation: If we call the number of offset coincidences "O" then, to use pseudoscientific notation, $SV = V / f(O)$; this translates as SV, the scientific value of a zero-delay coincidence, is equal to its value as affected by environmental monitors, wave form, etc, divided by some "function" (f) of the number of offset coincidences. This was greeted with such visceral scorn by at least some of the physicists that this footnote is the last place we will see such a quasi-formula.

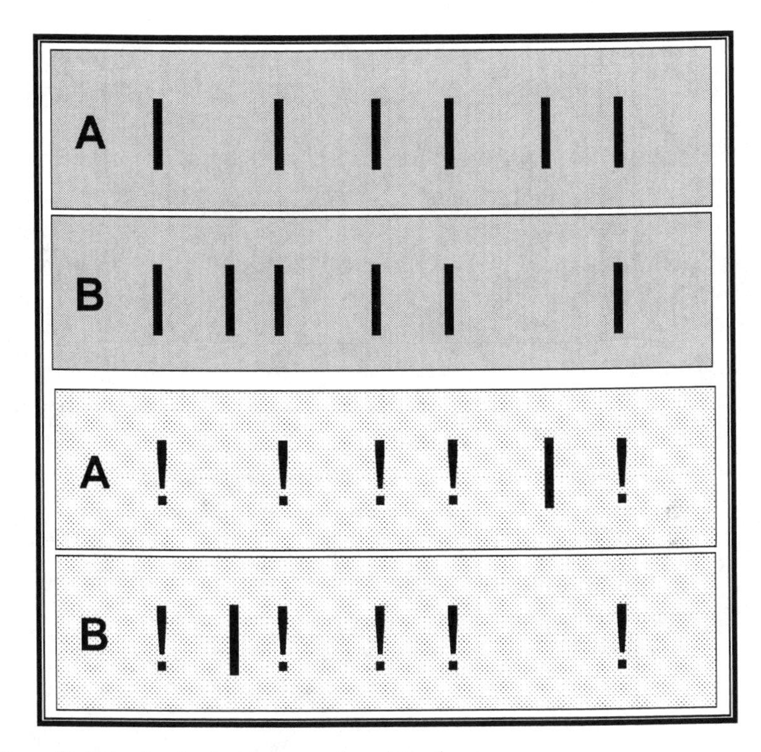

Figure 26. Lots of real coincidences and two bits of noise

what might be called "coincidenceness." I'll use the exclamation mark version in the rest of the argument.

Imagine, now, that one carries out the maximum number of time slides with the ribbons arranged as a continuous loop. If the zero-delay coincidences are left in, the number of offset coincidences is 36 – 5 = 31. This is because there are six components in each data stream and they can combine in 6 × 6 different ways to form coincidences but five of those are at zero delay and must be subtracted once the exercise is completed.

If, on the other hand, the zero-delay components—the exclamation marks—are removed at the outset, there will be only a single offset coincidence found in the time slides. So in the first case the zero-delay coincidences would be devalued by the existence of thirty-one offset coincidences and in the second case they would be devalued by the presence of only one offset coincidence. It is obvious that the right answer is the second one; there are lots of zero-delay coincidences and their value is hardly diminished at all by the fact that there is one offset coincidence in the data streams. This has nothing to do with gravitational waves—it is simply about the logic hunting for coincidences with "scientific value" and working out that scientific value by doing time slides and counting offset coincidences.

The next step is an "argument from induction." Remove one of the zero-delay coincidences from figure 26. The numbers change a bit but the argument does not. Then remove another and the argument still remains unchanged. Continue this way until there is only one zero-delay coincidence left and the argument still remains unchanged. Add some more offset noises and the argument still remains unchanged. At no point is there a sudden change in what is going on—nowhere does quantity change to quality. If this argument is correct, the right way to do the analysis must be to remove the components of the zero-delay coincidences before doing the time slides whether there are lots of zero-delay coincidences or only one. The argument, if it is right, works because it looks only at the logic of coincidence-hunting and does not slip into the language of real signals versus noise; it works because exclamation marks and solid bars are treated as identical.

In the six months before the envelope was opened I could get only one physicist respondent to take these ideas seriously.[3] That physicist agreed in principle with the point that we should not be asking whether the signal was real or not but merely working with the coincidences that we had in front of us. But, like Dogwood (see note 1), the respondent had a very different method, based on Bayes's theorem, for proceeding from there.

After the envelope was opened and the debate about Big Dog was over, one respondent who found time to look back at my argument questioned the "inductive" part of the procedure—where I remove coincidences one by one in order to work out what to do in the one-coincidence case by inference from the many-coincidence case of figure 26. This respondent argued that though there was no sudden change as exclamation marks were removed there was still a big difference between the beginning and the end point of the inductive procedure because of the large effect of a single extra offset coincidence in the actual, one-coincidence, case, as compared to the hypothetical, many-coincidence case of figure 26. This may well be a good

3. It would not be unreasonable to say that in trying to do physics I was acting, or at least opening myself up to being defined, as a "crank." We can define a crank as someone who believes they have a special insight into the physical world even though they are not properly connected into the specialist oral and institutional culture. It is this culture—the paradigm, or form-of-life of the field—that sets out the boundaries of what counts as a proper problem and the boundaries of what can count as a proper solution to existing problems. Cranks go for problems and solutions that lie outside these boundaries.

argument but the point is that it was an argument and to have an argument one must start with something that is worth arguing against rather than simply dismissing. So, right or wrong, it was encouraging.

A lengthier discussion of my argument about little dogs was to take place on 12 July 2011. It coincided with a big gravitational-wave conference that happened at my home university, Cardiff.[4] I was invited to give a public lecture on my work at the meeting. My talk was attended by many of the gravitational-wave scientists and I was able to persuade two of the most senior persons in the gravitational-wave community to come to my office on the following day to discuss the little dog business. For a couple of hours we argued the matter back and forth using PowerPoint slides. My conclusion was, once more, that my approach may not be the best possible but was not unreasonable. The basis of this claim was the good natured tone of our discussion, the fact that the meeting continued into the evening as a social occasion, and an e-mail I wrote the following day, which was not met with a rebuttal.[5]

> Dear [R] and [S],
> Thank you very much for yesterday's talk—it was very interesting and useful to me, not to mention very enjoyable. Sorry, S, that you could not make dinner.
>
> S, I attach the newly published paper I mentioned about interactional expertise.
>
> I thought I would try to sum up what I thought came out of the discussion to see if you agree this is fair.
>
> First, in earlier e-mail to me R you say, among other things:
>
> > *I believe that the problem here is that you have misunderstood (or poorly described) what we are trying to do. We are trying to discriminate between accidental coincidences of noise events which do not have a common origin and coincidences which do have a common origin, e.g., a gravitational wave. Your argument, that we are looking for coincidences, doesn't make this distinction clear, and will confuse readers. Turn one of the coincidences [in figure 26] into a coincidence between two noise glitches i.e., one pair of !'s into a pair of |'s and then see whether your reasoning still holds.*

4. 9th Edoardo Amaldi Conference on Gravitational Waves, Cardiff University, 10–15 July 2011.

5. These two scientists are frequently represented in the book, identified as anonymous trees, but here I'll use letters instead of tree names.

I hope that yesterday's discussion will have convinced you that any mistakes on my part do not emerge out of any lack of understanding of this crude nature. Certainly, nothing said yesterday indicated to me that I did not understand at this level.[6]

On the matter of time slides etc. I thought I learned two things that I had not grasped or fully grasped before:

1) I had not fully grasped your "eliminate one coincidence at a time" strategy and this was because, as you pointed out, there was no strength dimension in my ribbon-diagram. I now see that if the coincidences in my diagram had been ordered in terms of their strength then a strategy that started with the strongest one and based an estimate of its sigma value on time slides *containing only noncoincident glitches that were as strong or stronger* would be a reasonable one; it would do no worse in terms of risking a false negative than the Big Dog case where there was only one potential signal. I still can't work out whether this strategy would work if the components of the "strongest signal" were significantly asymmetric so that you have to include lots of weak glitches—including the components of other coincidences—but I can see that your approach is correct in principle (unless nature is capricious enough to give you a lot of coincidences of nearly equal strength). If you have any further thoughts on the asymmetry problem please let me know. Anyway, I'll try to explain this in the book.
2) Completely new to me was the suggestion that in a case of multiple coincidences (e.g., like the 5 on my ribbon diagram) the background one would look for in the time slides would be accidental sets of five coincidences rather than single coincidences. This seems to me to make a lot of sense—though, again, I would guess there is a lot of devil in the detail. Once more I will try to explain this in the book.
3) My overall conclusion from 1 and 2 above is that (a) my argument in the book does *not* lead to the best approach in this matter and that, unsurprisingly, you are well ahead of me in thinking out this stuff but (b) the arguments I put forward were not completely stupid and trying to think it through independently has been a worthwhile exercise in a number of ways not least because it re-

6. Readers who follow my argument that there is nothing to be seen but coincidence will realize that it makes no sense to replace an explanation mark with a solid bar.

> sulted in a clearer understanding and my gaining more knowledge of the thinking in the field . . .[7]

That is roughly how I think yesterday's discussion will affect the book so, if you still have time and patience for this, let me know if you think that is reasonable or where you think it is not.

Cheers

Harry

To explain, under what I refer as point 1, above, what R pointed out to me was that all the coincidences in my figure were of equal magnitude. He suggested that in the normal way they would vary in magnitude. If this was the case one could proceed by taking the biggest one first and running time slides with the equivalent of the little dogs left in. This way, the biggest one would not produce a spurious signal when combined with the other components of the coincident signals because they would not be large enough. If the biggest component was found to be a signal it could then be eliminated and the next biggest could be analyzed in the same way and so on. I had not thought of that way of proceeding and had to admit that it made good sense. On the other hand, my response, which I offered at the time of the discussion, was to ask what would happen if the two components of the largest coincidence (as measured by signal-to-noise ratio), were of different sizes. In that case, when the large one was "run against" the opposite ribbon, spurious signals with equal overall signal-to-noise ratio would be generated by coincidences with components equal in size to the smaller component of the original. This would make the whole procedure less reasonable. And it was the case that Big Dog itself was made up of components of very different sizes—one being between 1.5 and 3 times as big as the other.[8] Under such circumstances it does not seem to me that choosing the biggest signal to eliminate first works as well as it appears to at first sight. Readers should be getting the sense of the extent

7. A missing sentence here refers to the promise of one of the scientists to send me the e-mails, which didn't arrive, where similar issues had been discussed before. There were another couple of paragraphs on other topics, which will be brought back under another heading.

8. In response to my query, Peter Saulson explained: "Signal to noise ratio is the usual way to compare signals in two interferometers. In the CBC search, we introduced a variant (NewSNR) to take account of whether the signal was a good match to our templates. The Combined NewSNR is what was reported in the paper. For the Big Dog, H1 had a NewSNR = 10.29, and L1 had a NewSNR = 6.29. I don't know whether it makes more sense to talk about a ratio of these numbers, or a ratio of their squares, but those are the numbers to compare" (personal communication, 31 January 2012).

to which I was trying to go beyond interactional expertise into the realms of contributory expertise.

Under point 2, R suggested that the right way to analyze data of the kind I had invented was not to look at it one coincidence at a time, but to consider the likelihood of finding spurious sets of five coincidences at a time. This too was a new idea to me and seemed a good way to do the analysis; at first sight it seems to resolve the little dog problem for large numbers of coincidences and, using the same kind of inductive inference that I used, resolves the case of a single coincidence in favor of leaving the putative signal in. This was the first I had heard of these notions, though S intimated that some of them had been discussed in the streams of e-mails that went backward and forward and promised to forward the relevant mailings to me; unfortunately they did not come.

So, we can see that I am simply not as good as R at thinking out solutions to these problems but my approach was sensible enough to elicit new (or new to me who had been listening hard for 6 months), solutions. The point of this detailed exposition is to establish something claimed in the chapter 17—that my approach may not have been the best but it was not crazy.

ACKNOWLEDGMENTS

Gravity's Ghost, First Edition

My approach to the analysis of science is to try to understand as much of it as I can—something I call "trying to develop interactional expertise" (Collins and Evans 2007). I do my best to put this technical understanding into my analysis. In the case of the science I describe here, the statistical analysis of data, I have also had to depend on the scientists to help me out quite a lot on the technical details by checking bits of "proto-text." As always, my friend Peter Saulson, author of the best treatise on interferometric gravitational-wave detection, onetime spokesperson of the LIGO scientific collaboration, and the most honest and honorable person I know, has been my constant guide in the writing of this book. That we disagree quite heartily in respect of some of its more sociological analyses and conclusions only demonstrates still further the selflessness of his contribution. I have also had help in understanding certain specific technical concepts and procedures from Alan Weinstein, Sergey Klimenko, and Mike Landry, to each of whom I sent short passages of the draft for comment. My onetime, bitter academic enemy, and now valued academic colleague, Allan Franklin, has generously looked into the history of high-energy physics in ways useful for this book but which I was technically not equipped to do for myself, and his findings are used and acknowledged in the text. Franklin intends to publish more on the recent history of high-energy physics so as to make clear where precedents and previous discussion of the kind reported here can be found. Graham Woan looked over the statistics chapter from the point of view of a gravitational-wave physicist especially interested in statistics, while Deidre McCloskey looked at it from the point of view of an expert in their use in the social sciences. Robert Evans, my colleague from Cardiff, read the whole manuscript carefully and pointed out a number of places where the typical reader from science and technology studies could easily misunderstand what I was trying to say. This led me to put in more explicit pointers aimed at that community. Richard Allen worked assiduously at copyediting to smooth the text and remove infelicities. More diffuse thanks go to the entire community of gravitational-wave physicists and to my colleagues and department at Cardiff for providing an environment in which work like this can be done. Needless to say, all remaining faults are my own.

As always, Christie Henry, my editor at the University of Chicago Press, has encouraged me from the outset, passing the reins for this particular volume to Karen Darling. I cannot say how wonderful it is to have found a publisher who believes you should publish what you want rather than what they want; it was a lucky day for me when I got together with Chicago.

The fieldwork was supported by a small grant from the UK Economic and Social Research Council (ESRC): "The Sociology of Discovery" (2007–2009: RES-000-22-2384).

Big Dog

As always, the book stands in the debt of the community of gravitational-wave physicists. I continue to be proud and happy in their company, even more so as I become "part of the furniture"—though never, I hope, a comfortable sofa. Of these physicists my greatest debt is to my friend Peter Saulson. I've pointed to his importance of chapter 14, but the idea of it only gelled on the second day of the Arcadia meeting as I was talking to Peter about what I was doing. Conversation is an irreplaceable resource and my conversations with Peter are an essential part of my methodology. They remain great conversations, of course, when they stray from physics to wider topics. To be fair to both of us, it should be said that the process of writing *Big Dog* coincided with the first stint of Peter's chairmanship of the physics department at Syracuse University, and this stopped him giving as much time to discussion as he would have liked—though we made up for much of this after the initial draft was completed. As explained, the first draft of *Big Dog* was sent to a number of physicists and a number of very useful comments were made on it. The book would be much worse and would contain many more mistakes without the close readings of Peter Saulson, Gabriella Gonzalez, and Stan Whitcomb, but I want to make clear that none of what follows is their responsibility; I did not always take the advice I was offered; remaining mistakes are mine and mine alone.

My onetime enemy and now colleague Allan Franklin made a vital contribution to the section of the text that deals with the recent history of physics publishing. He scoured physics journals and uncovered their developing traditions in respect of the relationship between levels of statistical confidence and the confidence expressed by their titles. He will write this material up at greater length in a book of his own (*Experimental Particle Physics, Then and Now*, University of Pittsburgh Press). I also thank my long-term colleague, Trevor Pinch, who made a vital contribution that helped determine to how I wrote up the final version of the methodological section of the book, which, I hope, has rendered it a little less priggish and a little more readable. Rob Evans and Martin Weinel have been supportive academic companions throughout and helped to make Cardiff a good place from which to work.

Once more I owe a huge debt to my publishers for their adventurousness in publishing this unusual second edition; it is a rare publisher who would be prepared to make the paperback nearly twice as long as the hardback. That the book has three parts instead of the initially submitted two, and is, I think, a much better book as a result, is due to the assiduous comments of referees and board members, skilfully marshalled by Karen Darling, who has enthusiastically pushed the project forward. I am simply delighted that Chicago, a true academic publisher, continues to take at least some of the awkward-looking work I submit to them. Mary Gehl provided invaluable help at the copy editing stage.

The analysis of Big Dog and my continuing contact with the gravitational-wave community is now funded by US National Science Foundation grant PHY-0854812 to Syracuse University, "Toward Detection of Gravitational Waves with Enhanced LIGO and Advanced LIGO," P.I., Peter Saulson. A portion of this grant goes toward the project entitled "To Complete the Sociological History of Gravitational Wave Detection."

Finally, my debt to Susan is beyond measure.

REFERENCES

For an additional list of references covering both gravitational-wave physics and the sociology of science see Gravity's Shadow *(Collins 2004).*

Astone, P., D. Babusci, M. Bassan, P. Bonifazi, P. Carelli, G. Cavallari, E. Coccia, et al. 2002. "Study of the Coincidences between the Gravitational Wave Detectors EXPLORER and NAUTILUS in 2001." *Classical and Quantum Gravity* 19 (7): 5449–65.

Astone, P., D. Babusci, M. Bassan, P. Bonifazi, P. Carelli, G. Cavallari, E. Coccia, et al. 2003. "Comments on the 2001 Run of the Explorer /Nautilus Gravitational Wave Experiment." http://arxiv.org/archive /gr-qc/0304004.

Astone, P., G. D'Agostini, and S. D'Antonio. 2003. "Bayesian Model Comparison Applied to the Explorer–Nautilus 2001 Coincidence Data." *Classical and Quantum Gravity* 20 (17): S769–84.

Barnes, B., D. Bloor, and J. Henry. 1996. *Scientific Knowledge: A Sociological Analysis*, London: Athlone Press.

Barnes, S. B. 1974. *Scientific Knowledge and Sociological Theory*. London: Routledge and Kegan Paul.

Beller, M. 1999. *Quantum Dialogue: The Making of a Revolution*. Chicago: University of Chicago Press.

Berger, P. L. 1963. *Invitation to Sociology*. Garden City, NY: Anchor Books.

Bijker, W., T. Hughes, and T. J. Pinch, eds. 1987. *The Social Construction of Technological Systems: New Directions in the Sociology and History of Technology*. Cambridge, MA: MIT Press.

Bloor, D. 1973 "Wittgenstein and Mannheim on the Sociology of Mathematics." *Studies in the History and Philosophy of Science* 4:173–91.

Bloor, D. 1976. *Knowledge and Social Imagery.* London: Routledge and Kegan Paul.
Bloor, D. 1999. "Anti-Latour." *Studies in History and Philosophy of Science* 30:81–112.
Collins, H. M. 1974. "The TEA Set: Tacit Knowledge and Scientific Networks." *Science Studies* 4:165–86.
Collins, H. M. 1975. "The Seven Sexes: A Study in the Sociology of a Phenomenon, or The Replication of Experiments in Physics." *Sociology* 9 (2): 205–24.
Collins, H. M. 1982. "Special Relativism: The Natural Attitude." *Social Studies of Science* 12:139–43.
Collins, H. M. 1984. "Concepts and Methods of Participatory Fieldwork." in *Social Researching*, ed. C. Bell and H. Roberts, 54–69. Henley-on-Thames, UK: Routledge.
Collins, H. M. [1985] 1992. *Changing Order: Replication and Induction in Scientific Practice.* 2nd ed. Chicago: University of Chicago Press.
Collins, H. M. 1988. "Public Experiments and Displays of Virtuosity: The Core-Set Revisited." *Social Studies of Science* 18:725–48.
Collins, H. M. 1998. "The Meaning of Data: Open and Closed Evidential Cultures in the Search for Gravitational Waves." *American Journal of Sociology* 104 (2): 293–337.
Collins, H. M. 1990. *Artificial Experts: Social Knowledge and Intelligent Machines.* Cambridge, MA: MIT Press.
Collins, H. M. 2002. "The Experimenter's Regress as Philosophical Sociology." *Studies in History and Philosophy of Science* 33:153–60.
Collins, H. M. 2004a. *Gravity's Shadow: The Search for Gravitational Waves.* Chicago: University of Chicago Press.
Collins, H. M. 2004b. "Interactional Expertise as a Third Kind of Knowledge." *Phenomenology and the Cognitive Sciences* 3 (2): 125–43.
Collins, H. M. 2007. "Mathematical Understanding and the Physical Sciences." In "Case Studies of Expertise and Experience," ed. H. M. Collins, special issue, *Studies in History and Philosophy of Science* 38 (4): 667–85.
Collins, H. M. 2008. "Actors' and Analysts' Categories in the Social Analysis of Science." In *Clashes of Knowledge*, ed. P. Meusburger, M. Welker, and E. Wunder, 101–10. Dordrecht: Springer.
Collins, H. M. 2009. "We Cannot Live by Skepticism Alone." *Nature* 458:30–31.
Collins, H. M. 2010. *Tacit and Explicit Knowledge.* Chicago: University of Chicago Press.
Collins, H. M. 2011a. *Gravity's Ghost: Scientific Discovery in the Twenty-First Century.* Chicago: University of Chicago Press.
Collins, H. M. 2011b. "Language and Practice." *Social Studies of Science* 41 (2): 271–300.
Collins, H. M. 2012. "Performances and Arguments." *Metascience* 21 (2): 409–18.
Collins, H. M., and R. Evans. 2002. "The Third Wave of Science Studies: Studies of Expertise and Experience." *Social Studies of Science* 32 (2): 235–96.
Collins, H. M., and R. Evans. 2007. *Rethinking Expertise.* Chicago: University of Chicago Press.

Collins, H. M., R. Evans, and M. Gorman. 2007. "Trading Zones and Interactional Expertise." In "Case Studies of Expertise and Experience," ed. H. M. Collins, special issue, *Studies in History and Philosophy of Science* 38 (4): 657–66.

Collins, H. M., and M. Kusch. 1998. *The Shape of Actions: What Humans and Machines Can Do.* Cambridge, MA: MIT Press.

Collins, H. M., and T. J. Pinch. 1979. "The Construction of the Paranormal: Nothing Unscientific is Happening." In *Sociological Review Monograph. No. 27: On the Margins of Science: The Social Construction of Rejected Knowledge*, ed. R. Wallis, 237–70. Keele, UK: Keele University Press.

Collins, H. M., and T. J. Pinch. 1982. *Frames of Meaning: The Social Construction of Extraordinary Science.* London: Routledge.

Collins, H. M., and T. J. Pinch. 1998a. *The Golem at Large: What You Should Know About Technology.* Cambridge: Cambridge University Press.

Collins, H. M., and T. J. Pinch. 1998b. *The Golem: What You Should Know About Science.* Cambridge: Cambridge University Press.

Collins, Harry, and T. J. Pinch. 2005. *Dr Golem: How to Think about Medicine.* Chicago: University of Chicago Press.

Collins, H. M., and G. Sanders. 2007. "They Give You the Keys and Say 'Drive It': Managers, Referred Expertise, and Other Expertises." In "Case Studies of Expertise and Experience," ed. H. M. Collins, special issue, *Studies in History and Philosophy of Science* 38 (4): 621–41.

Collins, H. M., and S. Yearley. 1992. "Epistemological Chicken." in *Science as Practice and Culture*, ed. A. Pickering, 301–26. Chicago: University of Chicago Press.

Desrosières, A. 1998. *The Politics of Large Numbers*, Cambridge, MA: Harvard University Press.

Evans, R. 1997. "Soothsaying or Science: Falsification, Uncertainty and Social Change in Macro-economic Modelling." *Social Studies of Science* 27 (3): 395–438.

Evans, R. 1999. *Macroeconomic Forecasting: A Sociological Appraisal.* London: Routledge.

Festinger, L., H. W. Riecken, and S. Schachter. 1956. *When Prophecy Fails.* New York: Harper.

Finn, L. S. 2003. "No Statistical Excess in EXPLORER/NAUTILUS Observations in the Year 2001." *Classical and Quantum Gravity* 20:L37–44.

Franklin, A. 1990. *Experiment, Right or Wrong.* Cambridge: Cambridge University Press.

Franklin, A. 1997. "Millikan's Oil-Drop Experiments." *Chemical Educator* 2:1–14.

Franklin, A. 2004. "Doing Much about Nothing." *Archive for History of Exact Sciences* 58:323–79.

Fleck, L. 1979. *Genesis and Development of a Scientific Fact.* Chicago: University of Chicago Press.

Galison, P. 1997. *Image and Logic: A Material Culture of Microphysics.* Chicago: University of Chicago Press.

Garfinkel, H. 1967. *Studies in Ethnomethodology*. Englewood Cliffs, NJ: Prentice-Hall.

Geertz, C., 1973. *The Interpretation of Culture*. New York: Basic Books.

Gigerenzer, G., Z. Switjink, T. Porter, L. Daston, J. Beatty, and L. Kruger. 1990. *The Empire of Chance: How Probability Changed Science and Everyday Life*, Cambridge: Cambridge University Press.

Giles J. 2006. "Sociologist Fools Physics Judges." *Nature* 442:8.

Gieryn, T. 1999. *Cultural Boundaries of Science: Credibility on the Line*. Chicago: University of Chicago Press.

Godin, B., and Y. Gingras. 2002. "The Experimenter's Regress: From Skepticism to Argumentation." *Studies in the History and Philosophy of Science* 30A (1):137–52.

Greenberg, D. S. 2001. *Science, Money and Politics: Political Triumph and Ethical Erosion*. Chicago: University of Chicago Press.

Gross, P., and N. Levitt. 1994. *Higher Superstition: The Academic Left and Its Quarrels with Science*. Baltimore, MD: Johns Hopkins University Press.

Gross, P., N. Levitt, and M. W. Lewis. 1996. *The Flight From Science and Reason*. New York: New York Academy of Sciences.

Hacking I. 1975. *The Emergence of Probability*, Cambridge: Cambridge University Press.

Hendry, R., J. Donnelly, G. Delacôte, L. Wolpert, and B. Jurdant. 1994. "Multiple Book Review." *Public Understanding of Science* 3:323–37.

Hendry, R., J. Donnelly, G. Delacôte, L. Wolpert, and B. Jurdant. 1990. *The Taming of Chance*. Cambridge: Cambridge University Press.

Holton, G. 1978. *The Scientific Imagination*. Cambridge: Cambridge University Press.

Kaiser, D. 2011. *How the Hippies Saved Physics*. New York: W. W. Norton.

Kennefick, D. 2000. "The Star Crushers: Theoretical Practice and the Theoretician's Regress." *Social Studies of Science* 30 (1): 5–40.

Kennefick, D. 2007. *Traveling at the Speed of Thought: Einstein and the Quest for Gravitational Waves*. Princeton, NJ: Princeton University Press.

Knorr-Cetina, K. 1999. *Epistemic Cultures: How the Sciences Make Knowledge*. Cambridge, MA: Harvard University Press.

Krige, J. 2001. "Distrust and Discovery: The Case of the Heavy Bosons at CERN." *ISIS* 95:517–40.

Kuhn, T. 1961. "The Function of Measurement in Modern Physical Science." *ISIS* 52:162–76.

Kuhn, T. 1962. *The Structure of Scientific Revolutions*, Chicago: University of Chicago Press.

Labinger, J., and H. M. Collins, ed. 2001. *The One Culture?: A Conversation about Science*. Chicago: University of Chicago Press.

Lakatos, I. 1970. "Falsification and the Methodology of Scientific Research Programmes." In *Criticism and the Growth of Knowledge*, ed. I. Lakatos and A. Musgrave, 91–196. Cambridge: Cambridge University Press.

Lakatos, I. 1976. *Proofs and Refutations*. Cambridge: Cambridge University Press.

Latour, B. 1983. "Give Me a Laboratory and I Will Raise the World." In *Science Observed: Perspectives on the Social Study of Science*, ed. K. Knorr-Cetina and M. Mulkay, 141–70. London: Sage.

Latour, B. 1987. *Science in Action*. Milton Keynes, UK: Open University Press.

Latour, B. 2005. *Reassembling the Social—An Introduction to Actor-Network-Theory*. Oxford: Oxford University Press.

Latour, B., and S. Woolgar. 1979. *Laboratory Life: The Social Construction of Scientific Facts*. London: Sage.

Laudan, L. 1983. *Progress and Its Problems: Towards a Theory of Scientific Growth*. Berkeley: University of California Press.

Leclerc-Madlala, S. 2002. "On the Virgin Cleansing Myth: Gendered Bodies, AIDS and Ethnomedicine." *African Journal of AIDS Research* 1:87–95.

Lynch, M., S. Cole, R. McNally, and K. Jordan. 2011. *Truth Machine: The Contentious History of DNA Fingerprinting*. Chicago: University of Chicago Press.

Mackenzie, D. 1981. *Statistics in Britain, 1865–1930*. Edinburgh: University of Edinburgh Press.

Mackenzie, D. 1998. "The Certainty Trough." In *Exploring Expertise: Issues and Perspectives*, ed. R. F. W. Williams and J. Fleck, 325–29. Basingstoke, UK: Macmillan.

Mackenzie, D. 2001. *Mechanizing Proof: Computing, Risk, and Trust*. Cambridge, MA: MIT Press.

Medawar, P. B. 1990. "Is the Scientific Paper a Fraud?" In *The Threat and the Glory, Reflections on Science and Scientists*, ed. D. Pyke, 228–33. Oxford: Oxford University Press.

Mannheim, K. 1936. *Ideology and Utopia: An Introduction to the Sociology of Knowledge*. Chicago: University of Chicago Press.

Merton, R. K. 1942. "Science and Technology in a Democratic Order." *Journal of Legal and Political Sociology* 1:115–26.

Merton, R. K. 1973. *The Sociology of Science: Theoretical and Empirical Investigations*. Chicago: University of Chicago Press.

Mortara, J. L., I. Ahmad, K. P. Coulter, S. J. Freedman, B. K. Fujikawa, J. P. Greene, J. P. Schiffer, W. H. Trzaska, and A. R. Zeuli. 1993. "Evidence Against a 17 keV Neutrino from 35S Beta Decay." *Physical Review Letters* 70:394–97.

Mulkay, M., J. Potter, and S. Yearley. 1983. "Why an Analysis of Scientific Discourse Is Needed." In *Science Observed: Perspectives on the Social Study of Science*, ed. K. D. Knorr-Cetina and M. Mulkay, 171–203. London: Sage.

Nye, M. J. 2011. *Michael Polanyi and his Generation: Origins of the Social Construction of Science*. Chicago: University of Chicago Press.

Pinch, T. J. 1980. "The Three-Sigma Enigma." Paper presented at the ISA-PAREX Research Committee Meeting, Burg Deutschlandsberg, Austria, September 26–29, 1980.

Pinch, T. J. 1985. "Towards an Analysis of Scientific Observation: The Externality and Evidential Significance of Observation Reports in Physics." *Social Studies of Science* 15:167–87.

Pinch, T. J. 1986. *Confronting Nature: The Sociology of Solar-Neutrino Detection*. Dordrecht: Reidel.

Pinch, T. J. 1993. "'Testing—One, Two, Three . . . Testing!': Toward a Sociology of Testing." *Science Technology Human Values* 18 (1): 25–41.

Polanyi, M. 1958. *Personal Knowledge*. London: Routledge and Kegan Paul.

Popper, K. R. 1959. *The Logic of Scientific Discovery*. New York: Harper & Row.

Porter, T. 1996. *Trust in Numbers*. Princeton, NJ: Princeton University Press.

Shapin, S. 1994. *A Social History of Truth: Civility and Science in Seventeenth-Century England*. Chicago: University of Chicago Press.

Shapin, S. 2001. "How to be Antiscientific." In *The One Culture? A Conversation about Science*, ed. J. Labinger and H. M. Collins, 99–115. Chicago: University of Chicago Press.

Shapin, S. 2008. *The Scientific Life: A Moral History of a Late Modern Vocation*. Chicago: University of Chicago Press.

Shapin, S., and S. Schaffer. 1987. *Leviathan and the Air Pump: Hobbes, Boyle and the Experimental Life*. Princeton, NJ: Princeton University Press.

Sibum, H. O. 1995. "Reworking the Mechanical Value of Heat: Instruments of Precision and Gestures of Accuracy in Early Victorian England." *Studies in History and Philosophy of Science* 26 (1): 73–106.

Smith, R., and B. Wynne, eds. 1988. *Expert Evidence: Interpreting Science in the Law*. London: Routledge.

Snow, C. P. 1959. *The Two Cultures*. Cambridge: Cambridge University Press.

Stepanyan, S., K. Hicks, D. S. Carman, E. Pasyuk, R. A. Schumacher, E. S. Smith, D. J. Tedeschi, et al. 2003. "Observation of an Exotic S = +1 Baryon in Exclusive Photoproduction from the Deuteron." *Physical Review Letters* 91:252001–5.

Stigler, S. 1999. *Statistics on the Table: The History of Statistical Concepts and Methods*. Cambridge, MA: Harvard University Press.

Weisberg, J. M., J. H. Taylor, and L. A. Fowler. 1981. "Gravitational Waves from an Orbiting Pulsar." *Scientific American*, October, 74–82.

Winch, P. G. 1958. *The Idea of a Social Science*. London: Routledge and Kegan Paul.

Wittgenstein, L. 1953. *Philosophical Investigations*. Oxford: Blackwell.

Wolpert, L. 1992. *The Unnatural Nature of Science*. London: Faber and Faber.

Wright-Mills, C. 1940. "Situated Actions and Vocabularies of Motive." *American Sociological Review* 5 (13): 904–9.

Wynne, B. 1982. *Rationality and Ritual? The Windscale Inquiry and Nuclear Decisions in Britain*. Chalfont St. Giles: British Society for the History of Science.

INDEX